# Financial and
# Cost Concepts
# for Construction
# Management

# Financial and Cost Concepts for Construction Management

Daniel W. Halpin

Georgia Institute of Technology

John Wiley & Sons
New York   Chichester   Brisbane   Toronto   Singapore

*Library of Congress Cataloging in Publication Data:*

Halpin, Daniel W.

    Financial and cost concepts for construction management.

    Bibliography: p. 408
    Includes indexes.
    1. Building—Estimates.    2. Construction industry—
Accounting.   I. Title.

TH435.H32 1985    624.1'068'1    84-22178
ISBN 0-471-89725-6

Printed in the United States of America

10

# Preface

As in any business-related activity, one of the most important goals of engineering and construction management is to achieve a profit through the expenditure of time, funds, and effort. Money spent on resources must be recovered by passing expenses to the client. In addition, the entrepreneur marks up these expenses with the intent of recovering the costs of management as well as earning a profit or fee for his or her entrepreneurial skill. The subject areas of estimating and cost accounting have traditionally provided the skills necessary for recovering costs—both the direct costs incurred in constructing a facility and the management costs required to orchestrate the work. The subjects of the financial health of the firm and the overall ability of company-level management to achieve success in the marketplace are generally addressed in financial accounting and management courses.

The connection between company-level finance and project-level cost and resource management is fuzzy at best because these topics are not generally considered in an integrated fashion. Integrated systems that relate field management decisions to the balance sheet and income statement at the company level are becoming more prevalent. The advent of small computers capable of collecting and integrating data at both project and company levels in a form that is useful to project management has expedited this development.

The objective of this text is to present both company and project levels of revenue and expense management in an introductory but integrated format to give the student a coordinated overview of how data at both levels interact. In many cases, managers have engineering or technology backgrounds that provide excellent preparation for solving technical problems in the field. However, their exposure to the financial context within which engineering and construction occur (if covered at all) may be limited to accounting courses oriented to nonconstruction disciplines such as manufacturing. These courses often emphasize the rudiments of bookkeeping without providing the student with a good understanding of how familiarity with accounting techniques can aid the engineering manager in the field. Such courses prepare the student to be a bookkeeper rather than a manager.

The treatment of financial and cost concepts in this text is oriented toward engineering and technology students with no accounting background. The first six chapters introduce accounting and bookkeeping principles as they relate to construction-oriented companies. The intent is not to prepare the student to set up the books of a company or to actually perform accounting functions. Rather, the objective is to provide future managers an understanding of financial accounting processes and to give them enough information to communicate at an informed level with accountants and bookkeepers. Although most of the material is presented within the context of a construction company, the information discussed and the financial concepts developed are equally applicable to a firm specializing in engineering design.

Chapters 1 through 3 introduce the accounting conventions commonly used in engineering and construction, as well as the financial documents generated as part of the accounting process. Chapters 4 and 5 introduce basic bookkeeping activities. Chapter 6 briefly examines some methods of analyzing financial documents. In Chapter 4, the interplay between project level accounts and general ledger accounts is introduced. This interplay is central to an understanding of how project level expenses flow upward to the company-level financial documents.

Chapters 7 and 8 introduce the concepts of job costing and the cost account structure. As expenses are collected, the job-cost system can be implemented to organize these data in such a way as to be useful to management for control purposes.

The role of estimating as it relates to cost recovery is covered in Chapters 9 and 10. Chapter 11 discusses the importance of cash flow considerations in planning and control. Chapters 12 through 14 address project cost data collection and the generation of management reports. Chapter 15 discusses general and administrative overhead cost recovery and how to determine the break-even point. Chapter 16 presents the methods of analysis that are available to the manager for comparing actual field costs with the originally projected or estimated costs. The concepts of variance analysis and trend analysis are discussed.

The material in this textbook is appropriate to courses at the junior and senior levels in engineering and technology programs. It is relevant to management education courses that emphasize career preparation in the construction, architecture, engineering technology, and urban planning fields.

Many people have made valuable contributions to this book. Material presented throughout the text is based on research projects submitted over the past ten years by graduate students in the construction management program at the Georgia Institute of Technology. Research papers prepared by James Adamczyk, Eloy Benedetti, Luigi Blasi, Jose Hoyos, John Simpson, and Oliver Wager contributed excellent background material for several of the chapters within the text. Dr. R. Kangari used the original manuscript as a text in his classroom and made helpful suggestions regarding the method of presentation. Problems and questions throughout the text were enhanced by contributions from Dr. Kangari and Mr. Leonhard Bernold.

The original concept for the book came as a result of graduate work done by John Simpson, who has since established a successful construction firm and put many of the concepts presented here into practice.

Finally, I would like to recognize the support and patience of my wife during the long hours devoted to the preparation of this and other text books.

Daniel W. Halpin
Atlanta, Georgia

# Contents

*vii*

# Financial and
# Cost Concepts
# for Construction
# Management

# Chapter 1

# COMPANY-LEVEL FINANCIAL CONTROL

## 1.1  FINANCIAL ACCOUNTING CONCEPTS

The expression "the bottom line" is commonly used in characterizing the outcome of some undertaking. The concept of the bottom line comes from financial management. Specifically, it refers to the line or figure in a financial statement that sums up performance on the part of company management in reaching financial goals. It can be considered a grade or score reflecting how successful management has been in advancing the fortunes of the company.

Management of engineering and construction firms requires a wide variety of talent. Managers are somewhat like decathlon athletes. They must be good or preferably excellent in a wide range of activities. An individual may be an outstanding engineer, but if he has no knowledge of contracts or project control, he cannot be successful. Financial management and cost control are two areas in which the manager must have a thorough grounding. In most companies performance is evaluated in terms of the bottom line. It is in these terms that the manager's grade or score is calculated. Clearly, an athlete in competition must have a good understanding of the scoring system. Similarly, it is imperative that the manager understand the nature and function of financial documents and cost reports.

These documents not only reflect how well one is doing, but provide a basis for developing new strategies and obtaining feedback on the effectiveness of these strategies. The better the manager understands the accounting system both at company and project level, the more effective he or she will be at "reading" trends and taking appropriate actions. The objective of this text is to provide a basic grounding in these topics and to aid the engineering and construction manager in better understanding the framework within which financial and cost control can be exercised.

## 1.2  FINANCIAL DOCUMENTS

Two financial documents are key to the proper fiscal management of any firm. Many functions within both the financial and cost accounting systems are tied directly to the production of these two basic control documents. The two documents that are so

critical that they establish the structure of the financial and cost accounting systems
are

1. The balance sheet
2. The income statement

The basic equation that controls financial accounting is

$$\text{Assets} = \text{liabilities} + \text{net worth}$$

This equality is as fundamental to the accountant as $F = ma$ is to the engineer or
physicist. The balance sheet is a document that reflects the financial status of the firm
at a given time in terms of this basic equation. The conceptual layout of the balance
sheet is shown in Figure 1.1.

Figure 1.1 indicates that the summary of assets is listed on the left side of the
balance sheet. Two types of assets are of interest. Some assets are more readily
converted to cash and are called current assets. Others are more difficult to convert
to buying power, since they require sale or resale by another party. These are called
fixed assets. The following types of assets are considered to be quickly available and
therefore liquid:

1. Cash
2. Money in bank accounts
3. Accounts receivable (i.e., charges billed to a client and due within a short time)
4. Loans that can be recalled or transferred
5. Stocks and bonds that are negotiable

Some typical fixed asset categories are:

1. Real property
2. Construction equipment

**FIGURE 1.1   Layout of the Balance Sheet**

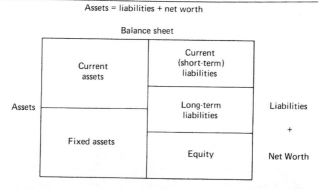

3. Office equipment
4. Long-term notes that cannot be readily recalled or transferred

On the right side of the balance sheet the sum of liabilities and net worth must equal assets on the left side. Liabilities are obligations that must be met either in the near term or the long term. Typical short-term liabilities are

1. Accounts payable
2. Subcontracts payable
3. Payroll payable
4. Taxes payable

Long-term liabilities are those for which payment is deferred. Notes with a bank to purchase equipment or to provide working capital are typical long-term liabilities. Property mortgages are also long-term obligations.

The basic function of an income statement is to measure the cash flow, or difference between revenues and expenses, over a defined period of time. Therefore it reflects the retained earnings for a period and becomes a companion document to the balance sheet. Its layout and relationship to the balance sheet are shown in Figure 1.2.

## 1.3  NET WORTH

Net worth is the value of the company. It consists of stockholders' equity and earnings. Simply put, if all liabilities are paid, what remains is the worth of the firm.

$$\text{Net worth} = \text{assets} - \text{liabilities}$$

If the assets of a firm are worth $1 million and the outstanding liabilities of the firm are $800,000, the net worth of the firm is $200,000. If 10,000 shares of stock are held by the stockholders, the value of the stock is approximately $20 per share. As the fortunes of the firm rise and fall, the apparent value of the stock fluctuates. In many cases, this "book" value of stock is not the traded value or the value stock brings

**FIGURE 1.2  Balance Sheet–Income Statement Relationship**

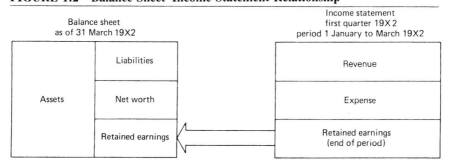

if it is sold. Book value is based on the assumed value of the firm's assets. The assumed asset value of a bulldozer may be $35,000. If, however, its owner is forced to sell it within the week, he may have difficulty in getting $20,000 for it. This would affect the stock value negatively. The $35,000 is, however, used in the calculation of net worth. Stock value is also strongly affected by the prospects of the firm. If it looks as if the firm will make a large profit in the near future, the stock buyer may pay more than the asset or book value per share to have ownership in an expanding firm.

To initiate a firm the principals normally contribute assets such as money, equipment, patents, and in some cases intangibles such as expertise. These assets are valued and become the initial net worth of the firm. These assets also constitute the owners' equity in the firm. As the firm operates, this equity either increases or decreases. The amount of increase or decline in value is a function of whether revenue generated is greater than the expenses accruing to the firm. This is measured periodically (e.g., monthly or quarterly), and becomes part of the balance sheet as *retained earnings*. This is a measure over time of the flow of income versus outgo (i.e., the cash flow). That is, for a given period,

$$\text{Retained earnings} = \text{revenue} - \text{expenses}$$

An expanded version of the basic accounting equation reflects this equality as follows:

$$\text{Assets} = \text{liabilities} + \text{net worth} + (\text{revenue} - \text{expenses})$$

If the company makes money during the period, the retained earnings are positive and the net worth increases. This leads to a compensating increase in assets (probably as cash in the bank) on the left side of the balance sheet. If the firm loses money, retained earnings are negative for the period and net worth decreases (i.e., liabilities increase in proportion to assets). A typical balance sheet for the Fudd Associates, Inc. is shown in Figure 1.3. The basic portions of the balance sheet are indicated.

Revenues are listed at the top of the income statement and derive from progress payments made by clients, revenues from equipment rental, and other income-generating activities. Expenses, including payroll, material, equipment, and other direct costs, overhead costs, and general administrative costs are listed below revenue items. Revenues minus these expenses yield income before taxes. Taxes must be subtracted, which further reduces income. The after-tax balance represents the retained earnings for the period. A simplified income statement is shown in Figure 1.4.

## 1.4   COMPONENTS OF THE BALANCE SHEET

Figure 1.5 graphically shows the various items that make up a balance sheet. Each of the slices of this pie-shaped diagram can be thought of as changing shape dynamically as a function of time. That is, each of the slices expands or contracts as time varies (causing, for instance, receivables to be converted to cash). From one instant to another, the contribution of any one of the components of the balance sheet will change. The objective of preparing the balance sheet is to make a "snapshot" of these components

**FIGURE 1.3  Balance Sheet, Fudd Associates, Inc. (31 December 19X2)**

| | | | |
|---|---|---|---|
| *Current Assets* | | | |
| Cash in bank and on hand | | | $ 56,753 |
| Accounts receivable | | | 153,112 |
| Inventories | | | 214,102 |
| Other current assets | | | 4,786 |
| Total current assets | | | $428,753 |
| | | | |
| *Fixed Assets (at cost)* | | | |
| Land and buildings | $ 54,123 | | |
| Equipment | 218,277 | | |
| Less accumulated depreciation on land and buildings | | 10,825 | |
| Less accumulated depreciation on equipment | | 152,205 | |
| Total fixed assets at book value | | $109,370 | |
| All other assets | | 1,247 | |
| Total assets | | | $539,370 |

| | | | |
|---|---|---|---|
| *Current Liabilities* | | | |
| Accounts payable | | | $ 49,882 |
| Notes payable—banks | | | 31,254 |
| Advance billings on uncompleted contracts | | $1,792,402 | |
| Less: Cost of contracts in process | | 1,639,441 | |
| Total current liabilities | | | $234,097 |
| | | | |
| *Long-Term Liabilities* | | | |
| Loans payable | | | $ 60,000 |
| | | | |
| *Net Worth* | | | |
| Capital stock | | | $200,000 |
| Retained earnings | | | 45,273 |
| | | | |
| Total liabilities and net worth | | | $539,370 |

5

**FIGURE 1.4   Simple Income Statement (Third Quarter, 19xx)**

| | |
|---|---:|
| Revenue from construction contracts | $412,510.00 |
| *Cost of contracts* | |
| Subcontracts and supervision of subcontractors | 347,891.00 |
| Field overhead | 2,063.00 |
| Gross profit | $ 62,556.00 |
| *Administrative and General Expenses* | |
| Office operating cost | $ 10,921.00 |
| Information costs | 650.00 |
| Bidding costs | 1,733.00 |
| Interest on existing loans | 900.00 |
| Earnings before federal income taxes | 48,352.00 |
| Federal income taxes | 17,411.00 |
| Net earnings | 30,941.00 |
| Retained earnings at beginning of period | 208,422.00 |
| Retained earnings (end of period) | $239,363.00 |

at a given time to reflect the financial condition of the firm. This snapshot is normally made at the end of some fiscal reporting period, such as the fiscal year.

It is desirable to have various fiscal-year-end statements for comparative analysis. One reason for requiring a fiscal-year-end statement is that one complete operational cycle may be compared with another. Moreover, if a contractor uses a period other than the fiscal year for his statement, he could present a statement prepared as of the

**FIGURE 1.5   Balance Sheet Items**

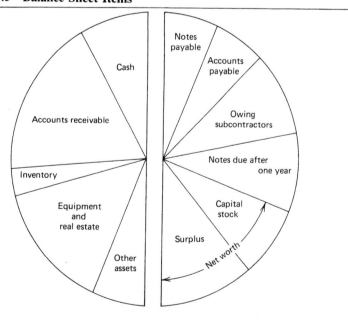

day most advantageous to him. The statement as of that date could reflect a much better position than usual, but in itself would not give a complete and accurate picture of the business.

The balance between assets, liabilities, and net worth is represented in Figure 1.5 by the two equal semicircles of the pie diagram. As noted earlier, the left-hand portion of the pie contains all of the assets of the business. They may grow or shrink in value, depending on current conditions. For example, a reappraisal of equipment may cause the assets to increase, whereas the uncollectibility of an account receivable would cause the assets to shrink.

The right-hand portion of the diagram represents what the business owes. That is, it reflects the claim on the assets by both the creditors and owners. The amount owed to creditors is a liability, and that portion which reflects the owners' claim on the assets is called net worth or owners' equity. In this instance, the capital stock and surplus (i.e., the accumulated retained earnings) are the net worth.

## 1.5   CURRENT ASSETS

Cash, earned estimates, accounts receivable, and inventory are *current assets* since, with the exception of cash itself, they are in varying states of being converted into cash.

According to the American Institute of Certified Public Accountants (AICPA), the definition of current assets is as follows:

> For accounting purposes, the term Current Assets is used to designate cash and assets or resources commonly identified as those which are reasonably expected to be realized in cash or sold or consumed during the normal operating cycle of the business.*

A current asset, therefore, would mean cash or an item convertible into cash within the normal operating cycle of the business. The operating cycle is defined as

> The average time intervening between the acquisition of materials or services entering [the production] process and the final cash realization . . .†

Another term that is used in place of current asset is *liquid asset*. In other words, a liquid or current asset is one that can be quickly liquidated into cash to discharge a liability of the business.

There is no question about cash being a liquid asset. Accounts receivable that are due and collectible within a short time are also considered good current assets. The portion of accounts receivable, however, that represents advance billings would not

---

* American Institute of Certified Public Accountants, Accounting Research Bulletin No. 43 (1953), p. 21.

† American Institute of Certified Public Accountants, *Audit and Accounting Guide—Construction Contractors,* prepared by the Construction Contractor Guide Committee (1981), p. 46.

be considered a current asset, since it represents that which must be used to liquidate a liability that is not yet shown on the balance sheet. It represents money held in trust by the contractor that was collected in advance of work that has yet to be performed. Receivables from completed contracts are current assets, if it is determined that the owner has the ability to pay.

Retainage is also considered a liquid asset if it is due and collectible within the operating cycle. A long-term contract retainage is not considered a current asset in the contractor's statement, since it will not be received during the normal operating cycle and will not be available during that time for the payment of current obligations.

Materials on hand (inventory) are not always considered current assets because they are the least liquid component of the current asset group.

## 1.6   FIXED ASSETS

*Fixed assets* are tangible assets of a permanent nature. Most companies find it necessary to make sizable investments of this kind that will result in incomes not only in the period of purchase but in future periods as well. Fixed assets are shown at purchase price less depreciation. That is, they should be shown at book value. Book value, in this context, means the depreciated value shown on the contractor's books and not the value used for appraisal or lending purposes.

If the property is appraised at a price greater than that shown on the balance sheet, then the reappraised value on the difference between the cost and the appraisal value may be shown separately on the asset side, provided that it is clearly shown as a reappraisal of the asset and an offsetting entry is shown on the net worth portion of the statement.

Fixed assets are important to the analyst to indicate a company's secondary cushion or borrowing power in the event that it is needed. A contractor who has a greater equity of fixed assets is a better risk than one who may have the same amount of working capital but who lacks borrowing power. Working capital is defined as the difference between current assets and current liabilities. That is,

$$\text{Working capital} = \text{current assets} - \text{current liabilities}$$

## 1.7   MISCELLANEOUS ASSETS

There are other assets that are shown on the balance sheet, in addition to the current and fixed assets. Some of the items usually shown on contractors' statements are

*1. Intangible Assets.*   Intangible assets are items such as goodwill, patents, copyrights, and organizational expense. These are items that have value to a going business but are not used in the normal sense to pay bills and can rarely be used as collateral for a loan. They are assets that are not available for the payment of debts of a going business. In analyzing the net worth ratios,* the intangible assets are excluded from the net worth figure.

*2. Stocks and Bonds.* On many occasions stocks and bonds on a contractor's financial statement are carried as current assets. To be properly classified as a current asset, the securities must represent invested surplus cash. Securities must be listed on one of the major stock exchanges and must be stocks that can be liquidated within a short time when cash is needed or stocks that may be used as collateral for a loan.

*3. Prepaid Expenses.* These items represent expenses that have been paid in advance. They may include such items as prepaid rent, taxes, insurance, interest, and many other items. They are not current assets in the usual sense that they will be converted into cash during the normal operating cycle. If they were not prepaid, they would require the use of the current assets, cash, during the accounting period. These amounts are usually small in relation to the total current assets and from a practical standpoint, it is agreed that prepaid expenses are a noncurrent asset.

## 1.8  LIABILITIES AND NET WORTH

The term *current liability* is used principally to designate an obligation whose liquidation is reasonably expected to require the use of existing resources properly classified as current assets, or the creation of the other current liabilities.

On the right-hand portion of the balance sheet, items such as notes payable, accounts payable, and subcontracts payable are listed as current liabilities. There are other current liabilities, such as taxes payable, accrued payroll,† advance payments and other accruals such as rent and insurance.

Some other accrued liabilities are

Payroll taxes and insurance
Union benefits
Sales tax
Accrued federal withholding
Accrued state withholding
Accrued vacation plan
Accrued federal unemployment insurance
Employers FICA
Rent
Payroll
Health and welfare fund

* See discussion of ratio analysis in Chapter 6.

† See Appendix B, Transaction No. 10.

Accrued liabilities are obligations that are incurred (often based on expenditure of resources or the passing of time) for which no formal billing document is received. All of the liabilities listed are those that the contractor admits to owing for various reasons, whether or not they are documented.

In the liquidation of current liabilities, the creation of another current liability could occur. This happens, for example, when a bank loan is obtained to pay off accounts payable to suppliers. The transaction does not involve the use of a current asset.

A *long-term liability* is an obligation that normally will not be paid within approximately one year of the balance sheet date. They range from unsecured notes to first mortgages. Long-term liabilities are incurred to finance additional plant, equipment, or land, to obtain additional working capital, to meet a current debt, or to pay off another long-term debt.

The capital accounts, which consist of capital stock and surplus, are the *net worth* of a corporation. Capital stock includes stock issued and outstanding. This is the original investment in the business in exchange for capital stock. The investment may be in the form of cash paid in or capital equipment. Surplus includes earned surplus (i.e., retained earnings), although this could sometimes be a negative figure. "Paid in" surplus represents money put in the business but for which no capital stock was issued or note taken for the money. In other words, it is money invested but not loaned.

## 1.9   THE INCOME STATEMENT

It can be seen that the balance sheet is a list of the assets and liabilities as of a given day. The balance sheet only shows the condition of the firm on that day and does not indicate whether the contractor is making or losing money. The measure of progress or lack of progress is shown by the income statement.

The income statement shows the actual net profit or loss for a given time period. In addition, it gives other information regarding revenues and expenses. Revenues measure the inflow of net assets (assets less liabilities) from selling goods and providing services. Expenses measure the outflow of net assets that are used up, or consumed, in the process of generating revenues. As measures of earnings performance, revenues reflect the services rendered by the firm and expenses indicate the efforts required or expended.

## 1.10   REQUIREMENTS FOR FINANCIAL STATEMENTS

Construction contractors are required by income tax laws to have annual balance sheets and income statements prepared for their firms. To supply required information to their bankers, bonding agents, and insurers, interim reports are often required. Larger firms have balance sheets and income statements prepared at regular intervals (quarterly or monthly) as a matter of course.

For the evaluation of any construction firm's financial position, the firm's bankers, bonders, and insurers, and in some cases even prospective clients, focus almost exclusively on the firm's financial statements. From an analysis of these statements, financial institutions decide the risk involved in doing business with the contractor and prospective clients judge his ability to undertake certain projects.

Industry ratios prepared from financial statements are the most accessible sources of information a construction manager has on the market in which he is competing. If his competitors are public companies, their financial statements are a readily available source of information.

The financial statements of a firm are nothing more than photographs of the firm's financial position at the point in time shown for the balance sheet or for the period just ended for the income statement. The construction manager can turn these still pictures into a moving picture by analyzing trends in consecutive financial statements. These trends can also be extrapolated and used for the prediction of the future. The construction contractor can learn a great deal about his firm by comparing its financial trends with industry standards and those of his competitors.

## REVIEW QUESTIONS AND EXERCISES

**1.1** At the end of the fiscal year 19X2, Rainer Construction Co., Inc. has the following financial profile:

| | |
|---|---|
| Total assets | $1,200,000 |
| Total liabilities | 850,000 |
| Total revenue before taxes | 300,000 |
| Total expenses | 265,000 |

(a) What is the total net worth of this company?
(b) Did the new worth increase, decrease or remain the same compared to FY 19X1? Why?
(c) What is the income after taxes? (Assume a 50% tax rate.)

**1.2** Define in your words the difference between a balance sheet and an income statement.
**1.3** The following balance sheet figures are available on the Cougar Construction Company:

| | Current | Long-Term | Total |
|---|---|---|---|
| Assets | $100,000 | $100,000 | $200,000 |
| Liabilities | 80,000 | 70,000 | 150,000 |

Find the total net worth and working capital of the company.
**1.4** The following data regarding Atlas Construction Company are available:

| | |
|---|---|
| Current assets | $300,000 |
| Current liabilities | 200,000 |
| Long-term liabilities | 500,000 |
| Total net worth | 200,000 |

(a) What is the value of the company's fixed assets?
(b) What is its working capital?

**1.5**  Name three accounts typical of the liability ledger.
In what ledger would you enter "discounts earned?"

**1.6**  How are the balance sheet and income statement related? Discuss the pattern of information flow between these two documents.

# Chapter 2

# CONVENTIONS IN ACCOUNTING

## 2.1 ACCOUNTING CONVENTIONS USED IN CONSTRUCTION

An understanding of the basic conventions used in the preparation of a financial statement is necessary to properly evaluate financial statements. The main conventions concern:

1. Bases of accounting used
2. Methods of income recognition used

The bases of accounting commonly used in the construction industry are:

1. Cash basis
2. Accrual basis

The methods commonly used for income recognition are:

1. Billings method
2. Percentage of completion method
3. Completed contract method

The relationships between these conventions are shown schematically in Figure 2.1. Within certain guidelines published by the Internal Revenue Service (IRS) and professional organizations such as the American Institute of Certified Public Accountants (AICPA), firms are free to use these conventions in the preparation of their financial statements. It is, however, required by law that a *convention of consistency* be used. As the name implies, the convention of consistency requires that once a business transaction is accounted for in a particular manner, it must be accounted for in this same way consistently thereafter. This doctrine makes it very difficult for a business to manipulate its figures by showing them on one occasion in a manner that sheds a favorable light on the result, and then, when it is convenient, changing to another approach.

*13*

**FIGURE 2.1   Accounting Conventions**

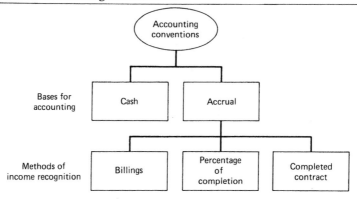

At various points in time, it is necessary to reflect the financial performance of the firm. As will be discussed in Chapters 4 and 5, at such times the company accounts are closed and a calculation is made of the income generated during the period since the last closing. The process of calculating the amount of income generated in a given period is called *recognition of income*. The way in which this calculation is made is a function of the method of income recognition. The method used for recognizing income is very important since it affects the amount of taxes that must be paid and the perception of the firm's success in making a profit.

Using the cash basis simplifies income recognition, since income is recognized on the basis of the actual transfer of monies. Monies actually received constitute revenue and monies actually disbursed constitute expense. The difference between physically transferred receipts and expenditures in a given period constitutes income. No attempt is made to match revenues with the costs that generated them. Therefore, the expenses reported in a given period are not directly linked to the revenues reported. That is, some of the revenues reported may have been generated by expenses reported in a previous period, and some expenses reported will create revenues in future periods. As a result, the value of this information for management purposes is limited.

Recognition of income when the accrual method of accounting is used is more complex, since a linkage is maintained between the expenses incurred and the revenues created by them. The times at which revenues are billed and expenses are incurred become the basis for income recognition. Good financial management requires the use of the accrual method, since this basis does attempt to match revenues with the expenses incurred in generating them. In order to gain a good understanding of the manner in which operations result in the production of income, it is essential that the link between revenue and expense be available for management review. Depending upon the needs of individual companies, any one or a combination of the three methods noted can be used to recognize income in accounting for project income under the accrual basis.

The conventions used have a very definite influence on the income of a company as reported for a given period and therefore merit considerable attention. The way in which these conventions are applied is also sensitive to the fact that much of the revenue and expense in construction occurs in a project format. Consequently, income recognition in the construction industry tends to be unique.

## 2.2 CASH BASIS

The *cash basis* accounts for expenditures at the time funds are actually received or disbursed. That is, a cash transaction is recorded in time when cash is transferred (received or paid out). This method can be used for both short-term contracts (contracts completed within one accounting period) and long-term contracts (contracts that run over more than one accounting period). The cash method is used in maintaining the balance of a personal checkbook account. When a check of payment is written, an entry is made showing the date and amount of the disbursement. A reduction of the checking account balance occurs. When a check or payment is received and paid into the account, it is recorded in the check register and the balance is increased. The method reflects increases and decreases in account balance at the time funds are transferred.

The cash method has the advantage of being simple and straightforward for many accounting situations (e.g., personal checkbook maintenance). A small contractor (who receives payment from customers and pays for materials and laborers as needed) will find the cash method adequate for his requirements. The cash method is, however, subject to manipulation and does not always reflect the true state of a firm's finances. By depositing incomes and deferring payments to workers and materials suppliers, the contractor can appear to be in a very strong position when in fact he may be in debt. Assume, for example, that income for work performed on three jobs in the amount of $6250 is deposited. The billings from suppliers and for labor amount to $7000. By deferring payment of the costs of the work, the contractor appears to have made a neat profit when in fact he has lost money on the work.

As a second illustration, assume that Fudd Associates, Inc. submits a billing to the client for work completed in December and receives a check on 10 January in the amount of $27,000 for work accomplished. Costs associated with work to include payroll and billings by subcontractors and vendors are $29,000. In fact, Fudd Associates is in a deficit position on the project at this point. However, if payment to the vendors and subcontractors is delayed, Fudd Associates will appear to be making a profit on the project. If the payroll for the month of December amounts to $10,000 and is paid while payment to subcontractors and vendors is delayed, the income from the project will appear to be $17,000. Until payments are actually made to cover expenses other than payroll expenses, these additional costs are not recognized. For this reason, financial statements prepared by this method are often termed inadequate for surety and credit purposes.

The cash method is commonly used by many small construction contractors because of its simplicity and because tax liability (for incomes from contracts) is recognized only when actual payment has been received. As will be seen later, under other methods tax liability is incurred at the time the income is recognized. This means that the contractor may be paying taxes on income that he has not actually received. The question of incurring revenues and expenses at such time as to minimize tax liability is central to the selection of accounting methods by large corporations. This is the province of the tax consultant. Firms with large amounts of revenue and expenditure (cash flow) are very sensitive to the impact of revenue and expense recognition.

In summary, cash basis is a good approach for small firms that operate on a pay-as-you-go basis. It has certain tax advantages but can be manipulated in such a way as to reflect a deceptive picture of the firm's true financial position. It is inadequate for management purposes, since it does not match revenues to the expenses incurred in generating them for income reporting.

## 2.3   ACCRUAL BASIS

The *accrual basis* is founded on the concept that an accounting transaction is recognized for accounting purposes at the time an obligation for payment is incurred. This means that an account reflects revenue from a billing to a client at the time the bill or invoice is prepared and presented (regardless of when payment is actually received). Similarly, the account reflects an expense arising from a bill received from a supplier at the time the invoice or request for payment is received. This means that obligations are recognized for accounting purposes at the time they are incurred rather than at the time of payment.

Properly administered, the accrual basis minimizes the kind of manipulation possible when the cash method is used. In the example of the contractor with $6250 in work performed for customers, the revenues from this work would be logged into the accounting system at the time the billings are sent to the three customers. The billings from the materials suppliers and the labor charges would be recognized when incurred; that is, the expense associated with the materials suppliers would be recognized upon receipt of billings from these vendors. The requirement to pay the labor force would be recognized at the end of the weekly pay period (whether or not paychecks are distributed). A better and more indicative linkage of revenues and expenses for income recognition purposes is achieved.

The accrual basis has the disadvantage of requiring a more extensive set of accounts and more bookkeeping time. For example, the accounts receivable and payable referred to in Chapter 1 are not required when the cash basis is used. These accounts are very active under the accrual approach, since they reflect revenues and expenses incurred but not yet received or paid. The differences (from a bookkeeping point of view) between the cash and accrual bases are illustrated in Chapter 4.

A second disadvantage inherent in the accrual method is tied to the concept of retainage commonly used with construction contracts. Clients typically hold back a certain percentage of the amount billed by the contractor until such time as the project is successfully completed. This retained amount is normally not received until the end of the job. Under the cash approach the retainage is not booked until received. Using the accrual basis, however, all of the billed amount is recognized as revenue at the time of billing. The retainage portion of billing is therefore recognized as revenue (and is subject to taxation) even though it may not be received until many months after it is recorded.

Finally, since the accrual basis results in a better linkage between revenues and expenses for income recognition, it is more difficult for the resourceful tax planner to manipulate cash flows so as to minimize tax liability.

In summary, the accrual basis reflects a more accurate picture of a firm's financial position. It can, however, result in a higher tax liability compared to the cash basis, and generally the accounts are more complex and require more effort to maintain.

## 2.4   BILLINGS METHOD OF INCOME RECOGNITION

When the accrual basis of accounting is used, any of the three methods of income recognition given in Section 2.1 is acceptable. Under the billings method, income is calculated for a given period by viewing the amounts billed as revenues and the costs incurred as period expenses. Project-related income is calculated by summing contract billings for the period and subtracting the period contract expenses incurred. This method provides a good reflection of income earned in certain product- and service-related companies. If an engineering design firm, for example, bills services to clients during a given period, the difference between the accumulated sales (i.e., billings) and the costs to the firm for the services provided establishes an acceptable stipulation of income earned.

The assessment of income earned on a construction project is more complicated, since the "sale" is, in effect, occurring across the life of the project. That is, the project itself is a large end-item that will be paid for in increments throughout the period of construction. If the project extends over several accounting periods, the question as to how much income should be recognized in a given period must be addressed.

If the accrual basis of accounting is being used, it is possible to sum billings to the client during a given period and view this as revenue for income recognition purposes. However, on longer-term contracts use of the billings method can lead to deviations from the actual revenues earned based on work performed. That is, the amount *billed* may not be the amount *earned*.

The billings that are submitted by the construction contractor are called *progress payments*. These partial payments for a given period (usually a month) of the total amount bid are calculated by estimating the percentage of the work performed in a given pay period. The client then pays the corresponding percentage of the total bid price for the pay period just completed. Many factors (including the accuracy of the percentage associated with the progress payment amount for a given period) can cause a condition referred to an *overbilling* or *underbilling*. That is, when revenue is based on progress payments billed to a particular point in project, the revenue (and associated income) billed may be either greater than or less than the revenue earned. If the cumulative amount billed is greater than the revenue earned, an overbilling has occurred. If the cumulative amount billed is less than the revenue earned, an underbilling condition exists. Therefore, the billings method does not reflect the true financial position of the project (and the firm) because the income earned has been misrepresented.

This problem develops on contracts that run over many progress payment periods. That is, the problem of over- and underbilling is typical of the long-term (multi-period) projects which are commonly found in the construction industry.

## 2.5  PERCENTAGE-OF-COMPLETION METHOD OF INCOME RECOGNITION

In order to better reflect the actual revenue earned on a particular project at a given time, the *percentage-of-completion* (POC) method can be used. In the percentage-of-completion method, gross revenue is recognized on each contract in proportion to the progress made on that contract. The revenue (and income) generated based on this progress is recognized whether or not it is billed to the client. The key criterion for recognizing it is that it has been *earned*. This gives a better indication in the financial statement of the firm's position and is therefore recommended for financial statement preparation.

Several methods are used in practice to establish the percentage completed in terms of value earned at a particular point in the project's life cycle. These are

1.  Cost-to-cost method
2.  Effort-expended method
3.  Units-of-work-performed method

A method called units-of-delivery is a modification of the POC method of income recognition.

The cost-to-cost or cost-completion method calculates the percentage complete on the basis of the following ratio:

$$\frac{\text{Cost incurred to date on project}}{\text{Estimated total cost of project}} \times 100$$

Another way of calculating this ratio is

$$\frac{\text{Cost incurred to date}}{\text{Cost incurred to date } + \text{ estimated cost to complete}} \times 100$$

In the effort-expended method, the percentage is based on the ratio of units of resource effort (e.g., man-hours or similar measure) to the total effort estimated to complete. This ratio can be expressed in terms of labor hours or labor cost. Based on man-hours, the calculation could be made as follows:

$$\frac{\text{Man-hours to date expended on project}}{\text{Total estimated man-hours on project}} \times 100$$

or, in terms of labor cost, as follows:

$$\frac{\text{Labor cost incurred}}{\text{Labor cost incurred } + \text{ estimated labor cost to complete}} \times 100$$

This method of calculation tends to offset distortions in the cost-to-cost method which occur when materials are delivered to the site (and therefore cost is incurred) but are not installed (i.e., the revenue is not yet earned).

The problems inherent in both the cost-to-cost and effort-expended methods relate to the fact that inefficient work during the early phases of the project (and the consequent higher expenses and/or resource expenditure) generates apparently higher revenues by increasing the numerator of the ratios above. This tends to inflate earned revenue amount.

To correct for inflated earned revenue, the percentage of completion can be calculated by using physical units of production as the basis for calculation. Using the *units-of-work-performed* method, we find that the POC ratio becomes

$$\frac{\text{Units of work performed to date}}{\text{Estimated total units of work in project}} \times 100$$

On a heavy construction earthwork job, this ratio might be

$$\frac{\text{Cubic yards of material excavated to date}}{\text{Estimated total cubic yards in project}} \times 100$$

This *physical completion* method of determining the POC is appropriate on projects where a single unit of work (e.g., yd$^3$, lineal foot) can be identified as representative of the work to be accomplished on the total project. It does, however, assume that revenue is generated and earned in proportion to the extent of physical completion of the selected work item. This method neglects earned revenue associated with mobilization and demobilization of the job. For instance, if extensive haul and access roads must be constructed before the first lineal foot of pipeline can be constructed, using the footage of pipeline as the work unit would result in no revenues earned during the extended period of access road construction.

The *units-of-delivery* method is described in the American Institute of Certified Public Accountants (AICPA) audit guide, *Audits of Government Contractors,* as follows:

> The units-of-delivery method recognizes as revenue the contract price of units of a basic production product delivered during a period and as the cost of earned revenue the costs allocable to the delivered units. . . . The method is used in circumstances in which an entity produces units of a basic product under production-type contracts in a continuous or sequential production process to buyer's specifications.*

This is a modification of the *units-of-work performed method.* It is appropriate to firms such as precasters providing architectural or structural elements for a given construction project.

Clearly all of the methods used to determine the percentage of completion are based on estimates. The accuracy of those estimates varies with the project and with different stages of completion of that project. Therefore, the accuracy of the estimated revenues will vary and must be updated.

The AICPA Guide for Construction Contractors' Statement of Position† states:

---

* *Audit and Accounting Guide—Construction Contractors,* prepared by the *Construction Contractor Guide Committee* (1981), p.111.

† See paragraph 23, Statement of Position 81-1, "Accounting for Performance of Construction-Type and Certain Production-Type Contracts," July 15, 1981, included as Appendix I in AICPA *Audit and Accounting Guide—Construction Contractors.*

The use of the percentage-of-completion method depends on the ability to make reasonably dependable estimates.

In general, the POC method is preferred and should be used unless the unique and risk-associated aspects of a particular project are so significant as to preclude development of reliable estimates (e.g., in some subsurface work, in unique high-technology facilities, etc.).

## 2.6   COMPLETED-CONTRACT METHOD OF INCOME RECOGNITION

On projects that are typical for a contractor and on which the scope is well defined, it is normal to expect that the costs of construction can be estimated within an accuracy of ±3 percent. For instance, a contractor specializing in the construction of school buildings and educational facilities has available data from previous projects and vendors which allow a very accurate estimate of costs if he has a full and adequate set of plans and specifications. As just noted, the percentage-of-completion method is the preferred method of income recognition for projects on which reliable estimates of cost and income can be made. More and more, however, contractors are required to bid on projects that have unique and unprecedented features. Even in the best case, cost data for such projects must be extrapolated from data on projects that are considerably different in terms of scope and size. In some cases, particularly in subsurface work such as tunneling and caisson construction, many unknowns regarding the belowground conditions exist, so reliable estimates of production and cost are extremely difficult. In cases where it is expected that actual costs and revenues may be difficult to estimate, the *completed-contract method* of account may be used.

The basic idea underlying the completed-contract method is that the recognition of income is deferred until the end of the project. At the end of the project, actual as-built information regarding the costs incurred and the revenues generated is available. Therefore, a "settling up" can occur, which gives the most accurate evaluation of income generated by the project. The determination of project income, therefore, is no longer dependent upon estimates. A precise calculation of income generated can be made by subtracting total project expense incurred from total revenues received. Since this recognition of income is deferred until the end of the project, tax liability is not established until the contract is completed.

Although income is not recognized until the end of the project, revenues and expenses are accounted for as they occur and are entered into appropriate accounts on an accrual basis. As will be seen in the illustration of Section 2.7, all accounting transactions on the job are recorded as they occur, but no calculation of income (i.e., income = revenue − expense) is made until the end of the job. This is in contrast to the billings and POC methods, which recognize income at the end of each fiscal or accounting period (typically end of the quarter or year). In the completed-contract method, revenues received by contractors from clients are considered as "advances" against the ultimate successful completion of the work. As such, they are booked in such a way as to indicate an obligation or liability on the part of the contractor. These accumulated advances are ultimately cleared at the end of the project when the contract is accepted as complete by the client.

The project expenses are considered to be contractor-owned assets until the project as a completed entity is transferred to the owner. The accumulation of expenses in appropriate accounts indicates the asset value accumulating in the project. Obviously, at any time (such as the end of an accounting period) a comparison of the revenues with accumulated expenses can be made to see whether the contractor appears to be making or losing money on the project. The final determination of income (either plus or minus) for tax purposes, however, is not made until the contract is completed.

## 2.7 THE SCULPTOR AS CONTRACTOR—COMPARISON IN A DIFFERENT CONTEXT

To illustrate the completed-contract (CC) method concept in a slightly different context, consider the following parallel. Assume that a famous sculptor has received a commission to do a bronze bust of the President of the United States. The government enters into contract with the sculptor to complete the project for $80,000. Assume that the bust is to be unveiled 18 months from the date of signing the contract. During the life of the project, the sculptor will receive partial payments towards the total contract amount. These are considered advances against completion of the work and establish an obligation or liability on the part of the sculptor to complete the bust as contracted. During the development of the work the sculptor's expenses for his time, labor, materials, and other items accumulate in an account that represents the cost associated with the project. Until the bust is presented in fulfillment of the contract, (from a completed-contract point of view) it is owned by the artist and the expenses reflect its value as an asset to him. On the other hand, the advances received represent liabilities on the part of the artist that will ultimately be offset by the presentation of the completed work.* At the time the work is transferred to the client (i.e., the government in this case), the total accumulated revenue of $80,000 is compared to the expenses incurred, and a one-time determination of income earned by the sculptor is made. If the $80,000 is greater than expenses incurred, the artist has a positive income, and he pays tax on this income. If expenses exceed the contract amount, the artist has experienced a negative income (loss) and pays no tax. In effect, he has produced something more valuable than the contract price to which he originally agreed. This one-time fixing of gain or loss is determined at the end of the contract.

It should be emphasized that the CC method enables the sculptor to know or to determine at any point in the contract what his revenues and expenses are. Therefore, he knows basically the value of his work to date and the revenue or advance received to date. However, a determination of final income occurs only when the work is transferred to the client at the end of the 18-month contract period.

Looking at this same situation if the POC method is used, assume that the contract

---

* The assumption is that the sum of the advances equals the original contract price of $80,000 plus any changes in the original contracted amount that occur.

is signed on 1 November of 19X4. The sculptor must pay taxes on his 19X4 income (income during the period 1 January to 31 December 19X4). Using the POC method, he will not defer recognition of income to the completion of the project in 19X6. He makes an estimate of his progress achieved during the period November to December 19X4. Assume that he estimates that he has completed 5 percent of the total contract.* On the basis of the 5-percent progress, he has earned $4000 of the total contract price during 19X4. Assume that his expenses to date (for travel, research, etc.) are $2000. He will report an income developing from this project of $2000 and pay tax on this amount.

Now assume that time has moved forward to 31 December 19X5. The sculptor is still working on the project and estimates his total progress towards completion is 75 percent. His expenses to date are $35,000. On the basis of 75-percent progress, he has earned $60,000 of the total contract. Therefore his cumulative income recognized on the project is $25,000. He recognized $2000 of this income during his tax year 19X4. Therefore, he recognizes $23,000 of income for tax purposes during 19X5.

He finishes the project on time in June of 19X6. The unveiling occurs at the White House on 1 July 19X6 and the President congratulates him on his outstanding achievement. He receives a $10,000 prize from a private foundation for his artistic achievement. His final expenses associated with the project amount to $50,000. He has a total income of $30,000 related to the project plus the $10,000 prize. He has paid taxes on $25,000 of this amount during 19X4–19X5. Therefore, his taxable income for the project during 19X6 is $15,000.†

This example indicates the income recognition differences inherent in the POC versus the CC method. Using the CC method, the artist would recognize an income of $30,000 on the project at the end of the contract and pay taxes only in 19X6. Under the POC method, income would be recognized at three individual points in time (i.e., for the three fiscal years during which the contract is in effect).

One method approaches income recognition on a period-by-period basis, whereas the other method recognizes income at a single point in time.

To summarize, when the completed-contract method is used, the income of each contract is recognized only when that contract is completed. Costs and billings are accumulated but not credited to the income statement until contract completion. The contractor has a choice of allocating his general and administrative expenses to contracts in progress or charging them directly to the income statement for the fiscal period in which they are incurred. As explained in Section 2.9, general and administrative expenses are costs not directly incurred in constructing the project, but required for running the firm. For the sculptor these costs might include fees paid to his agent for procuring work, his accountant's fee for preparing his income tax return, and the cost of the telephone in his studio.

The main advantage of this accounting method is that tax payments are deferred

---

* He may have used cost-to-cost or whatever estimating method he felt was appropriate. This is his estimate of *earned* revenue. The amount he has *billed* for the work may be and most probably is different from this estimate of earned revenue.

† Taxes, of course, are paid on the $10,000 prize as well as the earnings on the contract.

until the date of contract completion when the contractor receives his retainage, thereby augmenting his working capital. Profit calculation is based on actual contract results, thus eliminating estimates which may overlook unforeseen future costs and other unforeseen losses. This method also allows management to influence or manipulate taxable income by deliberately hastening or deferring the completion of contracts. The disadvantage of the method is that it may result in irregular recognition of income and, therefore, may generate a very high taxable income in one year and a very low taxable income in a following year.

## 2.8 TYPICAL PROJECT-RELATED ACCOUNTS

Since the end item in the construction industry is produced in the context of a project format, the revenues and expenses generated by a construction company are primarily project related. Some of these expenses are directly related to the physical elements of the project (e.g., concrete, electrical boxes, etc.) and others to the supervision or control of the field production. The costs directly associated with placement of a particular physical component of the project are called *direct field costs*. They consist of the labor, material, and equipment costs to include associated burdens (e.g., FICA, sales taxes, and the like), which are required to place a physical piece of construction. Costs that pertain to the direct supervision and control of the project at the job site are referred to as field or *project overhead costs*. They are also sometimes called *field indirect costs*. These are all costs associated with maintaining a field supervisory staff on site. They include such costs as the renting of a field office or maintenance of a trailer on site, site-related telephone and utility costs, the salary of supervisory staff, and so on.

Some typical direct cost accounts associated with a building construction project are shown as account items number 100 through 685 in Figure 2.2. Typical project overhead accounts are shown in the same figure as line items 700 to 950. All of these project direct and overhead expenses are accounted for by using one of the methods discussed previously.

## 2.9 ACCOUNTING FOR HOME-OFFICE SERVICE SUPPORT (GENERAL AND ADMINISTRATIVE COSTS)

In addition to the project expenses, other costs are associated with home office staff. In order to support ongoing projects, a staff is commonly maintained at regional and/ or home offices. This staff handles many centralized functions such as procurement, estimating, project scheduling, and other company level activities. In addition to this staff, the company officers can be thought of as support staff. Which functions are accomplished on a centralized basis is a management decision. Because of their nature it is more convenient to account for these company-level overhead functions separately from the project accounts. It would be tedious and complicated for the secretary to the president to distribute his or her time to the individual projects under construction. This is generally true for many centralized service support personnel. Therefore, such

## FIGURE 2.2 List of Typical Project Expense (Cost) Accounts

*MASTER LIST OF PROJECT COST ACCOUNTS*
*Subaccounts of General Ledger Account 80.000*
*PROJECT EXPENSE*

| | *Project Work Accounts* 100–699 | | *Project Overhead Accounts* 700–999 |
|---|---|---|---|
| 100 | Clearing and grubbing | 700 | Project administration |
| 101 | Demolition | .01 | project manager |
| 102 | Underpinning | .02 | office engineer |
| 103 | Earth excavation | 701 | Construction supervision |
| 104 | Rock excavation | .01 | superintendent |
| 105 | Backfill | .02 | carpenter foreman |
| 115 | Wood structural piles | .03 | concrete foreman |
| 116 | Steel structural piles | 702 | Project office |
| 117 | Concrete structural piles | .01 | move in and move out |
| 121 | Steel sheet piling | .02 | furniture |
| 240 | Concrete, poured | .03 | supplies |
| .01 | footings | 703 | Timekeeping and security |
| .05 | grade beams | .01 | timekeeper |
| .07 | slab on grade | .02 | watchmen |
| .08 | beams | .03 | guards |
| .10 | slab on forms | 705 | Utilities and services |
| .11 | columns | .01 | water |
| .12 | walls | .02 | gas |
| .16 | stairs | .03 | electricity |
| .20 | expansion joint | .04 | telephone |
| .40 | screeds | 710 | Storage facilities |
| .50 | float finish | 711 | Temporary fences |
| .51 | trowel finish | 712 | Temporary bulkheads |
| .60 | rubbing | 715 | Storage area rental |
| .90 | curing | 717 | Job sign |
| 245 | Precast concrete | 720 | Drinking water |
| 260 | Concrete forms | 721 | Sanitary facilities |
| .01 | footings | 722 | First-aid facilities |
| .05 | grade beams | 725 | Temporary lighting |
| .07 | slab on grade | 726 | Temporary stairs |
| .08 | beams | 730 | Load tests |
| .10 | slab | 740 | Small tools |
| .11 | columns | 750 | Permits and fees |
| .12 | walls | 755 | Concrete tests |
| 270 | Reinforcing steel | 756 | Compaction tests |
| .01 | footings | 760 | Photographs |
| .12 | walls | 761 | Surveys |
| 280 | Structural steel | 765 | Cutting and patching |
| 350 | Masonry | 770 | Winter operation |
| .01 | 8-in. block | 780 | Drayage |
| .02 | 12-in. block | 785 | Parking |
| .06 | common brick | 790 | Protection of adjoining |
| .20 | face brick | | property |
| .60 | glazed tile | 795 | Drawings |
| 400 | Carpentry | 796 | Engineering |
| 440 | Millwork | 800 | Worker transportation |
| 500 | Miscellaneous metals | 805 | Worker housing |
| .01 | metal door frames | 810 | Worker feeding |
| .20 | window sash | 880 | General clean-up |
| .50 | toilet partitions | 950 | Equipment |
| 560 | Finish hardware | .01 | move in |
| 620 | Paving | .02 | set up |
| 680 | Allowances | .03 | dismantling |
| 685 | Fencing | .04 | move out |

*Source:* R. Clough, *Construction Contracting*, 3rd ed., copyright © 1975 by John Wiley & Sons, Inc. Reprinted by permission of John Wiley & Sons, Inc.

**TABLE 2.1  Key Balance Sheet Accounts**

| Convention | Key Balance Sheet Accounts |
|---|---|
| Cash | 1. No accounts receivable |
|  | 2. No accounts payable |
|  | 3. No inventory |
| Accrual—billings | 1. Asset account—work in progress or |
|  | 2. Unbilled costs of work in progress |
| Accrual—percentage-of-completion | 1. Asset account—costs and estimated earnings in excess of billings |
|  | 2. Liability account—billings in excess of costs and estimated earnings |
| Accrual—completed-contract | 1. Asset account—costs on uncompleted contracts in excess of billings |
|  | 2. Liability account—billings in excess of costs on uncompleted contracts |

personnel and their associated costs are accounted for centrally and separately from the individual projects. The accounts maintained to accumulate service support expenses are often referred to as general and administrative (G&A) accounts. They are also called *home office* or *fixed overhead* expense accounts.

If the company management elects to prorate the accumulated G&A cost back to particular projects, this block allocation of home office overhead cost to individual projects is referred to as *absorption* costing. That is, the overhead costs are absorbed by the individual project. If this procedure is implemented when the completed contract method is used, actual declaration of these expenses for tax purposes is deferred until the end of the project. These costs will become part of the asset (work in progress) value maintained by the contractor until the project is transferred to the client.

The contractor may prefer to keep the G&A costs separate from the project cost activity. That is, he may prefer to keep these costs as period- or time-dependent fixed costs that accrue whether there is activity on the projects or not. General and administrative costs are called fixed costs, since they do not tend to vary significantly with the level of project activity (i.e., number of projects and size of project work load). This approach of expensing G&A costs to a given time period is referred to as *direct costing* of home office overhead. This simply means these costs will be taken as expenses in the period in which they occur and will not be associated with the project costs as a deferred expense to be taken at the time when project income is recognized. In the example calculations of the next chapter, the G&A expenses will be direct-costed.

## 2.10  KEY BALANCE SHEET ACCOUNTS (CASH BASIS)

The relationship of the key balance sheet accounts to the accounting convention used is summarized in Table 2.1. In the following sections the relationship between these balance sheet accounts and the accounting flows that they summarize will be discussed.

The use of the cash basis of accounting by a particular contractor can be detected by the absence of certain summary accounts in the balance sheet. There is no item

**FIGURE 2.3   Absence of Key Accounts on Cash Balance Sheet**

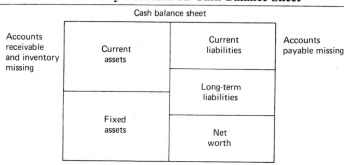

for accounts receivable in the current asset portion of the balance sheet, because entries to the accounting system are not made until funds are transferred. Therefore, there is no provision for accounts reflecting billings made but not received. Similarly, no accounts payable item is reported as part of the current liabilities portion of the balance sheet. Since an obligation is not recognized until it is paid when the cash basis is used, accounts reflecting billings received from creditors but not paid are not maintained. Further, materials purchased for installation against ongoing projects are not reflected as inventory under the current assets. Such inventories would be considered company property of a fixed nature under the cash method. As such, they would not be considered highly liquidable although they are purchased for project use. These missing key accounts are indicated diagrammatically in Figure 2.3. The Fudd Associates, Inc. balance sheet shown in Fig. 1.3 clearly shows that the company is not using the cash method of accounting.

## 2.11   KEY BALANCE SHEET ACCOUNTS USING THE BILLINGS METHOD

Accounts receivable, accounts payable, and inventory accounts are basic items reported in the balance sheet when using the accrual-billings method. In addition, an account is utilized to reflect work in progress that has not been billed. Such an account would embrace payrolls paid, materials installed, and other project costs that have not yet been billed to the client. These expenses are considered to be a current asset and are summarized in the account called work in progress. For example, assume that a large electrical manufacturer is fabricating an electrical transformer set to be delivered to a client. The set is partially complete, and $60,000 in materials and labor have been expended in its manufacture. Billing will not occur until the transformer is complete and shipped to the purchaser. However, the substantial expenditure of resources is reflected in the balance sheet as unbilled costs of work in progress. This account is also called unbilled costs of work in process. Inventory purchased against particular projects is carried as a current asset. These accounts and their positions in the balance sheet are shown schematically in Figure 2.4.

**FIGURE 2.4   Key Balance Sheet Accounts Using Billings Method**

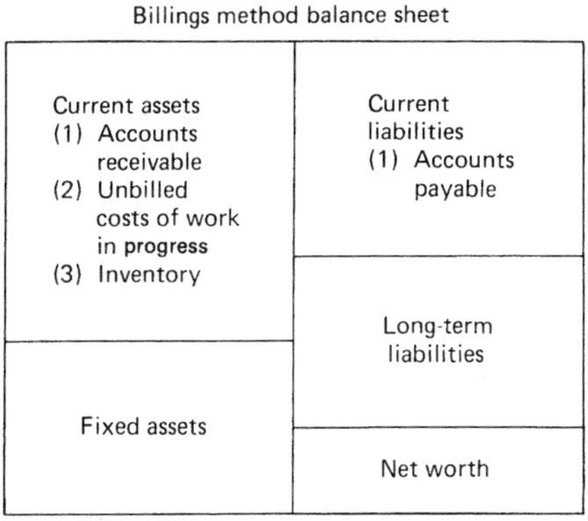

Billings method balance sheet

| | |
|---|---|
| Current assets<br>(1) Accounts<br>  receivable<br>(2) Unbilled<br>  costs of work<br>  in progress<br>(3) Inventory | Current<br>liabilities<br>(1) Accounts<br>  payable |
| | Long-term<br>liabilities |
| Fixed assets | |
| | Net worth |

## 2.12   KEY ACCOUNTS USING THE POC METHOD

In the case of the percentage-of-completion method, the key balance sheet accounts must account for the amount of underbilling and overbilling. *Underbilling* is a situation in which the value earned (based on POC) exceeds the amount billed, generating an asset. *Overbilling* is a situation in which the value earned (based on POC) is less than the amount billed, generating a liability.

The POC method provides for recognition of income (on the basis of earned value) at the end of each accounting period. Revenues are recorded in the project billings account and expenses in the work in progress account as work progresses.

At the end of each accounting period, income flows for all projects appear on the income statement and the balance sheet when the POC method is used. The accounts to reflect overbilling or underbilling on projects are updated at the end of each accounting period.

The aggregate of underbilling occurring on projects in progress leads to an asset, since this is work performed for which billing has not yet occurred. The contractor still "owns" this work, because ownership has not been transferred to the client in the form of a billing that obligates him to reimburse the contractor. The asset account reflecting the amount of underbilling on a POC balance sheet is called "costs and estimated earnings in excess of billings." At the end of the accounting period, the percentage complete is calculated for each project. For projects where the percentage complete exceeds the amount billed, the amount of excess is posted to this asset account.

In the case of overbilling, the percentage complete represents cost and earnings less than the actual billed amount. This means that the contractor is obligated to perform work for which he has already transferred the obligation for payment to client. In effect, he has billed for an "advance" against work to be performed. This obligation is recognized as a liability in the account "billings in excess of costs and estimated earnings." As with the underbilling, the amount of the overbilling is calculated at the end of the accounting period for each project. This is done by comparing the cost and

**FIGURE 2.5   Key Balance Sheet Accounts Using POC Method**

POC balance sheet

| Current assets | Current liabilities |
|---|---|
| (1) Accounts receivable<br>(2) Inventory<br>(3) Costs and estimated earnings in excess of billings | (1) Accounts payable<br>(2) Billings in excess of costs and estimated earnings |
| | Long-term liabilities |
| Fixed assets | Net worth |

earnings based on the percentage complete with the amounts billed. Excess billings are posted to this liability account to reflect the contractor's obligation. Income is calculated and recognized for all projects at the end of the accounting period. This is based on the percentage complete revenue level regardless of the amount billed. When accounts are closed, the expense accounts are compared to the cost plus estimated earnings based on percentage complete to determine the amount of income to be recognized on a project-by-project basis. The net amount of overbilling on all projects taken as a group is reflected in the balance sheet. These accounts are shown schematically in Figure 2.5.

## 2.13   KEY ACCOUNTS USING THE CC METHOD

In the case of the completed-contract method, the function of the key balance sheet accounts (updated at the end of the accounting period) is similar to that of the over- and underbilling accounts in the POC method. They reflect the amount of billings in excess of project costs (overbilling) and conversely costs in excess of billings *on uncompleted contracts*. Completed contracts are booked for income purposes to the income statement.

Uncompleted contracts, however, generate overbilling and underbilling amounts. In the Fudd Associates, Inc. balance sheet in Figure 1.3, it is clear that the completed contract method is being used, although the account titling is slightly different. Billings on uncompleted contracts are listed as $1,792,402. Cost of uncompleted contracts is $1,639,441. Therefore, an overbilling condition on uncompleted contract exists. The asset account "cost of uncompleted contracts" is also referred to as "work in progress" or "construction in progress." It has been listed in Figure 1.3 as a "negative" liability to be deducted from the aggregate billings on uncompleted contracts. It is an asset in the sense that all expense incurred on uncompleted contracts remains the "property"

**FIGURE 2.6   Key Balance Sheet Accounts Using CC Method**

CC balance sheet

| | |
|---|---|
| Current assets<br><br>(1) Accounts receivable<br>(2) Inventory<br>(3) Costs on uncompleted contracts<br>     in excess of billings | Current liabilities<br><br>(1) Accounts payable<br>(2) Billings in excess of<br>     costs on uncompleted<br>     contracts |
| Fixed assets | Long-term liabilities |
| | Net worth |

of the contractor until transferred to the client at completion of the project. The key balance sheet accounts are shown in Figure 2.6.

Although the accounts that contain the project-related assets and liabilities in both POC and CC methods have similar titles, the method used can be determined by the presence or absence of the words "and estimated earnings." For instance, if the balance sheet contains the asset account "costs *and estimated earnings* in excess of billings," the POC method is being used. In the case of the CC method, the words "and estimated income" do not appear and the phrase "on uncompleted contracts" is added. The corresponding account on the asset side of the completed contract balance sheet is then "costs on *uncompleted contracts* in excess of billings." The same test can be applied to the liability accounts. This fact provides a quick means of determining the accounting method used when one is picking up a balance sheet for the first time.

## 2.14   UNIQUENESS OF CONSTRUCTION ACCOUNTING SYSTEMS

The financial accounting systems used in construction have evolved from the peculiarities of the construction industry. Contracting by its very nature is a transient operation. Frequently, a firm's efforts are directed toward the production of a single unique product requiring a multitude of different materials, scores of various types of labor, and many expensive pieces of equipment. Upon completion of this operation, which may require weeks and sometimes years to complete, the contractor moves his production facility to a new site and begins work on a totally different product.

In manufacturing, completed goods are normally transferred to and stored in inventory until sold; the actual profit on a unit of production usually is unknown until some future date. Once a construction project is completed, however, the profit or loss for the company is a reality no longer subject to random variation. In manufacturing, annual income is calculated by using the straight accrual (i.e., billings) method. Thus net income is obtained by taking the difference between the period's sales and the sum of the cost of goods sold plus related costs.

When the accrual basis is used, the actual mechanics of accounting may vary from firm to firm; nevertheless, all systems of this type are based upon the following principles:*

1. Income is to be recognized when it is earned or billable.
2. Costs and expenses are to be recognized when incurred.
3. Income and the related costs and expenses are to be taken into operating results in the same accounting period.
4. All known losses are to be fully provided for in the period in which they become determinable with reasonable certainty.

In contrast to the straight accrual or billings method, which is based on a given accounting period, the normal duration for a construction project can run for several accounting periods. Consequently, the traditional accrual method is not an adequate method for measuring income on long-term construction projects.

Both the completed-contract and percentage-of-completion methods are used in accounting for income and costs associated with long-term construction contracts. Both methods are recognized as acceptable by the Internal Revenue Service for determining taxable income.

A contractor cannot legally use either of these methods for tax purposes unless the contract duration extends across two or more reporting periods. Once a firm elects to use either method for purposes of reporting taxable project income, it is bound to continue reporting on that basis, unless permission to change is authorized by the Internal Revenue Service.

When the percentage-of-completion method is used, the project's profit is accrued over the life of the contract in accordance with the progress each year. The percentage of completion represented by each year's work may be based upon an engineering estimate of the work performed relative to the total work required under the contract or the relationship between the cost incurred and the total estimated costs to complete the contract.

When the completed-contract method is used, income and expenses are recognized only when a particular job is substantially completed.

> Until a job is completed, revenues and expenses are shown as assets and liabilities, respectively. It is, however, permissible to offset related revenues against related costs and show the net as a liability or asset.†

Since income recognition attaches to the individual project, hybrid methods can be used in which income is recognized on some projects by the percentage-of-completion method and on others by the completed-contract method (see Sections 5.5 and 5.6). If a combination of both methods is utilized, the extent to which the percentage-of-

---

* William E. Coombs, *Construction Accounting and Financial Management* (New York: McGraw-Hill, 1958), p. 27.

† Felix P. Kollaritsch, "Operating and Financial Ratios of Ohio Highway Contractors," Bureau of Business Research Monograph No. 118, College of Commerce and Administration, The Ohio State University, 1966, p. 12.

completion method or the completed contract method is used will depend solely on the nature of the firm's work. As long as a firm applies these methods in a consistent manner and the books and financial statement reflect the actual facts, there should be no sound objection to this procedure.

## REVIEW QUESTIONS AND EXERCISES

**2.1**  Name the four methods of accounting commonly used in construction. Give two advantages and two disadvantages of each method.

**2.2**  Why do many contractors prefer to use the percentage-of-completion or completed-contract methods rather than the billings method?

**2.3**  What is the difference when handling the general and administrative costs between direct and absorption costing? What effect would each method have under (1) the percentage-of-completion method, and (2) the completed-contract method.

**2.4**  What is the difference between the percentage-of-completion method and the completed-contract method of accounting? Name advantages and disadvantages of both.

**2.5**  How would recognition of income vary between the billings method and the percentage-of-completion method? Could they show the same amount of income? Discuss.

**2.6**  Give the names of three acceptable ways of calculating the percentage of a contract complete using the percentage-of-completion method.

**2.7**  What advantages does the billings method have over other methods of income recognition?

**2.8**  Give an example of an overbilling situation. What accounting convention is implied?

**2.9**  What is meant by the term G&A expense?

**2.10**  When is a project considered complete when using the completed-contract method?

**2.11**  Given the data shown in Tables P2.1 and P2.2 for Pegasus International, Inc., what method of accounting is being used? How has the working capital of Pegasus International changed from 19X4 to 19X5? Interpret this development considering the increase in revenue from 19X4 to 19X5.

**TABLE P2.1  Pegasus International, Inc., Summary of Consolidated Operations (Dollar Amounts and Shares Are in Thousands, Except Per-Share Amounts)**

|  | *19 × 5* | *19 × 4* |
|---|---|---|
| Revenue | $1,325,423 | $801,322 |
| Costs and expenses |  |  |
| Cost of revenue | 1,211,402 | 726,937 |
| Corporate administrative and general expenses | 19,392 | 14,881 |
| Other (income) and expenses |  |  |
| Interest on indebtedness | 1,354 | 1,671 |
| Interest income | (6,565) | (7,089) |
| Provision for estimated losses on planned |  |  |
| disposition of assets | — | 400 |
| Total cost and expenses | 1,225,583 | 736,800 |
| Earnings before income taxes | 99,480 | 64,522 |
| Total income taxes | 52,429 | 31,277 |
| Net earnings | 47,411 | 33,245 |
| Preferred dividend requirements | 1,803 | 1,558 |
| Earnings applicable to common stock | $   45,608 | $ 31,687 |

**TABLE P2.2  Pegasus International, Inc., Consolidated Balance Sheet**

| Assets | 31 December, 19x5 and 19x4 | 19x5 | 19x4 |
|---|---|---|---|
| **Current assets** | Cash | $ 10,138,000 | $ 13,063,000 |
| | Short-term investments at cost, which approximates market | 101,310,000 | 49,298,000 |
| | Accounts and notes receivable | 90,526,000 | 83,442,000 |
| | Costs and earnings in excess of billings on uncompleted contracts | 65,745,000 | 44,728,000 |
| | Inventories at lower of cost (average and FIFO methods) or market | 30,011,000 | 24,210,000 |
| | Other current assets | 7,751,000 | 9,005,000 |
| | Total current assets | 305,481,000 | 223,746,000 |
| **Property, plant and equipment at cost** | Land | 11,162,000 | 11,744,000 |
| | Buildings and improvements | 39,622,000 | 26,909,000 |
| | Machinery and equipment | 50,650,000 | 51,597,000 |
| | Drilling and marine equipment | 107,617,000 | 71,300,000 |
| | Construction in process | 42,834,000 | 44,230,000 |
| | Subtotal | 251,885,000 | 205,780,000 |
| | Less accumulated depreciation and amortization | 75,441,000 | 76,732,000 |
| | Net total property, plant and equipment | 176,444,000 | 129,048,000 |
| **Oil and gas properties at cost** | Subtotal | 59,814,000 | 50,683,000 |
| | Less accumulated depletion and depreciation | 30,455,000 | 20,105,000 |
| | Net total oil and gas properties | 29,359,000 | 30,578,000 |
| **Other assets** | Excess of cost over net assets of acquired companies, less accumulated amortization | 15,105,000 | 16,351,000 |
| | Other | 12,058,000 | 12,542,000 |
| | Total other assets | 27,163,000 | 28,893,000 |
| **Total** | | $538,447,000 | $412,265,000 |

| Liabilities and Shareholders' Equity | 19x5 | 19x4 |
|---|---:|---:|
| **Current liabilities** | | |
| Accounts payable | $ 66,050,000 | $ 54,124,000 |
| Billings in excess of costs and earnings on uncompleted contracts | 69,082,000 | 53,375,000 |
| Accrued liabilities | 58,089,000 | 37,393,000 |
| Current portion of long-term debt | 1,395,000 | 1,385,000 |
| Income taxes currently payable | 24,895,000 | 8,209,000 |
| Deferred income taxes | 49,599,000 | 31,989,000 |
| Total current liabilities | 269,101,000 | 186,385,000 |
| **Long-term debt due after one year** | | |
| Total | 4,242,000 | 7,311,000 |
| **Other noncurrent liabilities** | | |
| Deferred income taxes | 11,623,000 | 11,792,000 |
| Other | 9,162,000 | 4,400,000 |
| Total | $ 20,785,000 | $ 16,192,000 |
| **Contingencies and Commitments** | | |
| **Shareholders' equity** | | |
| Capital stock | | |
| Series B preferred | $ 567,000 | $ 663,000 |
| Common | 9,368,000 | 9,203,000 |
| Additional capital | 116,636,000 | 115,184,000 |
| Retained earnings | 117,748,000 | 77,327,000 |
| Total shareholders' equity | 244,319,000 | 202,377,000 |
| Total | $538,447,000 | $412,265,000 |

# Chapter 3

# COMPARISON OF ACCOUNTING CONVENTIONS

## 3.1 PROJECT-LEVEL COMPARISON

In order to demonstrate the differences between various accounting conventions, consider the following bridge project with expenditures and revenues as described in Table 3.1. This project has an expected duration of 12 months, and the bid price is $1,342,000. Fudd Associates plans to begin this project 1 July 19X2 and complete the job 30 June 19X3. The expected cost of the project is $1,220,000. This cost does not include the home office expense to administer the project. The general and administrative cost will be $70,000. It is assumed for simplicity that the policy regarding the billing and payment of vendors, shown in the following table, is observed. Retainage of 10 percent is withheld during the entire duration of the project.

| Expense | Billing Received | Payment by Fudd Associates |
|---------|-----------------|---------------------------|
| Subcontract | End of month | 15 days after billing no retainage |
| Materials | As received through month from vendors | 10 days after beginning of month following billing |
| Payroll | — | 4 times in month |
| Equipment | Rented equipment billed end of month | Paid end of month |
| Overhead | Obligations accrue during month | Paid on last day of month |

Revenue (cost plus markup) is billed by Fudd Associates to the client on the first day of the month. The billed amount minus retainage is received from the client at the end of the month billed. Mobilization and demobilization payments are received 30 days after billing. Mobilization is billed at $44,000 on the first of the month, and a payment of $39,600 is received on the last day of the month. The mobilization expense is assumed to be incurred as a lump sum amount at the beginning of the first month. Figure 3.1 shows schematically the revenues and expenditures for this project

**TABLE 3.1   Table of Expenses**

| Month | Mobilization Demobilization | Sub-contractors | Materials | Payroll | Equipment | Field Overhead |
|---|---|---|---|---|---|---|
| 0 | $40,000 | $0 | $0 | $0 | $0 | $0 |
| 1 | 0 | 10,000 | 10,000 | 10,000 | 20,000 | 1,000 |
| 2 | 0 | 30,000 | 20,000 | 15,000 | 10,000 | 5,000 |
| 3 | 0 | 30,000 | 30,000 | 20,000 | 20,000 | 6,000 |
| 4 | 0 | 40,000 | 30,000 | 20,000 | 30,000 | 6,000 |
| 5 | 0 | 50,000 | 40,000 | 40,000 | 20,000 | 6,000 |
| 6 | 0 | 50,000 | 40,000 | 40,000 | 15,000 | 6,000 |
| 7 | 0 | 40,000 | 30,000 | 40,000 | 10,000 | 6,000 |
| 8 | 0 | 40,000 | 10,000 | 20,000 | 10,000 | 6,000 |
| 9 | 0 | 70,000 | 10,000 | 10,000 | 10,000 | 6,000 |
| 10 | 0 | 30,000 | 5,000 | 5,000 | 10,000 | 6,000 |
| 11 | 0 | 30,000 | 5,000 | 5,000 | 5,000 | 6,000 |
| 12 | 20,000 | 50,000 | 0 | 5,000 | 5,000 | 5,000 |
| Total | $60,000 | $470,000 | $230,000 | $230,000 | $165,000 | $65,000 |

Total cost = $60,000 + $470,000 + $230,000 + $230,000 + $165,000 + $65,000
    = $1,220,000
Profits + overhead @ 10% = $122,000
Bid price = $1,342,000

for the period July to December 19X2, calculated by the cash method. Figure 3.2 demonstrates the billings method for calculating revenue and expenses for the same project. Calculations for the project revenue calculated by using the percentage-of-completion method for the same period (July to December) are shown at the bottom of Figure 3.2. The amounts of income generated by the project, as calculated by the four methods discussed previously, are summarized in Table 3.2.

When the cash basis is used, the actual amounts received from client and paid for work and materials are recognized at the time payment is received or disbursed. Since the retainage is not received, it is not recognized at this time. Therefore, although the same number of progress payments is recognized with this method as with the billings method, the total revenue is less than that recognized by the billings method. Since expenses exceed revenue when the cash basis is used, the income from the project to this point is negative. That is, the gross end-of-year profit from this project is −$66,590. If a prorated portion of the G&A expense is allocated to the expenses, the net profit is −$101,590. It is obvious that the contractor can reduce this loss by deferring expenses. By delaying payment to subcontractors and material vendors and deferring equipment expense in December, the $61,590 loss is offset by the $105,000 in deferred expense and a net gross profit of $3,410 is realized. Clearly, the profit and loss picture for the project can be greatly manipulated by the timing (i.e., delay) of disbursements to cover expenses.

The revenue when the billings method is used for the period of consideration recognizes the retainage which, although earned, will not be received until the end of the job. Both the recognized revenue and expenses are greater with the billings method than with the cash basis. The income is even more negative, and the gross profit for

## FIGURE 3.1   Transactions Using Cash Method

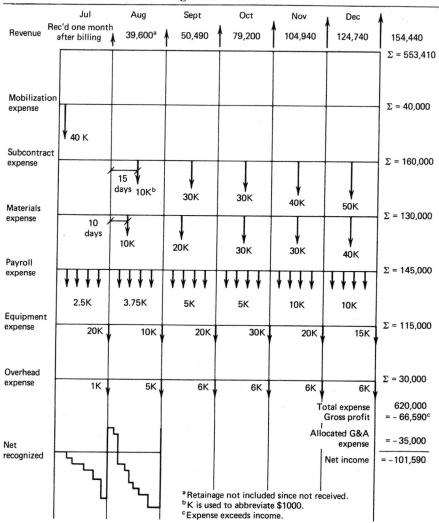

| | Jul | Aug | Sept | Oct | Nov | Dec | |
|---|---|---|---|---|---|---|---|
| Revenue | Rec'd one month after billing | 39,600[a] | 50,490 | 79,200 | 104,940 | 124,740 | 154,440 |
| | | | | | | | Σ = 553,410 |
| Mobilization expense | | | | | | | Σ = 40,000 |
| | 40 K | | | | | | |
| Subcontract expense | | | | | | | Σ = 160,000 |
| | | 15 days 10K[b] | 30K | 30K | 40K | 50K | |
| Materials expense | 10 days | | | | | | Σ = 130,000 |
| | | 10K | 20K | 30K | 30K | 40K | |
| Payroll expense | | | | | | | Σ = 145,000 |
| | 2.5K | 3.75K | 5K | 5K | 10K | 10K | |
| Equipment expense | 20K | 10K | 20K | 30K | 20K | 15K | Σ = 115,000 |
| Overhead expense | | | | | | | Σ = 30,000 |
| | 1K | 5K | 6K | 6K | 6K | 6K | |
| | | | | | Total expense | | 620,000 |
| | | | | | Gross profit | | = –66,590[c] |
| | | | | | Allocated G&A expense | | = –35,000 |
| Net recognized | | | | | | Net income | = –101,590 |

[a] Retainage not included since not received.
[b] K is used to abbreviate $1000.
[c] Expense exceeds income.

the period is –$95,100. The net profit accounting for prorated G&A expense is –$130,100. In this method, the ability to defer payment and thus manipulate the profit (loss) of the job is not available.

When the cost completion method is used, the revenue exceeds the incurred expenses, and a gross profit of $71,000 is recognized. With the POC method the revenue exceeds the expense until the total bid price is recognized. If expenses exceed the initial cost estimate, the maximum revenue that can be recognized is the bid price (unless adjustments for change orders and scope changes increase the original bid amount). If expenses exceed the original bid price or if projected expenses indicate

## FIGURE 3.2 Transactions Using Billings Method

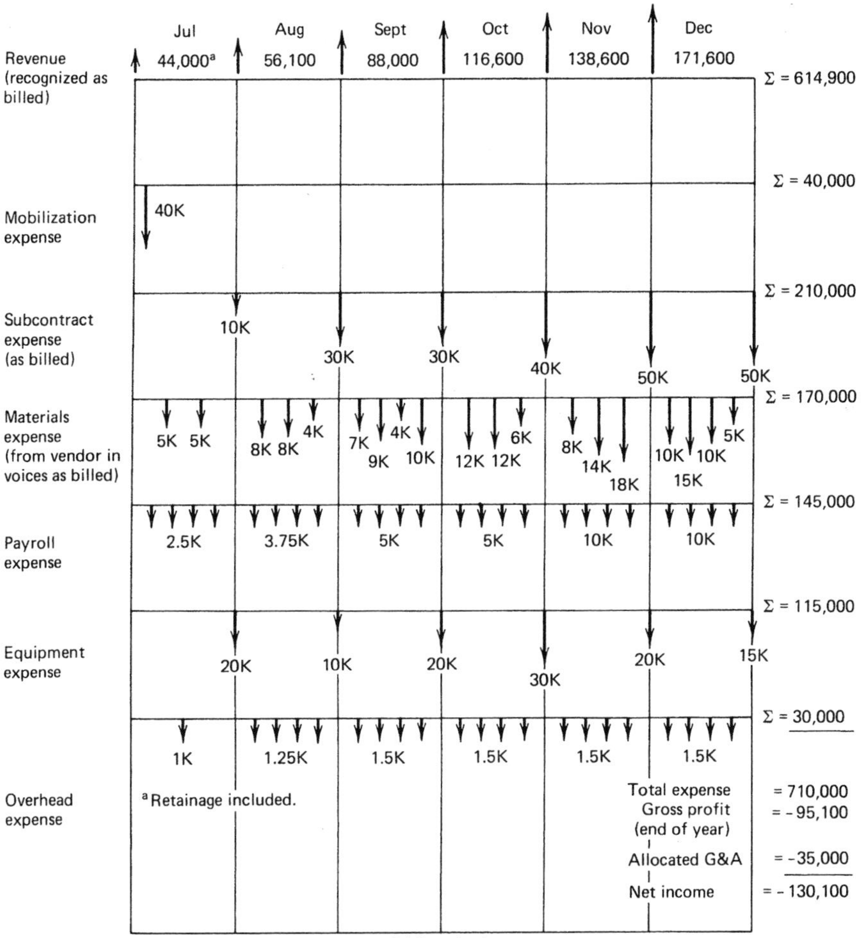

| | Jul | Aug | Sept | Oct | Nov | Dec | |
|---|---|---|---|---|---|---|---|
| Revenue (recognized as billed) | 44,000[a] | 56,100 | 88,000 | 116,600 | 138,600 | 171,600 | Σ = 614,900 |
| Mobilization expense | 40K | | | | | | Σ = 40,000 |
| Subcontract expense (as billed) | | 10K | 30K | 30K | 40K | 50K    50K | Σ = 210,000 |
| Materials expense (from vendor invoices as billed) | 5K  5K | 8K 8K    4K | 7K   4K   9K  10K | 6K  12K 12K | 8K   14K   18K | 10K 10K  5K  15K | Σ = 170,000 |
| Payroll expense | 2.5K | 3.75K | 5K | 5K | 10K | 10K | Σ = 145,000 |
| Equipment expense | 20K | 10K | 20K | 30K | 20K | 15K | Σ = 115,000 |
| Overhead expense | 1K | 1.25K | 1.5K | 1.5K | 1.5K | 1.5K | Σ = 30,000 |

[a] Retainage included.

| | |
|---|---|
| Total expense | = 710,000 |
| Gross profit (end of year) | = –95,100 |
| Allocated G&A | = –35,000 |
| Net income | = –130,100 |

POC method

Percentage complete (based on cost) $= \dfrac{(\$\ 710,000)}{(\$1,220,000)} = 0.58196.$

Revenue = 0.58196 ($1,342,000) = $781,000.

Income = revenue – expense = $781,000 – $710,000 = $71,000.

bid price overrun, a loss must be declared. The net profit for the period with POC is $36,000.

When the completed-contract (CC) method is used, revenue and expense recognized for the period July to December 19X2 are zero. Although these transactions are being accounted for, they will not be recognized until the project is complete in 19X3. Therefore, the net profit from the job is zero. As noted above, general and administrative expenses can be recognized at completion of the project or prorated to the

**TABLE 3.2   Summary of 19X2 Income Based on Different Accounting Conventions**

| Convention | Revenue | Expense | Gross Profit | Net Profit (Deducting G&A Expense) |
|---|---|---|---|---|
| Cash | $553,410 | $620,000 | $ − 66,590 | $ − 101,590 |
| Billings | 614,900 | 710,000 | − 95,100 | − 130,100 |
| POC | 781,000 | 710,000 | 71,000 | 36,000 |
| Completed contract | 0 | 0 | 0 | − 35,000 |

period in which they are actually incurred. They are shown in Table 3.2 being allocated to the July to December 19X2 period. The net profit is − $35,000.

## 3.2   PERCENTAGE-OF-COMPLETION METHOD CONCEPTS

The percentage-of-completion method is founded on the premise that income should be based upon revenue earned, a premise that may not be reflected by the billings. The billings may, in fact, represent a request to be paid for more value than has actually been placed. In this case, the client is overbilled, and the contractor is collecting in advance for work he will accomplish. This happens, for instance, when a contractor bidding in a unit price format unbalances his bid. That is, he overcharges for quantities to be placed early in the project and compensates by underpricing quantities on work to be placed late in the project. The process of unbalancing the bid is described as follows:

> Essentially, for those items that occur early in the construction, inflated unit prices are quoted. So, for instance, excavation that in fact costs $16 per cubic yard will be quoted at $24 per cubic yard. Foundation piles that cost $16 per linear foot will be quoted at $20 per linear foot. Since these items are overpriced, in order to remain competitive, the contractor must reduce the quoted prices for latter bid items. "Close-out" items such as landscaping and paving will be quoted at lower-than-cost prices. This has the effect of moving reimbursement for the work forward in the project construction period. It unbalances the cost of the bid items leading to front-end loading.*

This approach means that the billing submitted early in the project will tend to exceed the actual value of construction placed. In effect, the contractor is overbilling. In order to reflect a true income for financial accounting purposes, an estimate of the true value placed is made by taking a percentage of the total bid price. How reliable this true value figure is depends upon how accurate the estimate of percentage complete is. Mobilization payments lead to the contractor's receiving monies in excess of actual value placed. The mobilization costs will be "worked off" across the life of the job and represent a sort of prepayment by the client.

In contrast, an underbilling condition can occur in which the revenue received by

---

* D. W. Halpin and R. W. Woodhead, *Construction Management* (New York: Wiley, 1980), p. 73.

the contractor is actually lower than the value of the work placed. If, for instance, an unintentional underestimation of the value of the work to be accomplished in any month or time period has occurred, the amount that can be billed will be less than the value of work actually placed. This can occur during estimation of the project costs prior to bid. Such underestimates on portions of the work may be offset by compensating overestimates occurring later on other portions of the work. At certain points in the life of the project, however, the amount billed to the client will be less than the actual value of work placed to date.

Underbilling may also occur in cases where the contractor must prepay for goods or services that are received, but for which (under conditions of the contract) he cannot charge until some later point in the contract. For instance, the contractor may be required to pay a subcontractor for subdrains installed early in the job in connection with the landscaping bid item. Payment for these subdrains will not occur until the end of the job when the landscaping item is paid. The value of the drains is installed, but the billing for this value cannot occur until later in the project. This variation of "true" value versus billed value through the life of the project is illustrated schematically in Figure 3.3.

**FIGURE 3.3   Schematic of Overbilling and Underbilling**

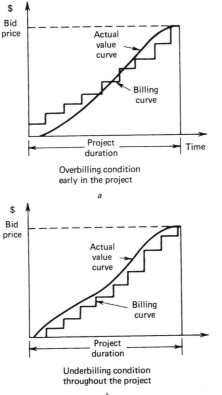

Similarly, if the contractor is underbilling, the work installed for which he has not been able to bill is carried as a current asset. As in the case of the landscaping drains, the work is "owned" by the contractor and becomes his asset until he can bill for the work.

## 3.3.   COMPLETED-CONTRACT METHOD CONCEPTS

The completed-contract (CC) method is meant to be used primarily for projects on which progress towards completion cannot be reliably estimated. In order to use this method, the project must extend across at least two accounting periods. That is, as a minimum it must begin in one period and end in the following period. Since revenue and expense are recognized only at the time the job is completed, the income reflected in the company will be irregular unless a relatively constant volume of work is being completed during each accounting period. To illustrate, consider Figure 3.4, which compares the work loads of Cougar Construction and Apex Construction.

Cougar Construction Company has five jobs completed in 19X2 and seven jobs completed in 19X3. Because of the relatively even number of completions in each period, recognition of income is relatively uniform. Let us assume for the purpose of illustration that each job has an end-of-job income of exactly $50,000. The financial statements for Cougar Construction would recognize exactly $250,000 in income for 19X2 and $350,000 in 19X3.

Apex construction does four larger jobs, all of which overlap two years. No income is recognized in 19X1, since none of the jobs are completed. Income from Job A is recognized in 19X2 because it is completed during this year. Income from Jobs B, C, and D is recognized in 19X3. Assume that cost information for Jobs A, B, C, and D constructed by Apex is as given in Table 3.3.

**FIGURE 3.4   Comparison of Job Volumes for Two Contractors**

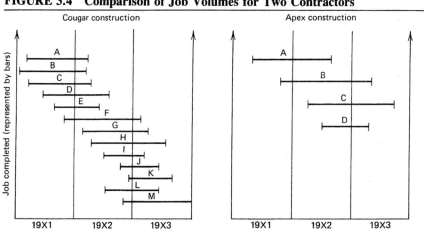

**TABLE 3.3  Project Information for Apex Construction Jobs, 19X1–19X3 (No Completed Contracts in 19X1)**

| Project | Contract Price | Total Cost | Cumulative Cost End 19X1 | Cumulative Cost End 19X2 | Cumulative Cost End 19X3 | Cumulative Billings End 19X1 | Cumulative Billings End 19X2 | Cumulative Billings End 19X3 |
|---|---|---|---|---|---|---|---|---|
| A | $ 720,000 | $ 620,000 | $370,000 | $620,000 | — | $460,000 | $720,000 | $1,000,000 |
| B | 1,000,000 | 800,000 | 100,000 | 580,000 | 800,000 | 125,000 | 850,000 | 1,800,000 |
| C | 1,800,000 | 1,480,000 | 0 | 815,000 | 1,480,000 | 0 | 792,000 | 1,800,000 |
| D | 800,000 | 720,000 | 0 | 400,000 | 720,000 | 0 | 420,000 | 800,000 |

If Apex Construction is using the completed-contract method, the amount of income reflected in Apex's financial statements for the years 19X1, 19X2, and 19X3 is as follows:

|  | Revenue | Expense | Income |
|---|---|---|---|
| 19X1 | $    0 | $    0 | $    0 |
| 19X2 | 720,000 | 620,000 | 100,000 |
| 19X3 | 3,600,000 | 3,000,000 | 600,000 |

If Apex is a publicly held company, these figures could make stockholders apprehensive about Apex's performance in 19X1 and 19X2, since the company reports only $100,000 of income on a volume of $4,320,000 of projects in progress.

## 3.4   A MULTIYEAR COMPARISON OF CC TO POC

If Apex were to use the percentage-of-completion method for reporting income, a better indication of the company's performance during the period would be reflected. On the basis of the percentage-of-completion method, Apex would recognize the amounts shown in Table 3.4.

The revenue on project A in 19X1 is calculated as $370,000/$620,000 \times ($720,000)$, which yields $429,677. Subtracting the incurred cost of $370,000 yields an income of $59,677. Similarly, for project B using cost completion to calculate the percentage complete gives $100,000/$800,000, or 12.5 percent complete. This amount is based on $100,000 incurred on a total cost of $800,000. Therefore, the revenue for 19X1 on project B is 12.5 percent of $1,000,000, or $125,000. Using the POC method gives an annual income before taxes of $84,677.

Similar calculations are made to obtain with incomes for 19X2 and 19X3. Gross profits for each year with POC versus CC are

|  | POC | CC |
|---|---|---|
| 19X1 | $ 84,677 | $      0 |
| 19X2 | 380,983 | 100,000 |
| 19X3 | 234,340 | 600,000 |

Clearly, the recognition of income is more uniform when the POC method is used. This method also presents a better indication of the performance of the company. If the general and administrative costs of home office operations for Apex Construction Co. are $100,000 annually and are direct-costed, the net profit for the period considered is as follows:

|       | POC          | CC          |
|-------|--------------|-------------|
| 19X1  | $ − 15,323   | − $100,000  |
| 19X2  | 280,983      | 0           |
| 19X3  | 134,340      | 500,000     |

This assumes that the G&A costs are taken in the accounting period (one year in this case) in which they are incurred.

The concept of overbilling and underbilling associated with the POC method can be seen in the context of the billings made by Apex Construction Co. In 19X1, Apex billed $460,000 on Project A, while the POC method indicates revenues of $429,677 actually earned. Therefore, Apex has overbilled on Project A. The billing on Project B during 19X1 exactly equals the recognized revenue, so no over- or under billing has occured on this project. The overbilling on Project A would be accounted for as a current liability in the liability account "billings in excess of cost and recognized income."

During 19X2 the billings and revenues earned on the four projects in progress are as shown below.

| Project | Billed Amount | Revenue Earned |
|---------|---------------|----------------|
| A       | $260,000      | $290,323       |
| B       | $725,000      | $600,000       |
| C       | $792,000      | $991,216       |
| D       | $420,000      | $444,444       |

An underbilling occurs on Project A (compensating for the overbilling in 19X1) to close out this project. Overbilling on Project B occurs because the earned revenue is only $600,000 while the billed amount is $725,000. On Projects C and D the amounts billed are less than the amounts earned as calculated by the POC method. Therefore, an underbilling condition exists on these projects.

**TABLE 3.4   Income Recognized by Apex (Percentage-of-Completion Method)**

| Year | Projects in Progress | % Complete[a] | Revenue | Expense | Project Income |
|------|----------------------|---------------|---------|---------|----------------|
| 19X1 | A      | 59.7  | $   429,677 | $   370,000 | $  59,677 |
|      | B      | 12.5  | 125,000     | 100,000     | 25,000    |
|      | Totals |       | $   554,677 | $   470,000 | $  84,677 |
| 19X2 | A      | 100.0 | $   290,323 | $   250,000 | $  40,323 |
|      | B      | 72.5  | 600,000     | 480,000     | 120,000   |
|      | C      | 55.0  | 991,216     | 815,000     | 176,216   |
|      | D      | 55.5  | 444,444     | 400,000     | 44,444    |
|      | Totals |       | $2,325,983  | $1,945,000  | $380,983  |
| 19X3 | B      | 100.0 | $   275,000 | $   220,000 | $  55,000 |
|      | C      | 100.0 | 808,784     | 665,000     | 143,784   |
|      | D      | 100.0 | 355,556     | 320,000     | 35,556    |
|      | Totals |       | $1,439,340  | $1,205,000  | $234,340  |

[a]End of year with cost completion method.

## 3.5   ACCOUNTING FOR PROJECTED LOSSES

Accounting guidelines published by the American Institute of Certified Public Accountants (AICPA) specify that when losses related to a project can be determined, they should be declared at the time they are recognized, regardless of what method of accounting is being used. The AICPA *Construction Contractors Guide* states:

> When the current estimates of total contract revenue and contract cost indicate a loss, a provision for the entire loss should be made. Provisions for losses should be made in the period in which they become evident under either the percentage-of-completion method or the completed contract method.*

To illustrate, let us assume that it becomes clear that due to an underestimation of costs and an overestimation of productivity on Project B, Apex Construction realizes during 19X2 that a loss of $100,000 will occur on this project. This loss should be reflected in the financial statements for the year 19X2. Doing so leads to a reduction in the income reported under the POC method from $280,983 to $180,983. If Apex is using the completed-contract method, they will report a gross profit for the year's operations of $ − 100,000. This is consistent with what is called the "cumulative catch-up" method, which states

> Account for the change in estimate in the period of change so that the balance sheet at the end of the period of change and the accounting in subsequent periods are as they would have been if the revised estimate had been the original estimate.†

Declaration of a loss on Project B revises the revenue and expense calculations for 19X2 as shown in Tables 3.5 and 3.6. The loss is carried in the general ledger accounts and on the balance sheet as a liability. In this way, losses are indicated in such a way

**TABLE 3.5   Impact of Loss Declaration for 19X2
(Percentage-of-Completion Method)**

|  | *Project A* | *Project B* | *Project C* | *Project D* | *Total* |
|---|---|---|---|---|---|
| Total revenue | $290,323 | $600,000 | $991,216 | $444,444 | $2,325,983 |
| Total expenditures | 250,000 | 480,000 | 815,000 | 400,000 | 1,945,000 |
| Additional estimated loss |  | (100,000) |  |  | (100,000) |
| Gross profit | $ 40,323 | $ 20,000 | $176,216 | $ 44,444 | $280,983 |
| G&A expense |  |  |  |  | 100,000 |
| Net income |  |  |  |  | $180,983 |

ᵃParentheses indicate a negative value.

* American Institute of Certified Public Accountants, *Construction Contractors Audit and Accounting Guide*, p. 148 (paragraph 85 of the Statement of Position 81-1) New York, 1981.

† Ibid.

**TABLE 3.6   Impact of Loss Declaration for 19X2 (Completed-Contract Method)**

|  | *Project A* | *Project B* | *Project C* | *Project D* | *Total* |
|---|---|---|---|---|---|
| Total revenue | $720,000 |  |  |  | $720,000 |
| Total expenditures | 620,000 |  |  |  | 620,000 |
| Additional estimated loss |  | $(100,000) | ———— | ———— | $(100,000) |
| Gross profit | $100,000 | $(100,000) |  |  | $     0 |
| G&A expense |  |  |  |  | 100,000 |
| Net income (loss) |  |  |  |  | $(100,000) |

that stockholders and others reviewing the financial statement are aware of a declared loss in operations.

## 3.6   SUMMARY

In Chapters 2 and 3, the bases of accounting and income recognition are

1.   Cash
2.   Accrual

When the accrual method is used, income can be recognized on the basis of

1.   Billings Method
2.   Percentage-of-Completion (POC) method
3.   Completed-contract (CC) method

When the percentage-of-completion method is used, several approaches to calculating the amount of work completed are available. These include:

1.   Cost-to-cost method
2.   Effort-expended method
3.   Units-of-work-performed method
4.   Units-of-delivery method

Use of the POC method reflects the amount of income *earned* rather than the amount billed. Since the amount billed may be greater than or less than the amount earned, an overbilling or underbilling condition typically develops. This over or underbilling is reported in the balance sheet. When the completed-contract method is used, over-billing or underbilling versus incurred cost is also reported in the balance sheet. Income recognition is deferred until individual projects are completed. Expenses on projects in progress are reported as assets in "work in progress" accounts. Revenues from projects in progress are reflected in liability accounts called "advance billings" accounts. Support expenses that cannot be readily assigned to a given project are booked to general and administrative costs.

General and administrative costs are either treated as period costs or allocated (absorbed) by individual projects. When the G&A costs are deducted from gross revenues at the end of the accounting period without allocation to individual projects, this is called *direct* costing. If the expenses are allocated to projects (and deferred when the completed-contract method is used), this is called *absorption* costing.

Each of the methods of accounting and income recognition has certain advantages and disadvantages. The methods being used can generally be detected by a close examination of the company balance sheet.

Just as the nature of the construction industry influences the type of accounting systems used in establishing a company's taxable net income, a company's chosen accounting method and the nature of its business operations contribute to the type, number, and nature of accounts present in its financial statements. Despite the complexity of the accounting methods, many companies consider it feasible to use one accounting method for tax purposes and another for financial reporting to its stockholders. The objectives of the accounting system when it is used for internal review and control purposes are different from those the company must consider when dealing with questions relating to tax liability. The POC method is often used for company management review purposes, whereas the CC method is used for tax liability.

## REVIEW QUESTIONS AND EXERCISES

**3.1**   You are the President of Dewey, Cheatum, and Howe Construction Co., Inc. You are reviewing the financial data for the first year of the company's operation ended 31 December 19X4. Your accounting records show the following status as of 31 December 19X4.

| | |
|---|---|
| Accounts receivable | $300,000 |
| Accounts payable | 100,000 |
| Accrued G&A costs | 20,000 |
| Uncompleted contract costs | 400,000 |
| Uncompleted contract billings | 500,000 |

You are also aware that the following has occurred during the 19X4 operational year:

| | |
|---|---|
| Cash collections | $1,700,000 |
| Cash paid—job costs | 1,500,000 |
| Cash paid—G&A costs | 280,000 |

You compare your billings with the actual value of construction placed (earned value) and realize that due to front-end loading on certain projects you have overbilled a total of $50,000 on the work.

(a) Calculate the net before-tax profit for 19X4 using cash basis.
(b) Calculate the net before-tax profit for 19X4 using accrual (billings) basis.

(c) Calculate the net before-tax profit using the percentage-of-completion method.

(d) Calculate the net before-tax profit using the completed-contract method.

3.2 Given the financial statement shown below, what method of accounting is Cyclone Construction using? Is Cyclone Construction overbilling or underbilling for the year 19X4? How much?

## Balance Sheet, Cyclone Construction
## (As of 31 December 19X4)

| | |
|---|---:|
| Assets | |
| Current assets | |
| Cash | $ 486,292 |
| Accounts receivable | 403,146 |
| Trade accounts | |
| Retainage | 84,296 |
| Total accounts receivable | 487,292 |
| Material inventory | 3,746 |
| Work in process (costs and estimated | 152,286 |
| earnings in excess of billings) | |
| Prepaid expenses | 12,296 |
| Other current assets | 2,566 |
| Total current assets | 1,143,622 |
| Fixed assets | |
| Machineary and equipment | 1,084,346 |
| Cars and trucks | 98,428 |
| Furniture and fixtures | 11,624 |
| Total depreciable assets | 1,194,398 |
| Less accumulated depreciation | 582,628 |
| Net fixed assets | 611,770 |
| Total assets | 1,755,392 |
| | |
| Liabilities | |
| Current liabilities | |
| Accounts payable | 472,288 |
| Retainage payable | 42,624 |
| Notes payable | 168,422 |
| Accrued liabilites | 31,654 |
| Work in process (billings in excess | 10,326 |
| of costs and estimated earnings) | |
| Other current liabilites | 2,842 |
| Total current liabilites | 728,156 |
| Long-term liabilities | 276,962 |
| Net worth | 1,005,118 |
| Capital stock | 30,000 |
| Retained earnings | 720,274 |
| Total net worth | 750,274 |
| Total liabilities and net worth | $1,755,392 |

3.3 The following data are available regarding the operations of the High Roller Construction Company for a given year.

## High Roller, Inc.

|  | Contract Price | Total Estimated Costs | Contract Costs Cumulative | Current Year | Cumulative Contract Billings |
|---|---|---|---|---|---|
| Completed contracts | $ 720,000 | $ 620,000 | $ 620,000 | $ 250,000 | $ 720,000 |
| Contracts in progress | 3,600,000 | 3,000,000 | 2,100,000 | 1,250,000 | 2,250,000 |

(a) Determine the value of the asset account *cost and estimated earnings in excess of billings on contracts in progress*.

(b) Compute the income of the company applying the percentage-of-completion method.

**3.4** Using the following data for the Michael Ryan Construction Company, calculate the annual income of the company using (1) the percentage-of-completion method and (2) the completed-contract method.

|  | Project A | Project B | Project C |
|---|---|---|---|
| Total contract amount | $820,000 | $1,400,000 | $1,000,000 |
| Cash expenditures to 12/31/X3 | 730,000 | 750,000 | 600,000 |
| Total expenditures to 12/31/X3 | 760,000 | 800,000 | 600,000 |
| Estimated expenditures to complete | 0 | 775,000 | 300,000 |
| Total estimated expenditures | 760,000 | 1,525,000 | 900,000 |
| Billings through 12/31/X3 | 800,000 | 900,000 | 750,000 |
| Cash collections through 12/31/X3 | 780,000 | 850,000 | 700,000 |
| Total G & A expense | $150,000 |  |  |

**3.5** What is meant by unbalancing a bid? What impact does unbalancing a bid have on recognition of income?

**3.6** What makes the accounting system in construction unique?

**3.7** When should losses be declared and why? What are the impacts of this rule on taxes, when using the different accounting methods?

# Chapter 4

# BOOKKEEPING FUNDAMENTALS

## 4.1 TRANSACTION PROCESSING

In order to understand how revenues and expenses are accounted for on a day-to-day basis, a basic knowledge of the bookkeeping process is necessary. A simplified diagram reflecting the sequence of actions that occurs during a typical bookkeeping operation is shown in Figure 4.1. The flow unit that is processed in the accounting sequence is called a *transaction*. A transaction occurs when a source document such as a billing from a subcontractor or a check to pay a supplier is generated. Transactions are also generated by actions required to close accounts at certain points in time. These transactions are not triggered by a specific source document but by the closing activity. Some typical types of source documents that lead to the generation of a transaction are listed in Table 4.1. By monitoring the flow of transactions through the accounting system, the manager can derive information regarding the company's balance of revenue versus expense as well as the balance sheet status of the company. Transactions are processed to reflect increase or decrease in the balances in the accounts maintained by the company. An account is an entity used to accumulate and monitor the balance fluctuations generated by some source of financial activity. The initial documentation of this activity is in terms of source documents.

Accounts are categorized in accordance with what they pertain to:

1. Assets
2. Liabilities
3. Net worth
4. Revenue
5. Expenses

Construction firms typically have a *chart of accounts* that lists all of the accounts maintained by the company. This is a kind of financial map of the company's income and expense centers and transfer points. This map determines to which area a given

**FIGURE 4.1   Transaction Processing**

Transaction ⟶ Journal ⟶ Ledgers ⟶ Trial balance ⟶ Financial statement

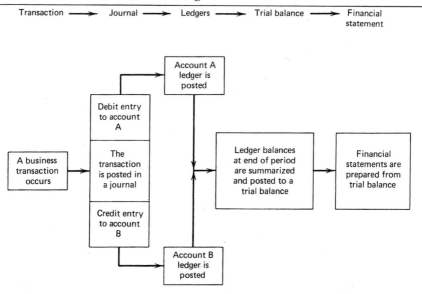

transaction is booked and how it flows through the accounting system. A typical chart of accounts for a construction company is given in Appendix A. Various trade and professional organizations publish standard *charts of accounts* that may be adopted by a particular construction company. One characteristic of construction accounting is that much of the revenue and expense activity is generated by projects and therefore expense and revenue accounts are maintained by project. A typical set of project-related expense accounts is shown in Figure 2.1.

Accounts may be maintained manually by using a standard form in a loose-leaf binder or similar physical document. Presently, computerized accounting systems are becoming so inexpensive that most companies of any size are adopting computerized systems for the financial accounting function.

**TABLE 4.1   Source Documents**

| *Examples* | *Provide Information About* |
|---|---|
| (A) Check stubs or registers | Cash disbursements |
| (B) Purchase invoices received from vendors | Purchases of materials or services |
| (C) Invoices from subcontractors for payment | Progress payment requests |
| (D) Equipment records | Rates of depreciation |

## 4.2 JOURNALIZING THE TRANSACTION

A new transaction is initially recorded in a *journal*. A journal is a chronological listing of transactions as they occur on a day-by-day basis. Journals establish the time at which transactions are generated and enter the accounting system. Journalizing involves recording the significant information regarding a transaction either when the transaction occurs, or in the chronological order in which the transaction occurs (if the record to journal is made at a time after the actual occurrence).

At the time a transaction is journalized, a decision must be made as to which accounts will be affected, and, whether the account will increase or decrease as a result of the transaction's being entered in the journal.

Accountants utilize the terms *debit* and *credit* to refer to the way in which the transaction will ultimately affect the individual account balance. Debit and credit refer to different effects, depending on what type of account is being addressed. Debits cause an increase in the balance of the asset and expense accounts, while causing a decrease in the balance of liability, net worth, and revenue accounts. Conversely, credits decrease the balance of asset and expense accounts, while increasing the balance of liability, net worth, and revenue accounts. These relationships are summarized as follows:

$$\text{Assets} = \text{liabilities} + \text{net worth} + \text{revenue} - \text{expense}$$

| Asset | | Liability | | Net Worth | | Revenue | | Expense | |
|---|---|---|---|---|---|---|---|---|---|
| *Debit* | *Credit* | *Debit* | *Credit* | *Debit* | *Credit* | *Debit* | *Credit* | *Debit* | *Credit* |
| + | − | − | + | − | + | − | + | + | − |

Each transaction causes one account to be debited and another account to be credited. The effect on the balance of the accounts affected varies in accordance with the types of accounts involved. That is, a transaction may cause the balance in one account to decrease while increasing the balance of the second account. However, some transactions cause the balances of both accounts to increase. Other transactions cause the balances of both accounts to decrease. For instance, if a transaction causes an asset account to be debited and a revenue account to be credited, the effect is to increase the balance in both accounts. If another transaction causes a liability to be debited and an asset account to be credited, the effect is to decrease the balance in both accounts. Transactions affecting accounts of the same type cause an increase in one and a decrease in the other. These effects should be kept in mind for the remainder of this chapter.

The bookkeeping process is, in fact, a method of maintaining balances between accounts. These balances are ultimately reflected in the balance sheet. Two counter-balancing entries are recorded for each transaction that is journalized. Because any transaction leads to two entries, this approach is referred to as *double-entry* book-keeping. As noted above, one of the entries debits an account, while the other credits an account. When the transaction is entered into the accounting system, the two

**FIGURE 4.2 Accounting Flows for Simple Transactions**

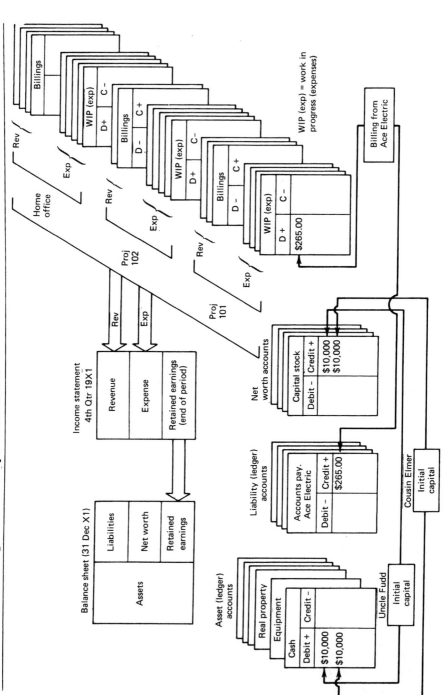

GENERAL LEDGER[a]

[a] Collectively the accounts are known as the *ledger* or *general ledger*.

52

accounts that are affected must be determined. Furthermore, it must be decided which account is debited and which is credited. A result of this double-entry system is that pairs of entries are recorded chronologically in the company journal(s).* Some typical transactions and their entry into the company journal are discussed below.

## 4.3   A TRANSACTION TO ENTER INITIALIZING CAPITAL

In order to gain an overview of the construction accounting flows, consider the schematic diagram shown in Figure 4.2. Let us assume that Cousin Elmer and Uncle Fudd have decided to form Fudd Associates, Inc. and have each contributed $10,000 cash as initial capital. This leads to entries in the asset accounts and the net worth accounts, as shown in Figure 4.2. The $20,000 contributed to form the company results in debits to the asset account called *cash* and balancing credit entries in the net worth account called *capital stock*. Here is a case where the transactions increase the balances in both accounts affected. The entry of these two transactions in the company journal would appear as shown below.

|   | Date | Description | Debit | Credit |
|---|------|-------------|-------|--------|
| | | *Journal* | | |
| 1 | Dec. 1 | Cash | $10,000 | |
| | | Elmer T. Fudd, capital | | $10,000 |
| | | (original investment in Fudd Associates) | | |
| 2 | Dec. 1 | Cash | 10,000 | |
| | | Ferdinand F. Fudd, capital | | 10,000 |
| | | (Original investment in Fudd Associates) | | |

## 4.4   A VENDOR BILLING TRANSACTION

Let us assume now that the first billing to arrive at Fudd Associates, Inc. is an invoice (request for payment) from Ace Electric Company for work done on wiring in the job site trailer on Fudd Associates Project 101. This charge of $265 must be booked against the "work-in-progress (expense)" account for this mobilization cost on Project 101.† That is, it is a project expense and must be recorded as such. Therefore, a debit (increase in balance) to the work-in-progress (expense) account must be made. The company now has an obligation to pay Ace Electric for the work performed. This obligation must also be reflected by the accounting system. Therefore, a balancing

---

* Larger companies have many centers through which source documents can enter the bookkeeping system. Therefore, journals are maintained at each source document processing center. In construction, for instance, journals could be maintained for each project as well as home office journals.

† As will be discussed later, mobilization is actually a subaccount or subsidiary account of the general ledger account work-in-progress expense—Project 101.

entry—a credit to the outstanding balance payable to Ace Electric—must also be made. This is made to the liability account area to an account payable called Ace Electric. Obligations to pay vendors, subcontractors, and other suppliers are reflected in the accounts payable area of the liability general ledger accounts. As noted in Figure 4.2, the total of all active accounts is referred to in accounting parlance as the general ledger. The entries to the company journal generated by the billing from Ace Electric are shown below. Convention in recording transactions requires that the debit be entered first and then the credit. This leads to a debit upper left, credit lower right appearance.

|   | *Journal* | | | |
|---|---|---|---|---|
|   | *Date* | *Description* | *Debit* | *Credit* |
| 1 | 5 December 19X1 | Work-in-progress expense—Project 101 | $265.00 | |
| 2 | | Ace Electric invoice # 101-1 | | $265.00 |

## 4.5  A BILLING TO THE CLIENT

Now let us assume that there is a mobilization payment clause in Contract 101 and that Fudd Associates sends a billing to the client, Donut Factory, Inc., requesting payment of the mobilization amount of $5000. As shown in Figure 4.3, this will cause the project billings account (Billings—Project 101) to be credited and the asset account—accounts receivable—to be debited. The request for payment is dated 7 December 19X1, and the journal entry for this transaction is shown below.

|   | *Journal* | | | |
|---|---|---|---|---|
|   | *Date* | *Description* | *Debit* | *Credit* |
| 1 | 7 December 19X1 | Accounts receivable—Job 101 | $5000 | |
| 2 | | Billings—Job 101 | | $5000 |
| 3 | | | | |
| 4 | | | | |
| 5 | | | | |

In the last two transactions, we have made entries based on the fact that obligations to make payment have occurred. In journalizing the invoice from Ace Electric Co., Fudd Associates, Inc. was recognizing an obligation to pay for something. However, payment has not been made; cash has not been transferred. The obligation to pay has been noted in the accounts payable designated Ace Electric. It is recognized effective 5 December 19X1.

In sending a billing to Donut Factory, Inc., for a mobilization payment, Fudd Associates is establishing an obligation on the part of the client to pay revenue. Again, no monies have been transferred. However, for accounting purposes, the impact on the various account balances of this transaction has been recognized effective 7 December 19X1.

FIGURE 4.3  Billing to Client (Billings Basis)

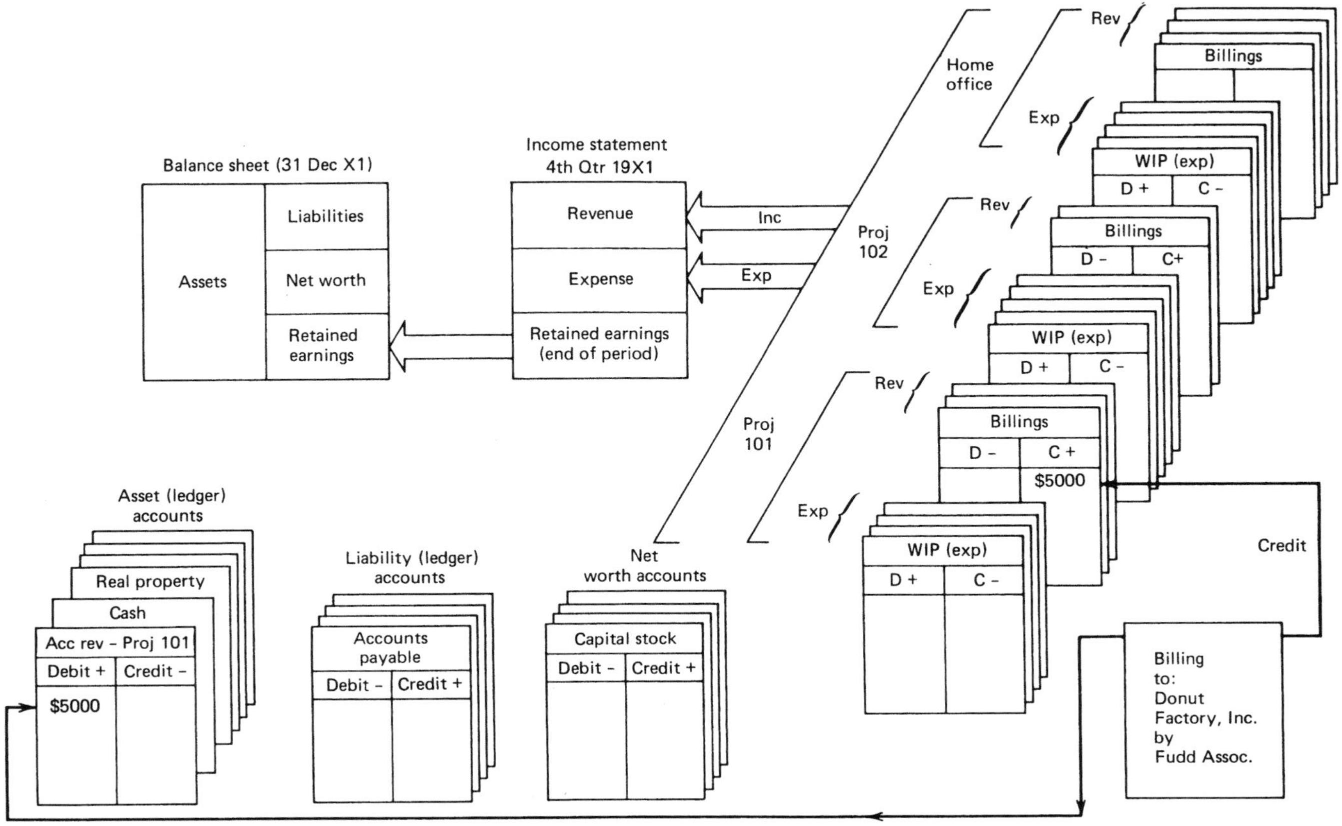

55

## 4.6   ENTRIES USING THE CASH BASIS

In Sections 4.4 and 4.5 we have been using the *accrual* basis. This basis provides that a transaction is recognized at the time that it occurs. That is, revenues and expenses are recognized as they are earned or incurred. Some firms prefer to use the *cash* basis of accounting. In this method, transactions are recognized only when payment is received (in the case of a revenue) or when payment is made (in the case of an expense). In other words, cash must be transferred before the transaction is journalized.

To illustrate, let us assume that Fudd Associates is using the *cash* basis of accounting during the first month of its operations (December 19X1). Under this method the obligation to Ace Electric Co. will be recognized only when actually paid by Fudd Associates. Similarly, the revenue from Donut Factory, Inc. associated with the billing for mobilization will be recognized only when payment is received. Let us assume that Fudd sends a check of payment to Ace Electric on 11 December. Further, assume that a check for the $5000 mobilization payment is received from the client on 21 December. Under the cash basis, the three journal entries we have discussed so far would appear as follows:

| | Date | Description | Debit | Credit |
|---|---|---|---|---|
| | | *Journal* | | |
| | 19X1 | | | |
| 1 | 1 December | Cash | $10,000.00 | |
| | | Elmer T. Fudd, Capital | | $10,000.00 |
| | | Cash | 10,000.00 | |
| | | Ferdinand F. Fudd, Capital | | 10,000.00 |
| 2 | 11 December | Mobilization Expense—Project 101 | 265.00 | |
| | | Cash | | 265.00 |
| 3 | 21 December | Cash | 5,000.00 | |
| | | Billings—Job 101 | | 5,000.00 |

When the cash method is used, the payment to Ace is recorded on 11 December and results in decreasing (crediting) the asset account—*cash*. Similarly, receipt of the payment from the client on 21 December results in a debit (increase) to the cash account balance on that date. Under this method, the accounts payable and accounts receivable have not been utilized. Further illustrations of typical construction transactions in the context of the flow diagrams of Figures 4.2 and 4.3 are given in Appendix B.

## 4.7   POSTING ENTRIES TO THE LEDGER

Periodically, data are transferred from the journal(s) to the appropriate accounts in the ledger. This process is called *posting*. Figures 4.2 and 4.3 illustrate schematically the accounts in the ledger that will be posted. The objective of posting is to record to the appropriate accounts the impact of various transactions so that at selected times the balance of entries in all accounts can be checked. As mentioned previously, accounting is the process of maintaining the balance implied by the basic equation assets = liabilities + net worth + (revenue − expenses.) From time to time, it is necessary to check this balance to make sure that all entries are consistent. The action of checking the balance is called taking a trial balance. This is usually done following the posting to verify the entries that have just been transferred. Posting may be done on a daily basis or at intervals as deemed necessary. Information reflected by the general ledger regarding the financial status of the company is not considered reliable until posting of all entries from the journal has been accomplished.

In order to illustrate posting, let us post the entries for the month of December journalized for Fudd Associates, Inc., using the cash basis. The flow of the entries from the journal to the various ledger accounts is shown in Figure 4.4. The accounts as shown in the figure can be thought of as pages in a loose-leaf ledger book.

Based on the eight entries in the journal (four transactions), posting results in four entries in the cash ledger and individual entries in (1) the net worth capital accounts for Elmer and Ferdinand Fudd, (2) the billings account for project 101, and (3) the mobilization expense account for Project 101. Manually, this transfer can be time-consuming and tedious. Therefore, bookkeeping operations are ideally suited for automation. Business machine companies over the years have developed partially mechanized, partially electronically automated machines to expedite both journalizing and posting. These bookkeeping machines evolved into total electronic automation with the advent of the computer. Initially large computer systems were used, and the accounting was accomplished on-line with computers at remote service bureaus or for larger companies with in-house computers. With the advent of the very small (micro) computers, now even small companies have access to completely computerized bookkeeping.

In order to verify the entries made up to any point in time, a trial balance can be made by summarizing the entries in all active accounts and checking to see that the sum of debits and credits in all accounts balance. Assume that a trial balance for the posting activity in Figure 4.4 is accomplished in connection with closing the books as of 31 December 19X1. The trial balance sheet for this operation is shown below.

**Trial Balances**
**Fudd Associates, Inc.**
**(31 December 19X1)**

| Account | Account No. | Debit Balance | Credit Balance |
|---|---|---|---|
| Cash | | $24,735.00 | |
| E. Fudd—capital | | | $10,000.00 |
| F. Fudd—capital | | | 10,000.00 |
| Billings—Project 101 | | | 5,000.00 |
| Expense—project 101 | | 265.00 | |
| | | $25,000.00 | $25,000.00 |

As can be seen, as of 31 December Fudd Associates has five active accounts. The account balances have been summarized and entered on the trial balance sheet. Since

FIGURE 4.4  Posting Journal Entries to Ledger Accounts (Cash Basis)

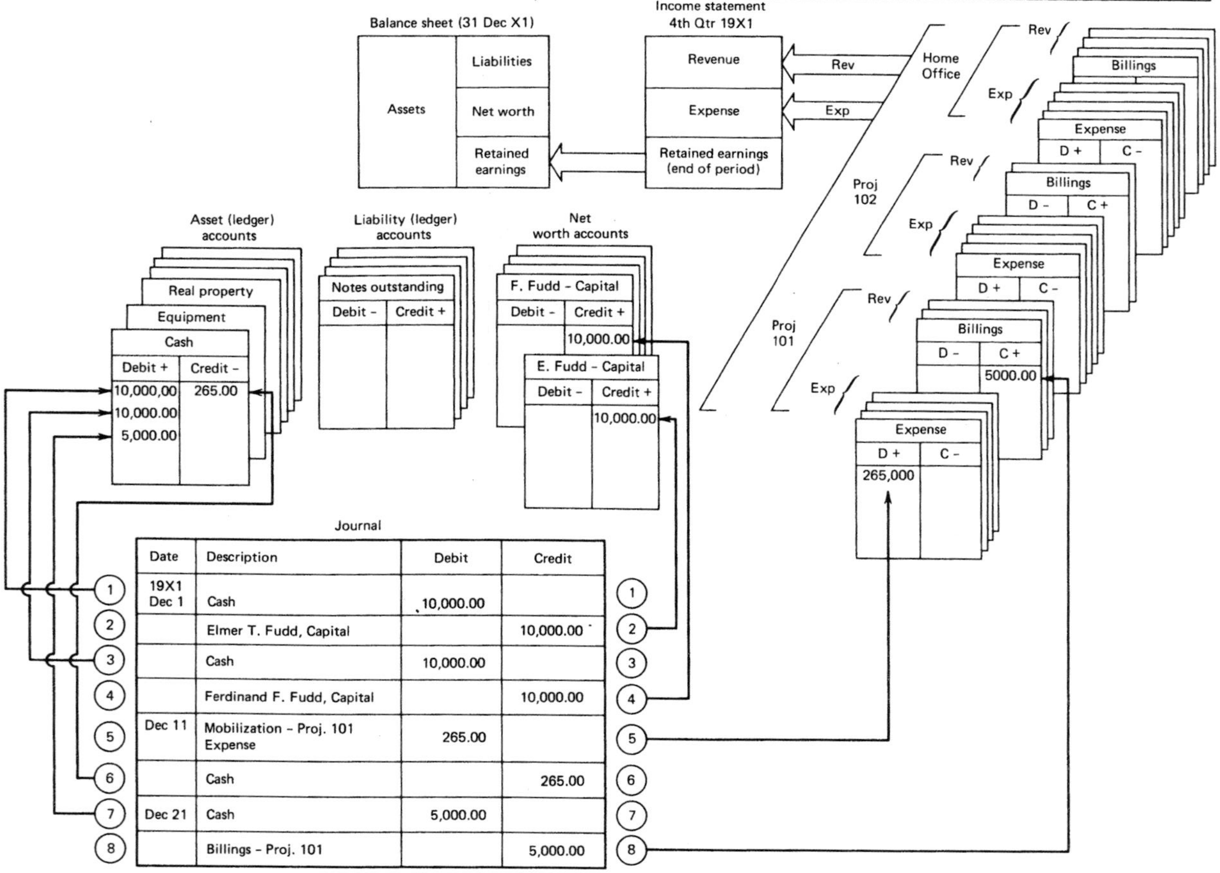

**FIGURE 4.5   Model Balance Sheet, Fudd Associates, Inc. (December 19X1 31)**

| Assets | | Liabilities | |
|---|---|---|---|
| Cash | $24,735.00 | *Owner's Equity* | |
| | | E. Fudd—capital | $10,000 |
| | | 1 December 19X1 | |
| | | F. Fudd—Capital | 10,000 |
| | | 1 December 19X1 | |
| | | Retained Earnings | $ 4,735[a] |
| Total assets | $24,735.00 | Total liabilities and owner's equity | $24,735 |

[a]Income is based on revenue minus expense, which in the example is the $5000 mobilization payment minus the $265 paid to Ace Electric.

four accounts have single entries, these are entered directly. The cash account balance is based on four entries, and the account has a debit balance (Dr balance)* of $24,735.00. The balance of accounts and of the basic accounting equation is verified in that the Dr balance and the Cr balance for all accounts are each $25,000. If the summation of Dr and Cr balances does not balance, the journalizing and posting must be reviewed to determine what error has occurred. If the balances are consistent, this in itself does not ensure that no errors have been made. Wrong accounts may have been debited and/or credited, but entries will balance. All that is checked is the basic requirement for a balance between debits and credits. Whether or not the debits and credits are posted to the proper accounts requires further review.

The ultimate purpose of bookkeeping operations is to produce financial statements that reflect the financial condition and health of the firm. Production of these documents typically occurs at the end of an accounting period. The operations required to organize the data contained in accounts for financial statement production are referred to as *closing* activities. Closing activities are discussed in detail in Chapter 5. Based on the month of December operations for Fudd Associates, the balance sheet and income statement produced by closing accounts on 31 December are shown in Figures 4.5 and 4.6.

* The abbreviation Dr is used for debit and Cr for credit.

**FIGURE 4.6   Model Income Statement, Fudd Associates, Inc. (For the Month Ended 31 December 19X1)**

| | |
|---|---|
| *Revenue* | |
| Income from construction contracts | $ 5000 |
| *Expenses* | |
| Project expenses | |
| | 265 |
| Net income | $ 4735 |

## 4.8   RELATIONSHIP OF IN-PROGRESS AND REVENUE/EXPENSE ACCOUNTS

Terminology regarding project revenues and expenses varies from firm to firm and project to project, depending upon the method of accounting used. However, the term "work in progress" is used broadly in accounting for billings and costs associated with projects under any of the accrual-based methods of accounting. In general, billings (revenue-generating documents) and costs are accounted for as work-in-progress (WIP) transactions until such time as revenue (based on billings) and expenses (based on invoices, etc.) are recognized in a formal sense for income calculation purposes. This recognition occurs at the end of the project when the CC method is used. It occurs for POC jobs at the time that percentage completion is calculated for income recognition purposes (typically at the end of an accounting period). Until the formal recognition of revenue and expense, the billings are *WIP-billings* and are considered liabilities pending declaration as revenue. The expenses are *WIP-cost* or *WIP-expense,* and as such are temporarily assets until they are transferred to the expense accounts for income recognition. When income recognition occurs, the WIP-billings (liability) become revenue and the WIP-costs (assets) become expenses. Figure 4.7 demonstrates this concept schematically.

The terminology "work in progress" (as a prefix) may appear in a company's books as construction in progress (CIP), contracts in process, or some similar phrase. The general idea is, however, that both billings and costs are "in progress" until the moment at which they are declared revenue and expense for purposes of income recognition. Therefore, at the time billings and costs enter the bookkeeping system, they are typically carried in work in progress accounts until the point at which clearing and closing of accounts occurs. This is a subtle point that has not been emphasized previously in discussing the flow of project-related transactions. A better understanding can be gained

**FIGURE 4.7   Concept of Work in Progress**

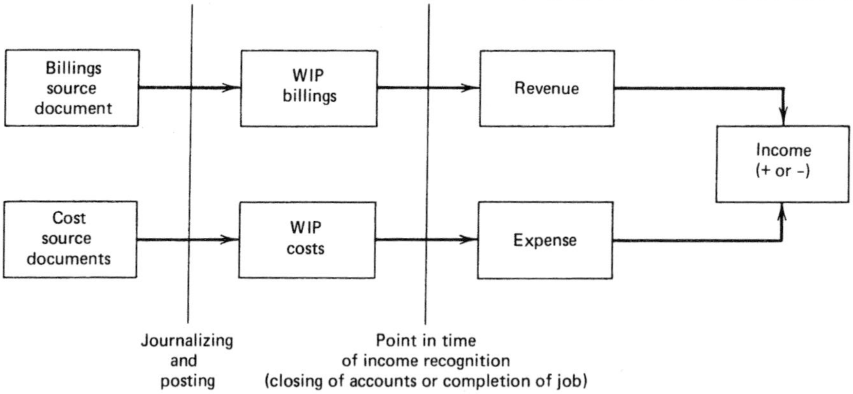

[a] Invoices and payroll.

by studying transactions 2, 3, 6, and 8 in Appendix B. This point should, however, be kept in mind when reading more advanced materials regarding financial accounting.

One additional point deserves comment. At the time of closing, some revenue earned has not been billed to the client, and some expenses incurred may not have been documented in terms of an invoice from the vendor or a billing from the appropriate subcontractor. Although these transactions have *accrued,* they have not been documented in terms of source documents. Since they have occurred, it is important to include them in the financial statements. Payroll is a good example. If the office staff is paid monthly, half of their pay will have accrued as an obligation by the fifteenth of the month. Similarly, half of a month's rent will accrue and must be reflected if a financial report is prepared as of the fifteenth of the month.*

The key idea is that even though a transaction is not documented in terms of a source document, if in fact a revenue or expense has accrued, this transaction should be accounted for at the time of closing accounts. Therefore, revenue earned but not billed should be recognized at closing, just as expenses incurred or accrued for which invoices or billings have not been received should also be recognized. Methods of handling these *accruals* are discussed in detail in bookkeeping and accounting texts.

## 4.9 A BASIC ACCOUNTING SYSTEM†

The level of complexity of a particular company's accounts varies, depending upon the level of information that the system must produce and the method by which the company management desires to have this information presented. The minimum level of complexity is defined by the need to present documents to the Internal Revenue Service to support payment of taxes. Banks and other lenders require certain information concerning a company before they will lend money. Depending upon the types of financial detail lenders require, management may decide that a higher level of detail (greater than required for basic tax purposes) is required.

Most construction companies are project-oriented and desire to have information regarding how revenues are generated on a particular project and how expenses are progressing project by project. Consequently, there is a need to select, sort, and present financial information on a project-by-project basis. Sorting of data on a project basis is not required by law; it is done to provide company management information for controlling work progress and bidding future projects. This project orientation leads to a hierarchy of detail, which in turn determines the level of complexity involved in journalizing and posting accounts.

---

* Some other types of accrued liabilities are listed in Section 1.8.

† This section is based on material presented by Felix P. Kollaritsch, *Job Order Systems for Highway-Heavy Contractors,* (Columbus: College of Administrative Science, The Ohio State University, 1974) Chapter 2.

Let us assume that Fudd Associates, Inc. wants to simplify its accounting activity so that statements can be prepared for tax purposes but that no segregation of data is required by project. Assume that the company's business activity for a given period is as shown below.

| Contract Status | Contract Costs | Contract Revenues |
|---|---|---|
| Completed | $1,300,000 | $1,600,000 |
| Uncompleted | 900,000 | 1,000,000 |
| Total | $2,200,000 | $2,600,000 |

A minimum number of accounts can be used to collect this information and summarize it for the purpose of statement development. The accounts required and the process of income determination and financial position development are shown schematically in Figure 4.8.

Figure 4.8 indicates that the four general ledger accounts required for this most simple of systems are:

1. Contract revenues
2. Contract expenses
3. Accounts receivable
4. Accounts payable or cash

An income summary account is utilized as a clearing account to feed information to the income statement. Four basic accounting flows occur during the accounting cycle

### FIGURE 4.8   Basic System Number 1

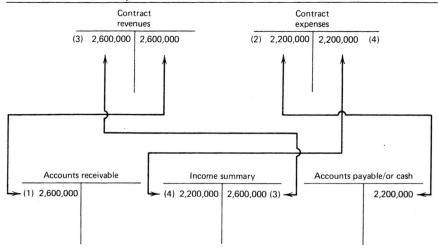

and are captured by these accounts. These four flows are indicated by labels in the diagram in Figure 4.8.

1.  As billings occur, this activity is ultimately accumulated in the contract revenue account and the accounts receivable. Balances in both accounts increase. The total revenue of $2,600,000 is shown as a single entry in both accounts, although this entry represents all billings to clients (i.e., multiple entries).
2.  The expenses incurred result in credit entries to the accounts payable (or cash) account and ultimately as debit entries in the account "contract expense." Again, the single entries represent many entries occurring during the period under consideration. These contract costs are expenditures for labor, materials, subcontracts, and the like.
3.  At the end of the accounting cycle (or period), closing entries are made to calculate income and establish financial position. The accumulated contract revenues on the work completed are consolidated and credited to the income summary account. This action closes the contract revenue account to zero (by the offsetting debit entry).
4.  Similarly, at the end of the accounting cycle, a closing entry to move the accumulated expense to the income summary account is also made. This causes the contract expense account to be credited in the amount of $2,200,000 (thus closing it to a zero balance). The income summary account is debited in the amount of the accumulated expense.

The balance in the income summary account at the end of the period will be a credit balance of $400,000. This $400,000 is recognized as income and ultimately moved to the retained earnings portion of the balance sheet.

|  | Dr | Cr |
|---|---|---|
| Income summary | $400,000 | |
| Retained earnings | | $400,000 |

This simple system effectively satisfies the need to report income and establish financial position. It does not, however, yield any information about which projects generated revenue and the level of expense incurred project by project. All revenues and expenses are lumped together. No summarization of data by project is available.

## 4.10   PROJECT- OR JOB-ORIENTED SYSTEM

A more detailed system can be used to break down costs and revenues on a project-by-project basis. Assume the following project-by-project distribution of revenue and expense for the period under consideration:

|  |  | Project Expenses | Project Revenues |
|---|---|---|---|
| Project 101 | Completed | $700,000 | $ 825,000 |
| Project 102 | Completed | 600,000 | 775,000 |
| Project 103 | Uncompleted | 900,000 | 1,000,000 |

### FIGURE 4.9    Basic System Number 2

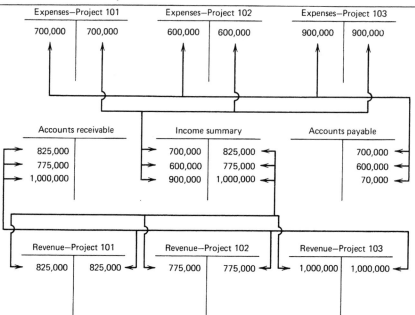

In this case the contract revenue account is divided into an account for each project. Similarly, accounts for revenues (WIP-billings) on each project are maintained. The flow of entries to these accounts during the accounting period and at the end of a period is shown in Figure 4.9.

As expenses are accrued on each project, they are posted to each project expense (WIP-expense) account and to accounts payable. The accumulated costs (consisting of many individual entries) on Project 101 are represented here by a single entry of $700,000 to the Project 101 expense account and an entry of $700,000 to accounts payable.* Similarly, as revenues are generated and billed they are ultimately posted to the account *revenue—project* and accounts receivable. On Project 101 this results in debiting (increasing) accounts receivable by a total amount of $825,000† and crediting (increasing) the account revenue—Project 101. Comparable entries are made on Projects 102 and 103. At the end of the accounting period, these entries are "closed" to the income summary account by crediting (decreasing to zero in this case) the expense accounts and debiting (decreasing to zero) the revenue accounts. This results

---

* Actually, many entries generated by invoices from vendors, billings from subcontractors, and the like would make up the single entry shown here. For this illustration the multiple entries are consolidated into a single entry.

† Again, this figure is a consolidation of many billings during the period.

in transfer of the expenses on a project-by-project basis to the income summary account (debits to this account). Compensating movement of the revenue amounts, resulting in credits to the income summary, is also made at this time. This allows a project-by-project comparison of revenues versus expense. The closing activity "clears" or brings to zero the balance in revenue and expense accounts.

## 4.11   A HIERARCHY OF ACCOUNTS

Most construction companies need more precise information regarding where and how *expenses* are being incurred. This information is essential (1) to control the project by comparing actual field costs with originally estimated costs and (2) to use in estimating future projects. It is relatively simple to aggregate all expenses into a single general ledger account as described in Section 4.9 or simply to post them to individual project expense accounts as done in Section 4.10. In both cases, however, no information is available on the relationship of costs to individual parts of the project. Therefore, information as to the cost of placing a particular steel beam or excavating a particular trench is lost in the consolidation of data into a relatively small number of general ledger accounts.

To achieve a better level of detail and a sorting and presentation of data that is more useful to management for control purposes, a system of *subsidiary ledger accounts* that are subordinate to the general ledger accounts can be established. Subsidiary accounts are simply subaccounts of general ledger accounts. They are maintained to provide greater detail regarding the aggregate information presented in the general ledger accounts. It is possible to have many levels of subsidiary accounts.

Typically, a project is subdivided into physical entities such as excavation, concrete work, and so on. These physical subentities in a project can be addressed as expense "line items" associated with the project. Assume that each of the three projects in Section 4.10 consists of five major physical subentities. Let us assume that these subentities are

1.   Excavation
2.   Foundation work
3.   Structural system
4.   Exterior finish
5.   Interior finish

Each of these subentities can be considered "line items" within the project. Subsidiary accounts to collect expense data on each of these subdivisions can be established to gain a better detail regarding expense activity within the project. Figure 4.10 shows schematically how the subsidiary ledgers collect data and support the entries in the general ledger account expense—Project 103 for the months of November and December 19X2.

**FIGURE 4.10  Subsidiary Accounts Supporting a General Ledger Account**

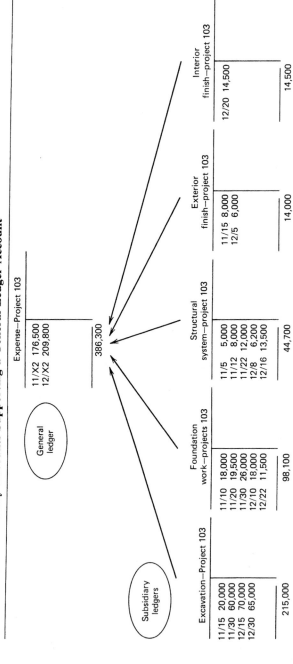

**FIGURE 4.11   Hierarchy of Expense Subsidiary Accounts**

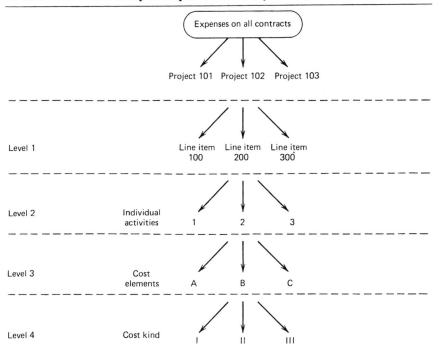

The entries in the subsidiary ledger accounts are carried forward to the general ledger account by "double debiting." That is, a debit to a subsidiary account generates a corresponding debit to the general ledger account, so a single entry results in a "double debit."* Posting to a subsidiary account such as excavation—Project 103 will normally occur throughout the accounting period as required. In this illustration, the debits associated with subsidiary accounts are consolidated and posted as a consolidated debit to the general ledger account at the end of each month. The general ledger account is essentially brought into balance with the subsidiary accounts at the end of each month. Obviously, the information in the subsidiary accounts provides greater detail.

The number of subsidiary accounts used by management can be expanded into a hierarchy of many levels. One hierarchy useful for construction projects is proposed by Kollaritsch (1974) and is shown in Figure 4.11.

As shown in Figure 4.11, the individual contract expense general ledger accounts are supported by four subordinate levels at which subsidiary ledgers can be maintained.

---

* The normal transaction generates a debit and a credit. This transaction would generate a single credit to offset a hierarchical set of two debits—one to the subaccount and one to the general ledger account. The amount of the debits at each level is equal.

The choice of level depends upon the desires of management for detail. For instance, at level 1 (line item) the excavation subsidiary ledger might be supported by second-level (activity) subsidiary accounts such as:

1. Stripping over burden
2. Blasting
3. Hauling

Within the level 2 account, hauling, a further subdivision to monitor labor, equipment, and subcontracts at level 3 can be maintained. Within the subcontracts level 3 subsidiary account, separate level 4 accounts for each of the subcontractors could be maintained. Figure 4.12 shows this hierarchy.

The four levels of subsidiary ledgers below contract level imply that debiting occurs in the four subsidiary ledgers and at the contract and general ledger levels. In this case, rather than having a "double debiting" situation, debits at six different levels must be posted. Manually posting is, of course, tedious, and considerable thought should be given to the time and effort that must be expended to maintain this kind of accounting control. In manual systems, lags occur between posting the lower-level subsidiary accounts and the corresponding parent accounts, which lead to the possibility of temporary imbalances between upper- and lower-level accounts. These difficulties can be eliminated to some degree by using computerized accounting systems, which can post all levels at virtually the same time.

A comprehensive schematic of the multilevel general and subsidiary ledger accounting approach just discussed is shown in Figure 4.13. This multilevel approach will be the basis for discussing expense coding systems and other cost-control-related subjects in later chapters. The level of complexity can be very great. If ledger accounts are maintained at "line item" level for four projects consisting of 20 line items, the required number of accounts is 80. If each of the line item accounts is supported by four activity level accounts, the number of total accounts expands to 320.

Adding four "cost element" accounts for each activity account and five "cost kind" accounts for each cost element would lead to a system of 6400 accounts to be posted to support the four contract level expense summaries ($4 \times 20 \times 4 \times 4 \times 5$). Again, posting this number of accounts manually is very tedious, a fact that raises the question as to whether the information captured for management control justifies the effort.

## 4.12  SUMMARY

A contractor must understand the basic nature of the bookkeeping process in order to follow the flow of revenue and expenses in connection with a project and to control costs properly. This chapter has presented some of the basic operations that are performed as part of the bookkeeping process. It has also introduced the concept of general and subsidiary accounts, which establish the context within which accounting flows occur. The next chapter deals with operations required to close accounts, produce financial documents, and recognize income for taxation purposes and performance evaluation.

**FIGURE 4.12 Four Levels of Subsidiary Ledgers**

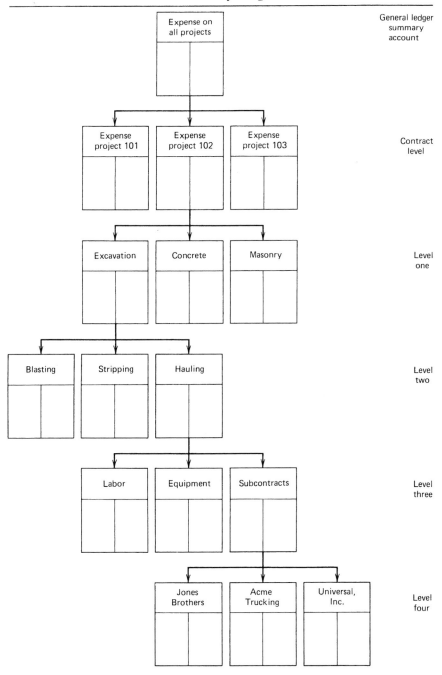

## FIGURE 4.13  A Comprehensive Multilevel Cost Accounting System

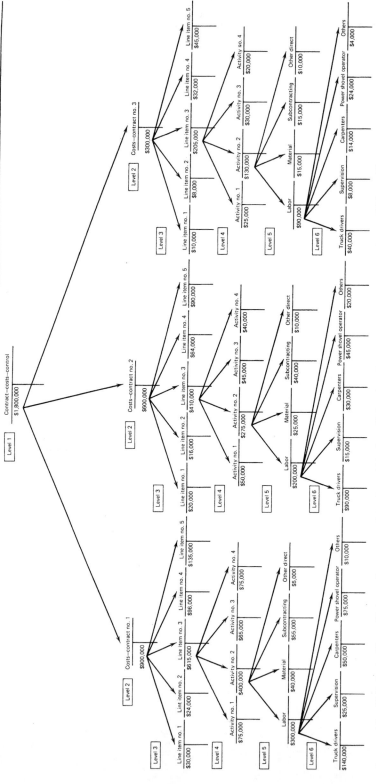

*Source:* After Kollaritsch, *Heavy-Highway Contractors* (Columbus: Ohio State University, College of Administrative Science, 1974), p. 20.

## REVIEW QUESTIONS AND EXERCISES

**4.1**  What is the difference between journalizing and posting?

**4.2**  Is it possible to post into an asset account and a revenue account? Give two examples with debiting and crediting an asset account.

**4.3**  What is meant by double debiting? When would it be used?

**4.4**  Today is 3 February 19X6. You are using the accrual method of accounting. You have received an invoice from Halco Industries in the amount of $565 for work performed on Project 101. You also mailed an invoice to Ajax International for payment of work performed by your company in the amount of $6,238 on Project 101. On 10 February you send a check to Halco to cover their billing and receive a check from Ajax on 12 February in the amount of your billing to them.

Make the proper journal entries on the sheet captioned as below.

### Journal

| Trans # | Date | Description | Debit | Credit |
|---------|------|-------------|-------|--------|
|         |      |             |       |        |
|         |      |             |       |        |
|         |      |             |       |        |
|         |      |             |       |        |
|         |      |             |       |        |

**4.5**  Journalize the following transactions for Tip Top Associates, Inc., incurred during the month of July 19X4 (assume accrual basis):

| | | |
|---|---|---|
| 6 July | Initializing capital deposited in bank | $12,000 |
| 10 July | Office supplies purchased | 250 |
| 12 July | Purchase computer | 3,500 |
| | (paying $1000 with balance on account) | |
| 15 July | Office rent | 425 |
| 20 July | Billing to Jones County Board of Education | 1,255 |
| 25 July | Salaries—employee payroll | 655 |
| 28 July | Paid on account for computer | 800 |
| 30 July | Cost of expended supplies | 145 |

(a) Prepare a balance sheet as of 30 July 19X6 for a period of one month.

(b) Prepare an income statement for the month of July.

**4.6**  Using the completed-contract method, make appropriate journal entries to document the following transactions:

| 15 January 19X3 | Contractor buys $300,000 of construction materials to be used on a project. |
| 10 February 19X3 | Engineers on project are paid $20,000. |
| 20 March 19X3 | Payroll for labor on project—$170,000. |
| 5 April 19X3 | Billings to client in amount of $700,000 (invoice 83-106). |
| 10 August 19X3 | 630,000 received from client—$70,000 held as retainage. |
| 20 December 19X3 | $160,000 is disbursed in payment of rental charges for project equipment. |
| 15 February 19X4 | Billings to client in the amount of $100,000 (invoice 84-23). |
| 15 March 19X4 (end of project) | Contractor receives $100,000 previously billed plus $70,000 held as retainage. |

**4.7**   Transactions during the period 1 December 19X5 to 30 November 19X6 are given below for ABC Construction:

| | | |
|---|---|---|
| (1) | Total income (fees) from construction contracts | − $400,000 |
| (2) | Material inventory purchased on account | 80,000 |
| (3) | Total office rent paid | 20,000 |
| (4) | Salaries paid to engineers | 90,000 |
| (5) | Accounts receivable collected | 60,000 |
| (6) | Accounts payable paid | 30,000 |
| (7) | Subcontractor accounts pair | 70,000 |
| (8) | Bidding expenses | 20,000 |
| (9) | Dividends paid to stockholders | 10,000 |
| (10) | Building depreciation | 40,000 |
| (11) | Construction equipment depreciation | 30,000 |

Using the transactions presented in Appendix B as a guide and the billings method:

(a) Journalize all of the above transactions.
(b) Post journal entries to appropriate accounts.
(c) Develop income statement for period 1 December 19X5 to 30 November 19X6.
(d) Develop the balance sheet for ABC Construction as of 30 November 19X6.

# Chapter 5

# FINANCIAL STATEMENT PREPARATION

## 5.1 DEFINING THE ACCOUNTING CYCLE

At certain times, normally at the end of accounting periods, it is necessary to check on the financial health of the firm and its operations. Accounting periods are determined on the basis of management needs and requirements of outside agencies for information concerning the financial strength of the firm. Accounting periods may be as long as one year. A shorter period, such as a quarter or a month, may be selected as required. The so-called accounting cycle is terminated at the end of the accounting period by closing accounts and generating financial statements. In this chapter, actions required to close accounts and generate statements under the percentage-of-completion and completed-contract methods are discussed.

## 5.2 CLOSING THE ACCOUNTING CYCLE

At the end of the accounting period, accounts are closed so as to recognize income and expense flows that should be picked up in the balance sheet, income statement, and other relevant financial documents. The process of closing accounts causes incomes and expenses to be transferred from *nominal* accounts to *real* accounts. Real accounts remain open from period to period and their period end balances are reflected in the financial documents (e.g., balance sheet and the like). Nominal accounts are closed at the end of the accounting period by bringing their balances to zero. Revenue and expense accounts, such as those associated with projects and other profit centers, are referred to as nominal accounts. At the end of the accounting period, the balances of these accounts are cleared out or transferred to real accounts, which link to reported items of summary information in the balance sheet and income statement. This process of clearing the nominal accounts brings the balance in these accounts to zero. The summarization of this "clearing" action appears in the real accounts. Real account balances are reflected in the income statement and balance sheet. This act of closing nominal accounts results in income recognition (based on earned value) if the percentage-of-completion method is used. If the completed-contract method is used, income is recognized during this closing process provided that a project (or projects) is completed during the accounting period being closed. The key balance sheet accounts indicative of project activity vary in accordance with the method of accounting being used.

## 5.3   TRANSACTIONS DURING A PERIOD

In order to understand this process of closing and statement preparation it is necessary to consider the activity during an accounting period. During a given period revenues and expenses are generated on a random or as required basis. These revenues and expenses are journalized and ultimately transferred to the general ledger accounts. The chart of accounts (see Appendix A) being used by a firm is a listing of all of the ledger accounts. In the following discussion, the general ledger accounts and the chart of accounts can be thought of as synonymous. The chart of accounts is organized into

1.   Revenue accounts
2.   Expense accounts
3.   Asset accounts
4.   Liability accounts
5.   Net worth (or equity) accounts

In other words, the chart of accounts is organized so as to provide feeder information to the financial statements. By examining the accounts in Appendix A, it is possible to recognize this breakdown of accounts into the five categories listed above. The movement of transactions to the general ledger accounts is shown in Figure 5.1.

Since in construction the focus of production is job- or project-oriented, the generation of revenue and expense transactions is generally associated with a particular job or project.* As can be seen in Figure 5.1, most of the expenses associated with a project are generated in the form of payroll transactions or from the accounts payable area. These expense transactions are journalized in the payroll and accounts payable journals and are ultimately posted to the appropriate general ledger accounts in the chart of accounts. Similar billings to the client are journalized in the accounts receivable journal and to the appropriate general ledger accounts.

Posting of these transactions to the general ledger typically occurs as a bulk or batch processing operation, which lumps different expenses together without trying to sort them for purposes of identification. That is, the weekly expenses on Job 101 will be consolidated and batch posted to the single GL account (i.e., work-in-progress–expense—Job 101). This procedure means that no attempt to relate the particular

---

* Nonproject-related expenses are posted to the appropriate G&A accounts in the general ledger.

**FIGURE 5.1 General Accounting Flows**

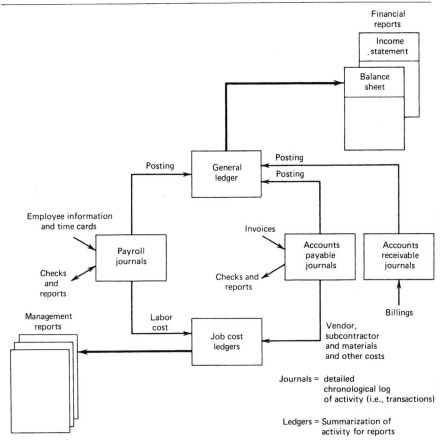

Journals = detailed
chronological log
of activity (i.e., transactions)

Ledgers = Summarization of
activity for reports

expense with a physical or cost component of the job is made. Of course, if subsidiary ledgers as described in Chapter 4 are used, a certain degree of linkage between cost and a physical subdivision can be achieved by multiple-level posting.

One alternative to the multilevel posting approach is to maintain a separate set of *job cost accounts*. These job cost accounts can be thought of as special parallel accounts established for each project. The specific purpose of these project accounts is to capture information regarding the linkage between various expenses and the part of the project with which they are associated. This approach is based upon feeding information to the detailed cost accounts at the same time that posting to the GL account occurs. A typical listing of project cost accounts was given in Figure 2.1. The simultaneous flow of information to the chart of accounts (GL accounts) and to the job cost accounts is shown schematically in Figure 5.1. The operation of the *job cost accounts* is discussed in detail in Chapter 7. It is sufficient at this point to understand that these accounts are defined to obtain detailed information relating expenses to particular cost centers in the job or project.

If a multilevel posting system is used to transfer expense information to these job-cost-related accounts, they may be considered subsidiary ledgers within the GL system. As such, they are subject to the debit-credit balancing requirements of a double-entry system. As has been noted, the posting process may be very tedious.

If the accounts are maintained separately from the GL system, they operate to collect costs within a detailed breakdown of the project's subelements. As such, they are not subject to the balancing requirements and multilevel posting described in Chapter 4. In this configuration, the job cost system can be thought of as a *single-entry* system designed simply to log or register costs against appropriate cost centers or subdivisions of the project. Depending upon the desires of management, this system can be maintained so as to remain in balance with the appropriate GL summary project expense account in the general ledger, or it can be allowed to become unbalanced as a result of its being used to maintain data for purposes other than balance sheet preparation. It is common to use the term *expense account* to refer to general ledger accounts listed in the chart of accounts. The term *cost accounts* is used to refer to job cost system (detailed) accounts whether they are considered subsidiary accounts of the general ledger or free-standing single entry accounts. The remainder of this chapter will discuss GL posting requiring balanced entries.

## 5.4  CONSOLIDATION ACCOUNTS BASED ON METHOD USED

The accounts within the general ledger used to aggregate period revenues and expenses differ in function, depending upon whether the percentage-of-completion or completed-contract method is used on a given project. It is, of course, possible to use the POC method on some projects and the CC method on others. The method of income recognition (POC or CC) attaches to the individual project.

The percentage-of-completion method recognizes income on an incremental or period-by-period basis. When using the POC method, the bookkeeper establishes an expense account within the expense area of the chart of accounts to aggregate costs associated with a particular job (e.g., work-in-progress–expense—Job 101). Similarly, when the POC method is used on a project, a billings account for the particular job is established within the revenue area of the chart of accounts (e.g., billings—Job 101).

When the CC method is used, expense and revenue, as such, are not recognized until the end of the job. Therefore, costs associated with a project are owned by the contractor and are posted to the asset account—*Construction in progress* for the job under consideration. Billings on a completed contract project are considered an obligation or advance and are posted to the liability account—*Advanced billings* for the particular job. These accounts are established within the chart of accounts.

Terminology for revenue and expense accounts under both POC and CC methods varies widely. In an effort to use a broad terminology to cover both methods, revenue accounts are sometimes called

1. Billings
2. Advanced billings
3. Work in progress billings

to name just a few of the designations in use. Similarly, expense accounts are called

1. Work in progress
2. Construction in progress
3. Work in progress (expense)
4. Work in progress (cost)

Any of these designations is acceptable. For the discussion in this and the following sections, the terminology in Table 5.1 is used.

## 5.5 POSTING THE GENERAL LEDGER DURING AN ACCOUNTING PERIOD

In order to understand the operation of the general ledger during a given accounting period, consider the situation presented in Figure 5.2. It is assumed that Apex Construction has five projects in progress during the period under consideration. These projects are numbered from 101 to 105. The POC method is being used on Projects 101 and 102. The CC method is being used on the remaining projects. Project 105 is completed during the period. This situation is summarized in Table 5.2.

As can be seen from Figure 5.2, billings associated with Projects 101 and 102 are posted to revenue accounts within the chart of accounts (i.e., billings—101 and billings—102). Similarly, expenses associated with these jobs are posted to the appropriate expense accounts (i.e., WIP–expense—101 and WIP–expense—102).* The accu-

**TABLE 5.1  General Ledger Accounts Posted When Using POC and CC Methods**

| Method of Income Recognition | G/L Account Under POC | G/L Account Under CC |
|---|---|---|
| *Type of Transaction* | | |
| Revenue | Billings—Job X | Advanced billings—Job Y |
| Cost | Work-in-progress–expense—Job X | Construction-in-progress—Job Y |

* If a job cost system is being maintained, parallel or subsidiary entries would be made to allocate costs as required. These entries are not considered in this discussion (see Chapter 7).

**FIGURE 5.2   During-the-Period Transactions**

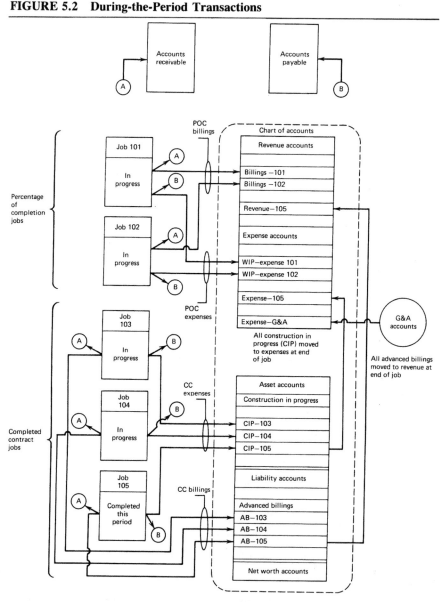

mulation of these billings and expenses will be used at the end of the period to calculate the incremental amount of income to be recognized.

The billings on completed-contract jobs are not recognized at this time as revenues. Recognition is deferred until the project is completed. Therefore, the billings on CC projects are posted to the appropriate advanced billings accounts (i.e., AB 103, AB 104, AB 105). The costs and expenditures associated with these jobs are not recognized

**TABLE 5.2   Example Summary Data**

| Project | Method of Income Recognition | Status | Bid Price | Estimated Expense |
|---|---|---|---|---|
| 101 | POC | In progress | $100,000 | $ 75,000 |
| 102 | POC | In progress | 200,000 | 160,000 |
| 103 | CC | In progress | 350,000 | 150,000 |
| 104 | CC | In progress | 400,000 | 200,000 |
| 105 | CC | Completed | 250,000 | 225,000 |

at this time as expenses. They are, therefore, posted to the appropriate construction-in-progress accounts (e.g., CIP—103, CIP—104, CIP—105), as shown in Figure 5.2.

The revenue postings when the POC method is used are balanced by appropriate entries to the asset account–accounts receivable. The advanced billings postings when the CC method is used are also offset by balancing entries to accounts receivable, as shown by the arrow labeled *A* in Figure 5.2. The expense postings under the POC method result in balancing entries to the appropriate accounts payable ledger accounts (e.g., payroll, Jones Lumber, and the like). Similarly, the construction-in-progress amounts posted on the CC projects are offset by entries to the appropriate accounts payable. These balancing entries are shown by the arrows labeled *B* in Figure 5.2. The accounts receivable and accounts payable themselves are general ledger accounts (that is, they are listed in the chart of accounts) and are shown in Figure 5.2 "pulled out" of their positions as asset and liability accounts for clarity of presentation.

One of the completed-contract projects (Project 105) is completed during the period under consideration. Since it is completed during this period, income will be recognized at the end of the period. When the project is completed, the accumulated expenses in the asset account construction-in-progress—105 are recognized as expenses and transferred to the account expense—105.

Similarly, the accumulated billings in advanced billings—105 are recognized as revenues and transferred to the account revenue—105. These actions essentially consolidate entries and clear these "holding" accounts preparatory to income recognition at the end of the period. These flows are shown by arrows connecting the appropriate accounts in Figure 5.2.

The general and administrative costs are assumed to be recognized incrementally (on a period-by-period basis).* Postings of these nonproject-related costs are shown as entries to the expense account G&A expenses.

## 5.6   CLOSING ACTIONS AT THE END OF THE PERIOD

At the end of the period, actions must be taken to "settle up" and recognize the amount of income for the period. These are the closing actions that clear the revenue and expense accounts and lead to calculation of income. In the case of Jobs 101 and 102, income is being recognized period by period (i.e., on percentage-of-completion basis).

---

* They are not prorated to or "absorbed" by the projects, but are direct-costed.

On Jobs 103 and 104, no income is recognized, since they are still in progress and being accounted for on a completed-contract basis. Income recognition will occur on Job 105, since it was completed during the period. Actions occurring at the end of the period for these five projects are shown schematically in Figure 5.3.

Income calculation for Projects 101 and 102 requires the calculation of the percentage complete. The cost-to-cost basis is used in this example. The period expenses for Project 101 amounts to $25,000 versus the total estimated expense of $75,000. This means that the percentage complete is 33.3 percent and the earned revenue is $33,333. The amount of income recognized is $8333. This is recognized in the account *Income summary* and ultimately in the income statement. Comparing the amount billed, $20,000,

## FIGURE 5.3    Closing Actions at End of Period

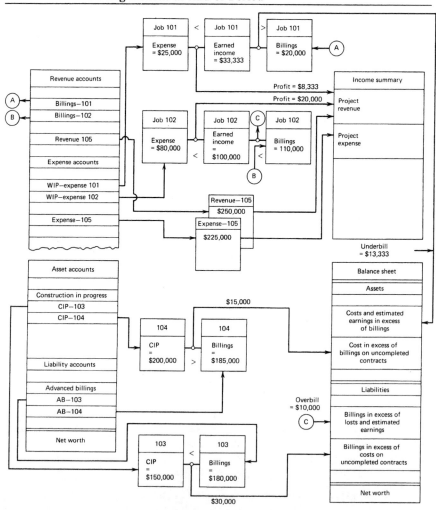

with the earned revenue, $33,333, indicates that an underbilling has occurred. This underbilling is posted to the asset account *Costs and estimated earnings in excess of billings*, and amounts to $13,333. Similar actions relating to Project 102 result in the recognition of $20,000 in income. The total estimated cost of $160,000 versus $80,000 expenses incurred indicates that the percentage complete is 50 percent. The earned revenue is $100,000 and a profit of $20,000 is recognized in the income summary. On this project, an overbilling has occurred, since the amount billed ($110,000) is greater than the amount earned ($100,000). This overbilling of $10,000 is reflected in the liability account billings in excess of costs and estimated earnings.

Activity on Projects 103 and 104 is recorded in the appropriate advanced billings and construction in progress accounts. The expenses on Project 104 exceed the billings and this results in an entry to the asset account *Costs in excess of billings on uncompleted contracts*. On Project 104 the billings exceed the expenses, and this overbilling is reflected in the liability account *Billings in excess of costs on uncompleted contracts*.

The American Institute of Certified Public Accountants recommends that these imbalances be reflected in excess accounts as described. Some accountants prefer to simply reflect the total amount of advanced billings (i.e., for all projects) as a liability on the right side of the balance sheet and the total amount of construction in progress as an asset—on the left side of the balance sheet. This approach (in principle) was taken in the balance sheet shown in Figure 1.3.* The preferred method, however, is to calculate the excess of billings versus expense (i.e., construction in progress) on a project-by-project basis for all contracts under the completed-contract method that are still in progress. For those projects where billings exceed expenses, the amount of excess is shown in the liability account *Billings in excess of costs on uncompleted contracts*. When expenses exceed billings, the amount of the excess is reflected in the asset account *Costs in excess of billings on uncompleted contracts*.

The excess of Project 104 is an asset (underbilling) of $15,000 and is recorded in the appropriate excess account. The excess on Project 103 is a liability (overbilling) of $30,000 and is so reflected in Figure 5.3.

Project 105, which was completed during the period and on which the deferred income will now be recognized, is handled in a manner similar to the two percentage-of-completion contracts. The exception is that the expenses are compared to the total project bid price (that is, the project is 100 percent complete). This is, of course, the advantage of the completed-contract method. That is, we know the expenses and we know the total amount bid, so the calculation of income is not based on an estimate, but rather on final amounts. In this case, the bid price (modified by any change order amounts) of $250,000 is compared to the actual expenses (previously transferred from the construction in Progress account) of $225,000. The earnings of $25,000 are transferred at this time to the income summary. Earnings recognized on Projects 101, 102, and 105 for this period amount to $53,333.

---

* Since billings exceeded construction in progress, the construction in progress (CIP) is shown as a negative entry on the right side of the balance sheet.

## 5.7   GENERATING FINANCIAL STATEMENTS— COMPLETED-CONTRACT METHOD

In order to better understand the process of generating financial statements, consider the following example. Fudd Associates, Inc. is using the completed-contract method of accounting on all six projects that are in progress during 19X3. A time-scaled bar chart of projects being constructed by Fudd Associates is shown in Figure 5.4. Before a balance sheet or income statement is prepared, two subsidiary schedules must first be prepared:

1. Schedule of work in progress as shown in Table 5.3. This provides a basis for determining the project-by-project advanced billings and construction in progress amounts.
2. Schedule of *completed contracts,* which supports the income statement, the completed contract sales, costs of completed contracts, and total gross profit earned on contracts. This schedule is shown in Table 5.4. As the schedule of work in progress for the previous financial year is required for the analysis, it is given in Table 5.5.

The gross billings represent the contractor's requisitions (e.g., requests for payment) to his clients during the fiscal period or the period of the contract. The other columns (direct labor, other direct costs, and total costs) are derived from the individual subsidiary job ledger accounts. The totals of these columns (total costs) should correspond with the "construction in progress" item in the general ledger. If the total billings on an uncompleted contract exceed the total cost of project, the excess is presented as a current liability. If the total cost of the project exceeds the total billings, the excess is shown as a current asset in the balance sheet.

Before the financial statements are prepared, the job superintendent should submit a detailed listing of the estimated future costs required to conclude each uncompleted contract. Added to the costs already incurred, this will make it possible to judge which contracts have incurred or will incur losses. As noted in Section 3.5, it is conservative accounting practice to record currently a full provision for these estimated future losses in the balance sheet. Doing so, of course, also constitutes a tax benefit.

**FIGURE 5.4   Time Scale Plot of Fudd Associates Jobs, 1982–1983**

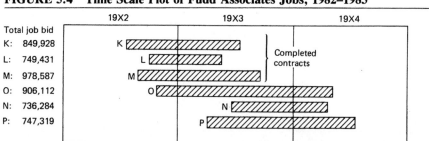

**TABLE 5.3    Schedule of Contracts in Process (Completed-Contract Basis), Fudd Associates, Inc. (Fiscal Year Ended December 19X3)**

| Contract | Contract Awarded | Gross Billings | Direct Labor | Other Direct Costs | Total Costs |
|---|---|---|---|---|---|
| N | $ 736,284 | $ 217,162 | $ 64,020 | $130,101 | $ 194,121 |
| O | 906,112 | 714,811 | 230,250 | 420,150 | 650,400 |
| P | 747,319 | 352,433 | 123,054 | 210,055 | 333,109 |
| | $2,389,705 | $1,284,406 | $417,324 | $760,306 | $1,177,630 |

The completed contracts schedule must be prepared before the preparation of the income statement can commence. As can be seen from Table 5.4, the columns used are similar to those used for the schedule of contracts in process. After the completion of a contract, the data are transferred from contracts in process to the completed contracts schedule. The detailed income statement is shown in Figure 5.5. A composite balance sheet is shown in Figure 5.6 for 19X2 and 19X3. As shown in Table 5.4, Fudd's revenue based on Jobs K, L, and M completed in 19X3 is 2,596,005. Expense for these jobs was $2,365,156. The income recognized is $230,849.

Direct contract overhead expenses represent direct job costs, which cannot be allocated easily to individual jobs. Such expenses include office engineering salaries and job travel expenses. These are carried separately on the income statement. Such payroll charges as payroll taxes, worker's compensation, liability insurance, and union welfare and pension funds have been applied directly to the individual job costs on the basis of a direct percentage of each direct labor dollar incurred. Therefore, these costs are considered part of the cost of completed contracts.

The statement showing the analysis of retained earnings is given below.

**Analysis of Retained Earnings (Completed-Contract Basis), Fudd Associates, Inc. (Fiscal Year Ended December 19X3)**

| | | |
|---|---|---|
| Retained earnings as of 1 January 19X3 | | $45,273 |
| Net income for fiscal year under review as per Figure 5.5 | $20,974 | |
| Less dividends paid | 12,000 | 8,974 |
| Retained earning as of 31 December 19X3 | | $54,247 |

**TABLE 5.4    Schedule of Completed Contracts (Completed-Contract Basis), Fudd Associates, Inc. (Fiscal Year Ended December 19X3)**

| Contract | Contracts Awarded or Contract Billings | Direct Labor | Other Direct Costs | Total Contract Costs | Excesses of Contract Billings Over Contract Cost |
|---|---|---|---|---|---|
| K | $ 867,987 | $251,432 | $ 551,433 | $ 802,865 | $ 65,122 |
| L | 749,431 | 259,205 | 501,642 | 760,847 | (11,416) |
| M | 978,587 | 237,122 | 564,322 | 801,444 | 177,143 |
| | $2,596,005 | $747,759 | $1,617,397 | $2,365,156 | $230,849 |

**TABLE 5.5   Schedule of Contracts in Process (Completed-Contract Basis), Fudd Associates, Inc. (Fiscal Year Ended December 19X2)**

| Contract | Contract Awarded | Gross Billings | Direct Labor | Other Direct Costs | Total Costs |
|---|---|---|---|---|---|
| K | $ 849,928 | $ 818,682 | $263,001 | $ 536,181 | $ 799,182 |
| L | 749,431 | 325,989 | 125,233 | 175,313 | 300,546 |
| M | 978,587 | 444,827 | 132,502 | 222,425 | 354,927 |
| O | 906,112 | 202,904 | 57,647 | 127,139 | 184,786 |
| | $3,484,058 | $1,792,402 | $578,383 | $1,061,058 | $1,639,441 |

Under the completed contract method, only three of the six projects generate income in 19X3, since only K, L, and M are completed during the year. Projects N, O, and P are still in progress; their status is reflected in the balance sheet. In all cases, billings on these projects have exceeded expenses, so that a liability has been generated to the excess account—*Billings in excess of costs on uncompleted contracts*—in the amount of $106,776. This liability is summarized below:

| Contract | Advance Billings | Work in Progress | Excess (Liability) |
|---|---|---|---|
| N | $217,162 | $194,121 | $23,041 |
| O | 714,811 | 650,400 | 64,411 |
| P | 352,433 | 333,109 | 19,324 |
| | | | $106,776 |

It should be reemphasized that the point at which a contract is completed is critical in establishing when income is generated. The income associated with the sale is recognized at the time the client accepts the project as fulfilling his requirements as established in the contract documents (i.e., at the point of substantial completion).

The definition of substantial completion is critical, since it establishes when income is recognized and tax liability is established. The AICPA *Construction Contractors Audit and Accounting Guide* establishes the following guidelines for substantial completion:

> As a general rule, a contract may be regarded as substantially completed if remaining costs and potential risks are insignificant in amount.*

Obviously by interpretating when substantial completion of the project occurs, the contractor can either shorten or lengthen the project moving tax liability from one period to another. Pending completion, the revenues are held as liabilities and the expenses are accumulated as assets.

---

* American Institute of Certified Public Accountants, *Construction Contractors Audit and Accounting Guide* (New York: AICPA, 1981), p. 132 (paragraph 52 of the Statement of Position 81-1).

**FIGURE 5.5   Statement of Income (Completed-Contract Basis), Fudd Associates, Inc. (Fiscal Year Ended 31 December 19X3)**

| | | |
|---|---:|---:|
| Completed contract sales | | $2,596,005 |
| Less cost of completed contracts | | 2,365,156 |
| Gross profit earned on contracts | | $   230,849 |
| | | |
| *Less Direct Contract Overhead Expenses* | | |
|    Engineering salaries | $51,082 | |
|    Supervision salaries | 46,415 | |
|    Travel expenses | 8,927 | $   106,424 |
|       Gross profit earned on operations | | $   124,425 |
| | | |
| *Less Selling Expenses* | | |
|    Bidding expenses | $ 6,428 | |
|    Travel expenses | 9,572 | |
|    Sales promotion | 5,324 | $     21,324 |
|       Selling profit | | $   103,101 |
| | | |
| *Less General and Administrative Expenses* | | |
|    Depreciation on land and buildings | $ 5,412 | |
|    Power | 2,423 | |
|    Office supplies, telephone, telegraph | 10,785 | |
|    Office salries | 49,088 | |
|    Interest on loans | 7,213 | $     74,921 |
|       Operating profit | | $     28,180 |
|       Profit realized on sale of equipment | | $       2,742 |
|          Total profit before federal tax | | $     30,922 |
|          Provision for federal income tax | | $       9,948 |
|             Net income for fiscal year | | $     20,974 |

## 5.8   GENERATING FINANCIAL STATEMENTS— PERCENTAGE-OF-COMPLETION METHOD

For purposes of comparison, assume that Fudd Associates, Inc. has used the percentage-of-completion method for income recognition rather than the completed-contract method. In this case, all six projects generate income (or loss) during the period of consideration. Table 5.6 is a worksheet that summarizes calculations required for recognition of income under the POC method. The percentage complete on each of the projects is calculated by using the cost-to-cost method. The calculations in the table on a column-by-column basis are described below.

1. Column 1: The cumulative cost of work as of the end of 19X2, based on Table 5.5.
2. Column 2: The cumulative cost of work as of the end of 19X3, based on Tables 5.3 and 5.4.
3. Column 3: Expense incurred for 19X3, calculated as column 2 minus column 1.

**FIGURE 5.6  Balance Sheet (Completed-Contract Basis), Fudd Associates, Inc. (31 December 19X2 and 19X3)**

| Assets | | | 19X2 | 19X3 |
|---|---|---|---|---|
| *Current Assets* | | | | |
| Cash in bank and on hand | | | $ 56,753 | $ 18,111 |
| Accounts receivable | | | 153,112 | 92,801 |
| Inventories | | | 214,102 | 166,024 |
| Other current assets | | | 4,786 | 3,102 |
| Cost in excess of billings on uncompleted contracts | | | 0 | 0 |
| Total current assets | | | $428,753 | $280,038 |
| *Fixed Assets (at Cost)* | | | | |
| Land and buildings | $ 54,123 | $54,123 | | |
| Equipment | 218,277 | 299,527 | | |
| Less accumulated depreciation on land and buildings | 10,825 | 16,237 | | |
| Less accumulated depreciation on equipment | 152,205 | 114,202 | | |
| Total fixed assets at book value | | | $109,370 | $223,211 |
| All other assets | | | 1,247 | 2,428 |
| Total assets | | | $539,370 | $505,677 |

| Liabilities and Net Worth | 19X2 | 19X3 |
|---|---|---|
| *Current Liabilities* | | |
| Accounts payable | $ 49,882 | $ 41,126 |
| Notes payable—banks | 31,254 | 23,528 |
| Billings in excess of costs on uncompleted contracts | 152,961 | 106,776 |
| Total current liabilities | $234,097 | $171,430 |
| *Long-Term Liabilities* | | |
| Loans payable | $ 60,000 | $ 80,000 |
| *Net worth* | | |
| Capital stock | 200,000 | 200,000 |
| Retained earning | 45,273 | 54,247 |
| Total liabilities and net worth | $539,370 | $505,667 |

## TABLE 5.6 Calculations for Percentage-of-Completion Method

| | Col. 1 Cumulative Cost End of 19X2 | Col. 2 Cumulative Cost End of 19X3 | Col. 3 = Col. 2 − Col. 1 Expense for Period 19X3 | Col. 4 Total Estimated Cost | Col. 5 = Col. 3 ÷ Col. 4 % for 19X3 | Col. 6 Contract Price | Col. 7 = 5 × 6 Period Contract Revenues Earned | Col. 8 = 7 − 3 Period Income Earned | Col. 9 Gross Billings for 19X3 | Col. 10 = 9 − 7 for Contracts in Progress Billings in Excess of Cost and Estimated Earnings[a] |
|---|---|---|---|---|---|---|---|---|---|---|
| K | $799,182 | $802,865 | $ 3,683 | $802,865 | 0.46 | $867,987 | $ 3,981.74 | $ 298.74 | $ 49,305 | $ 0 |
| L | 300,546 | 760,847 | 460,301 | 760,847 | 60.5 | 749,431 | 453,394.49 | (6,906.51) | 423,442 | 0 |
| M | 354,927 | 801,444 | 446,517 | 801,444 | 55.7 | 978,587 | 545,210.56 | 98,693.56 | 533,760 | 0 |
| O | 184,786 | 650,400 | 465,614 | 847,815 | 54.9 | 906,112 | 497,630.30 | 32,016.30 | 511,907 | 14,276.70 |
| N | 0 | 194,121 | 194,121 | 665,175 | 29.2 | 736,284 | 214,873.06 | 20,752.06 | 217,162 | 2,288.94 |
| P | 0 | 333,109 | 333,109 | 694,640 | 47.9 | 747,319 | 358,370.79 | 25,261.79 | 352,433 | (5,937.79) |
| | | | $1,903,345 | | | | $2,073,460.94 | $170,115.94 | | |

[a] Entries in parentheses are "costs and estimated earnings in excess of billings."

4.   Column 4: Total estimated cost to complete. For contracts K, L, and M this is the total contract cost given in Table 5.4. For the contracts in process (N, O, and P) this is based on estimated cost to complete added to total costs to date, given in Table 5.3. The estimated total costs on contracts in process and estimated costs to complete are summarized as:

|   | Cost to Date | Estimated Total Cost | Estimated Cost to Complete |
|---|---|---|---|
| O | $650,400 | $847,815 | $197,415 |
| N | 194,121 | 665,175 | 471,054 |
| P | 333,109 | 694,640 | 361,531 |

5.   Column 5: Percentage completed during 19X3 based on cost-to-cost calculation. An implicit assumption here is that the total estimated cost to complete Project O has not changed during the 19X3 period. Therefore, the percentage completed is the ratio of column 3 to column 4.

6.   Column 6: Total contract price including change orders. Note that the original contract price on Project K has been increased.

7.   Column 7: The earned revenue (cost and estimated earnings) for 19X3 is calculated by multiplying column 6 by column 5.

8.   Column 8: The recognized income on each project for 19X3 is calculated by subtracting the incurred expense (column 3) from the earned revenue (column 7). A loss is recognized on Project L.*

9.   Column 9: Gross billings for completed contracts and Project O are calculated by subtracting values in Table 5.5 from those in Table 5.4. Gross billings for N and P are taken from Table 5.3.

10.  Column 10: Both Projects O and N have billings in excess of cost and estimated earnings. An underbilling of $5937.79 has occurred on Project P.

The income statement for 19X3 under the POC method is given in Figure 5.7. The gross amount of recognized income is $170,116.

Because of the overhead, selling, and G&A costs, the operating profit when the POC method is used is a negative $32,553. That is, the company has an operating loss on the year versus a positive operating profit of $28,180 when the CC method is used. The before-tax income is negative, and therefore no tax is paid for 19X3.

The balance sheet for Fudd Associates, Inc. under the POC method is given in Figure 5.8. It is interesting to note the effect on the balance sheet accounts of using the POC method versus the CC method. When the POC method is used, the amount held in the liability excess accounts is considerably smaller and the net worth accrued from recognized income in previous periods is considerably larger. This statement

---

* The total loss on Project L is $11,416, so part of this loss (i.e., $11,416 − $6,906.51) was recognized in the previous period.

**FIGURE 5.7   Statement of Income (Percentage-of-Completion Basis), Fudd Associates Inc.**
**(Fiscal Year Ended 31 December 19X3)**

| | | |
|---|---:|---:|
| Contract Revenues Earned | | $2,073,461 |
| Cost or Revenues Earned | | 1,903,345 |
| Gross Profit | | $ 170,116 |
| | | |
| Less Direct Contract Overhead Expenses | | |
| | | |
| Engineering salaries | $51,082 | |
| Supervision salaries | 46,415 | |
| Travel expenses | 8,927 | $ 106,424 |
| Gross profit earned on operations | | $ 63,692 |
| | | |
| Less Selling Expense | | |
| | | |
| Bidding expenses | $ 6,428 | |
| Travel expenses | 9,572 | |
| Sales promotion | 5,324 | $ 21,324 |
| Selling profit | | $ 42,368 |
| | | |
| Less General and Administrative Expenses | | |
| | | |
| Depreciation on land and buildings | $ 5,412 | |
| Power | 2,423 | |
| Office supplies, telephone, telegraph | 10,785 | |
| Office salaries | 49,088 | |
| Interest on loans | | $ 74,921 |
| Operating profit | | $ (32,553) |
| Profit realized on sale of equipment | | $ 2,742 |
| Total profit before Federal tax | | $ (29,811) |
| Provision for Federal Income Tax | | 0 |
| Net income for fiscal year | | $ (29,811) |

gives stockholders a better picture of the value of the company. The income statement also reflects a more balanced picture of the year's operations. Where the CC method indicated a profit for the year, this profit was actually generated in the previous year's activity.

Although many companies utilize the percentage-of-completion method for recognition of income on long-term contracts, there are, in fact, some differences in the way in which the percentage-of-completion method is applied in practice. These subtle differences are not always described in the notes to a company's financial statements, even though they can have material effects when compared to operating results reported by other companies also utilizing the percentage-of-completion method.

A few companies defer a portion of income recognition under the percentage-of-completion method until a contract reaches a specific completion level. For instance, some companies do not report any income on a percentage-of-completion basis until

**FIGURE 5.8   Balance Sheet (Percentage-of-Completion Basis), Fudd Associates, Inc. (31 December 19X3)**

| Assets | | 19X3 |
|---|---|---|
| Current Assets | | |
| Cash in bank and on hand | | $ 18,111 |
| Accounts receivable | | 92,801 |
| Inventories | | 166,024 |
| Other current assets | | 3,102 |
| Cost and estimated earnings in excess of billings | | 5,937.79 |
| Total current assets | | $285,975.79 |
| Fixed Assets (at cost) | | |
| Land and buildings | $ 54,123 | |
| Equipment | 299,527 | |
| Less accumulated depreciation on land and buildings | 16,237 | |
| Less accumulated depreciation on equipment | 114,202 | |
| Total fixed assets at book value | | $223,211 |
| All Other Assets | | 2,428 |
| Total assets | | $511,614.79 |

| Liabilities and net worth | | 19X3 |
|---|---|---|
| Current Liabilities | | |
| Accounts payable | | $ 41,126 |
| Notes payable—banks | | 23,528 |
| Billings in excess of cost and estimated earnings | | 16,565.64 |
| Total current liabilities | | $ 81,219.64 |
| Long-Term Liabilities | | |
| Loans payable | | $ 80,000 |
| Net Worth | | |
| Capital stock | | $200,000 |
| Retained earning | | 150,395.15 |
| Total liabilities and net worth | | $511,614.79 |

a contract is at least 20 percent completed, and others wait until contracts are 30 percent, 40 percent, or 50 percent completed.

Fortunately, over the long term the effects of the innumerable ways by which the percentage-of-completion method is applied are washed out or at least greatly reduced. Over the short term, however, the subtle differences in the method of application of the percentage-of-completion method can, in specific instances, have a significant effect, one that might not be readily ascertainable from the information available solely from audited financial statements.

## 5.9. SUMMARY

In this chapter, the closing actions that must be taken when the completed contract and the percentage-of-completion methods are used have been presented and compared. The intent is not to prepare the reader as an accountant to actually perform these operations. Rather, the objective is to provide an introduction to the nature of these activities so that the reader can understand (1) their effect on the company performance as shown in the balance sheet, and (2) the way in which expenses and cost accounts are reflected in the income statement and balance sheet at the time of closing. The topics discussed in these first five chapters pertain to the province of "financial accounting" and focus on company-level operations. Later chapters (following the next one on financial analysis) concentrate on "cost accounting" and the capturing of data for cost control and future estimating purposes.

## REVIEW QUESTIONS AND EXERCISES

**5.1** What is a "real" account? How does it differ from a nominal account?

**5.2** What is the reason for closing accounts?

**5.3** Give at least five objectives of financial reporting.

**5.4** What is meant by "The statement of financial position presents a snapshot of the firm's financial position as of a given date."

**5.5** How are the income statement and balance sheet linked when closing accounts?

**5.6** Rework Problem 3.4 assuming Projects A and C recognize income using the completed-contract method and Project B is accounted for using the percentage-of-completion method.

**5.7** The balance sheet of CPM Construction Company as of 31 December 19X3 is given below. Assume that this company is using the POC method of income recognition. Further, assume that 65 percent of the projects with total bid price of $850,000 have been completed in 19X4.

(a) Journalize the following transactions. Identify each transaction as asset, liability, revenue, expense, etc.

**TABLE P5.1    Balance Sheet, CPM Construction Company (31 December, 19X3)**

| Assets | | Liabilities | |
|---|---|---|---|
| Cash | $ 75,000 | Accounts payable | $ 85,000 |
| Accounts receivable | 110,000 | Notes payable | 50,000 |
| Buildings | 300,000 | Long-term loans | 60,000 |
| Less accumulated depreciation on the buildings | (150,000) | Total liabilities | $195,000 |
| Construction equipment | 240,000 | Net worth | |
| Less accumulated depreciation on equipment | (80,000) | Capital stocks | 250,000 |
| | | Retained earnings | 70,000 |
| Other assets | 20,000 | Total net worth | 320,000 |
| Total assets | $ 515,000 | Total net worth and liabilities | $515,000 |

(b) Establish relevant accounts for posting. Divide them into categories as assets, liabilities, net worth.
(c) Close accounts as of 31 December 19X4.
(d) Develop the income statement for 19X4.
(e) Develop the balance sheet as of 31 December 19X4.

| Transaction Number | Date | Transaction |
|---|---|---|
| 1 | 1/2/X4 | CPM Co. bought construction equipment for $130,000 for which the company paid $15,000 cash and remaining on account. |
| 2 | 2/4/X4 | CPM Co. was billed $20,000 by Smith Material Supplier for cost of material. |
| 3 | 3/4/X4 | CPM paid $20,000 to Smith Material Supplier related to transaction #2. |
| 4 | 3/8/X4 | CPM billed client for $320,000 (bill #1 on Job 101). |
| 5 | 4/7/X4 | CPM was billed $60,000 by National Rental Co. for renting construction equipment. |
| 6 | 5/8/X4 | CPM received $290,000 from client for bill #1 on Job 101. |
| 7 | 6/7/X4 | CPM paid $60,000 to the National Rental Co. for the bill received on 4/7/X4. |
| 8 | 7/3/X4 | CPM paid $70,000 cost of labor. |
| 9 | 8/16/X4 | CPM was billed $45,000 by subcontractor. |
| 10 | 9/16/X4 | CPM paid $45,000 to the subcontractor for the bill received on 8/16/X4. |
| 11 | 10/1/X4 | CPM billed client for $280,000 (bill #2 on Job 101). |
| 12 | 10/20/X4 | Accounts receivable of $20,000 are collected. |
| 13 | 11/15/X4 | CPM received $265,000 from client for bill #2. |
| 14 | 12/15/X4 | Accounts payable of $40,000 are paid. |
| 15 | 12/25/X4 | CPM paid $145,000 in payroll expense. |
| 16 | 12/30/X4 | Dividends paid in the amount of $20,000 to stockholders. |
| 17 | 12/30/X4 | Building depreciation of $30,000 recognized for the year. |
| 18 | 12/30/X4 | CPM depreciates its construction equipment for the total of $65,000 each year (this includes also the depreciation of the equipment bought on 1/2/X4.) |

# Chapter 6

# ANALYZING COMPANY FINANCIAL DATA

## 6.1  REASONS FOR FINANCIAL ANALYSIS

Financial data are required so that interested parties can establish the financial status of a company. The data generated in preparing the financial statements are presented in the statements reflecting various aspects of the company's operation and financial health. Lenders and investors are particularly interested in analyzing these data to determine the level of risk inherent in participating as an equity partner or shareholder, or in providing credit or lending cash to the company. Bonding companies are also interested in these data so that they can determine the financial capacity of the firm to undertake certain projects.

The equity investor (i.e., stockholder or principal) supplies the basic risk capital. This capital is exposed to all the risks of ownership and provides a cushion or shield for the preferred or loan capital that is senior to it. This is why the equity interest is referred to as the *residual interest*. The information needs of equity investors are among the most demanding and comprehensive of all users of financial data. Their interest in an enterprise of which they own a share is the broadest, because their interest is affected by all aspects and phases of operations, profitability, financial condition, and capital structure.

The viewpoint of credit grantors or lenders differs from that of the equity investor in the way they analyze future prospects and in the objective they seek. The equity investor looks for his reward primarily in future prospects of earnings and in changes in those earnings. The lender, on the other hand, is concerned primarily with specific security provisions of his loan (such as the fair market value of assets pledged) and the repayment of principal and interest. He looks for the existence of assets and collateral, projections of future flows of funds, and the reliability and stability of such flows.

The techniques of financial analysis used by lenders, as well as the criteria of evaluation, vary with the term, the security, and the purpose of the loan. In the case of short-term credit extension, the creditor is concerned primarily with current financial condition, with the liquidity of the current assets, and the rate of their turnover.

Since the profitability of an enterprise is a major element in the lender's security, the analysis of profitability is an important criterion to the lender. Profit is viewed as the primary source of interest payments and as a desirable source for principal repayment.

Management derives a number of important advantages from a systematic monitoring of financial data and the basic relationship that they display. Financial data analysis can be undertaken by management on a continuous basis because it has unlimited access to internal accounting and other records.

Financial statement analysis can serve the needs of many other groups of users. Labor unions, for example, can use the techniques of financial analysis to evaluate the financial statements of enterprises with which they engage in collective bargaining.

## 6.2   TOOLS OF FINANCIAL ANALYSIS

In the analysis of financial statements, the analyst has a variety of tools from which he can choose those best suited to his specific purpose. The following principal tools of analysis will be described briefly:

1.   Comparative financial statements
2.   Index-number trend series
3.   Common-size financial statements
4.   Ratio analysis

The most important factor revealed by comparative financial statements is *trend*. The comparison of financial statements over a number of years will also reveal the direction, velocity, and the amplitude of trend. Further analysis can be undertaken to compare trends in related items. A comparison of financial statements over two to three years can be undertaken by computing the year-to-year change in absolute amounts and in terms of percentage changes.

Balance sheet amounts over a five-year period for the Apex Construction Co. are shown in Table 6.1. The annual percentage change is also given in this table. Information of this nature lends itself to a graphical presentation. A plot of the total assets, current assets, current liabilities, and net worth over the five-year period is shown in Figure 6.1. As can be seen from the plot and the table, current assets and the total assets increase sharply during the 19X3 to 19X5 period. The gain in total assets in this period is mainly due to the gain in current assets. From 19X5 to 19X7, the current assets remain constant and the total assets increase at a rate of approximately 2 percent per year. During the five-year period in review, the fixed assets increase approximately at an annual rate of 8.8 percent.

**TABLE 6.1  Five Years in Review, Apex Construction Company (in $1000)**

| | 19X3 | 19X4 | % Change | 19X5 | % Change | 19X6 | % Change | 19X7 | % Change |
|---|---|---|---|---|---|---|---|---|---|
| Current assets | $79,738 | $112,506 | 41.0 | $128,115 | 13.8 | $129,372 | 1.0 | $128,786 | (0.4)[a] |
| Fixed assets | 16,817 | 17,882 | 6.3 | 20,425 | 14.2 | 20,774 | 1.6 | 24,243 | 16.7 |
| Total assets | 99,631 | 133,301 | 33.8 | 152,385 | 14.3 | 155,897 | 2.3 | 159,018 | 2.0 |
| Current liabilities | 66,448 | 80,361 | 20.9 | 91,510 | 13.9 | 92,408 | 1.0 | 91,990 | (0.4) |
| Long-term liabilities | 16,333 | 32,915 | 101.52 | 39,970 | 21.4 | 36,470 | (8.7) | 31,287 | (14.2) |
| Net worth | $16,850 | $20,025 | 18.8 | $22,905 | 14.4 | $27,019 | 18 | $35,741 | 32.3 |

[a]Parentheses indicate downward % change.

**FIGURE 6.1   Balance-Sheet Trend Chart**

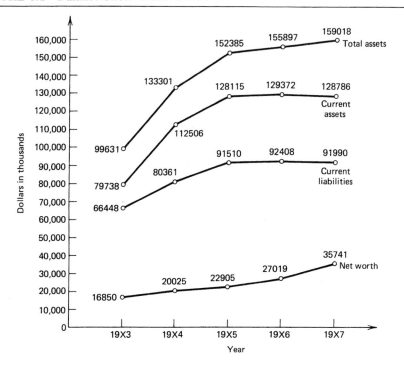

Inspection of the firm's assets over the 19X5 to 19X7 period illustrates that the fixed assets are increasing at a faster rate than current assets. This could be due to the purchase of land, buildings, and/or construction equipment. If a steeper increase in acquisition of fixed assets occurs, working capital could become limited.

Current liabilities and long-term liabilities both increase during the 19X3 and 19X5 period by an annual average of 18.9 percent and 72.4 percent, respectively. Then during the 19X5 and 19X7 period the current liabilities level off, and the long-term liabilities are decreasing at an average annual rate of 13.9 percent.

The firm's net worth shows a positive increase over the five-year review period. The percentage of increase from one year to the following year increases over each year in the five-year review period. This is obvious when we review the index numbers (see Table 6.2).

Other positive trends revealed upon review of the company's balance sheet figures are a steady decrease in the firm's long-term liabilities during the 19X5 to 19X7 period along with a steady increase in the firm's net worth over the same review period. This is a favorable indication. Net worth expanded at an average rate of approximately 28 percent during the period 19X3 to 19X7. Inspection of the current liabilities over the

**TABLE 6.2  Five Years in Review, Apex Construction Company**

|  | *19X3* | *19X4* | *19X5* | *19X6* | *19X7* |
|---|---|---|---|---|---|
| Current assets | 100 | 141 | 161 | 162 | 162 |
| Fixed assets | 100 | 106 | 121 | 124 | 144 |
| Total assets | 100 | 134 | 153 | 156 | 160 |
| Current liabilities | 100 | 121 | 138 | 139 | 138 |
| Long-term liabilities | 100 | 202 | 245 | 223 | 192 |
| Net worth | 100 | 119 | 136 | 160 | 212 |

19X5 to 19X7 period reveals stability in the firm's work load. Examination of these trends aids the manager in monitoring performance and identifying potential problems.

## 6.3  INDEX-NUMBER TREND SERIES

When a comparison of financial statements over a longer multiyear period is undertaken, the year-to-year method of comparison described above becomes cumbersome. The best way to effect such longer-term comparisons is by means of index numbers.

In planning an index-number trend comparison, it is not necessary to include in it all of the items in the financial statements. Only the most significant items need to be included in such a comparison.

The computation of a series of index numbers requires the choice of a base year that will, for all items, have an index amount of 100. Since such a base year represents a frame of reference for all comparisons, it is best to choose a year that, with regard to business conditions, is as typical or normal as possible. If the earliest year in the series compared cannot fulfill this function, another year is selected.

The balance sheet data in Table 6.1 are presented in index-number format with 19X3 used as the reference year in Table 6.2. A plot of the index numbers over the five-year period is shown in Figure 6.2. This chart highlights the sharp increase in net worth during the period. In this sense it is helpful to reflect management report data in a manner that can be readily grasped by management and stock holders.

The most well-known economic index system is the Dow Jones index of major stocks on the New York Stock Exchange. In construction, the *Engineering News Record* Building and Construction Cost Index is a typical trend indexing system. (See Figure 10.12.)

## 6.4  COMMON SIZE FINANCIAL STATEMENT

In the analysis of financial statements it is often constructive to find out the proportion of a total group or subgroup that a single item within them represents. In a balance sheet, the assets as well as the liabilities and capital are each expressed as 100 percent,

**FIGURE 6.2   Trend Chart based on Index Numbers**

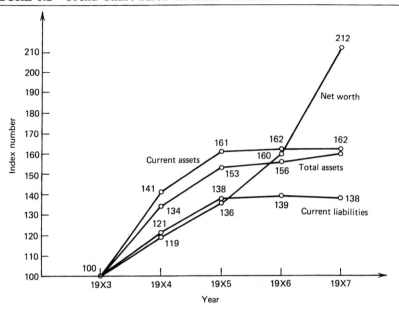

**FIGURE 6.3   Balance Sheet Structural Breakdown**

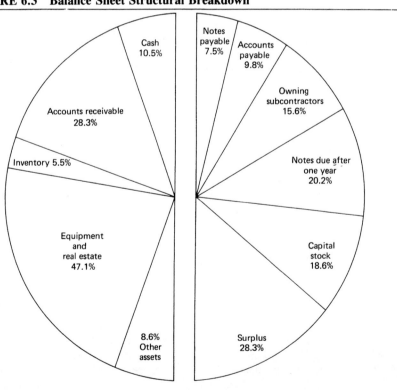

and each item in these categories is expressed as a percentage of the respective totals. Figure 6.3 is a pie chart representation of the percentage associated with major subgroups on each side of the balance sheet. Similarly, in the income statement net sales are set at 100 percent, and every other item in the statement is expressed as a percentage of net sales. Figure 6.4 shows an income statement in which this approach has been used. Since the total is always 100 percent, this common reference of size or normalizing to a common reference has resulted in such statements being referred to as "common size" statements. Similarly, following the eye as it reviews the common size statement, this analysis is referred to as "vertical," for the same reason that the trend analysis is often referred to as "horizontal" analysis.

**FIGURE 6.4   Income Statement, Newman Construction (Year Ended 31 December 19X6)**

| | | % of Sales |
|---|---|---|
| *Net Sales* | $2,143,761 | 100.0 |
| Cost of Sales | | |
| Materials | 1,013,913 | 47.30 |
| Labor (includes all payroll taxes and union fringes) | 548,271 | 25.57 |
| Subcontracts | 51,450 | 2.40 |
| Total direct costs | $1,815,432 | 84.68 |
| *Gross Profit* | $ 328,329 | 15.32 |
| Operating Expenses | | |
| Variable Overhead | | |
| Auto and truck | 12,089 | .57 |
| Communications | 5,575 | .26 |
| Interest | 1,275 | .05 |
| Miscellaneous | 213 | .01 |
| Office supplies | 1,943 | .09 |
| Travel and entertainment | 871 | .04 |
| Total variable overhead | $ 21,966 | 1.02 |
| *Fixed Overhead* | | |
| Contributions | $ 675 | .03 |
| Depreciation | 8,238 | .38 |
| Dues and subscriptions | 514 | .02 |
| Insurance | 12,475 | .58 |
| Legal and audit | 2,315 | .10 |
| Licenses and taxes | 175 | .01 |
| Payroll taxes (office only) | 51,790 | 2.42 |
| Rent | 2,506 | .12 |
| Repairs and maintenance | 375 | .02 |
| Salaries—office | 23,418 | 1.10 |
| Salaries—officers | 38,410 | 1.79 |
| Total fixed overhead | $ 140,891 | 6.57 |
| *Total Overhead* | $ 162,857 | 7.59 |
| Net Profit Before Taxes | $ 165,472 | 7.73 |
| State and federal taxes | 77,076 | 3.60 |
| Net profit | $ 88,396 | 4.13 |

*Source:* I. J. Jackson II and M. H. Gilliam, eds., *Financial Management for Contractors,* The Fails Institute (New York: McGraw-Hill, 1981).

**FIGURE 6.5   Comparison of Asset Structure of Various Industries**

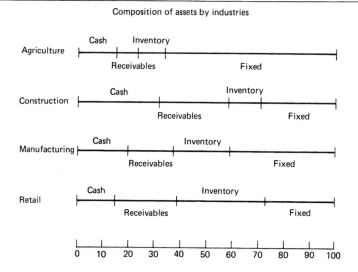

Composition of assets by industries

The analysis of common size financial statements may best be described as a *structural analysis* of a balance sheet. Structural analysis focuses on two major aspects:

1.  What are the sources of capital of the enterprise?
2.  Given the amount of capital from all sources, what is the mix of assets with which the enterprise has chosen to conduct its operations?

A comparison of asset distribution in various industries is shown in Figure 6.5.

Comparison of the common size statements of companies within an industry or with common size composite statistics of that industry can alert the analyst to variation in account structure and distribution.

## 6.5   RATIO ANALYSIS

The relationship of one item to another expressed in simple mathematical form is known as a *ratio*. Ratios are tools of analysis that, in most cases, provide the manager with clues and symptoms of underlying conditions or alert him or her to items requiring further investigation and inquiry. The analysis of a ratio can disclose relationships as well as bases for comparison which reveal conditions and trends that cannot be properly detected by mere inspection of the individual components of the ratio.

A single ratio in itself is meaningless, since it does not furnish all the information required to reflect financial position of an enterprise. A ratio becomes meaningful when compared with past ratios of the same company, some predetermined standard, or ratios of other companies in the same industry. The range of a ratio over time is also significant, as is the trend of a given ratio over time.

Many ratios can be developed from the multitude of items included in the balance sheet and the income statement. Some ratios have general application in financial analysis; others have specific use only in certain circumstances or in particular industries.

Some sources provide average ratios for groups of companies. These averages are generally desired for an entire industry, but they may be restricted to enterprises of certain size, or to certain groups within the specific industry. Reporting services such as Dun & Bradstreet and Robert Morris Associates prepare standard ratios for various lines of business in which the different branches of the construction industry are also listed. A listing of ratios for contractors in the commercial building area is given in Table 6.3. The ratios are presented by size of the firm in terms of annual revenues and provide a range of ratios indicating median, low, and high values.

There are four categories of ratios that are commonly used to gain an overview of a firm's financial position. These are

1. Liquidity ratios
2. Leverage ratios
3. Activity ratios
4. Profitability ratios

Liquidity ratios relate a firm's ability to meet its short-time obligations with readily liquidable assets. These are the most widely accepted ratios and relate some form of *current assets* to *current liabilities.*

Leverage ratios relate the firm's manner of incurring debt to a number of aspects of the company's operation. The company gains leverage by borrowing. Therefore, these ratios typically relate *debt* to some other parameter. These ratios are sometimes called *debt management* ratios.

Activity ratios are designed to relate *revenues* to other balance sheet items in such a way as to indicate the level of turnover of sales. Therefore, revenues are typically the numerator of these ratios. One exception is the fixed-asset-to-net-worth ratio. These ratios are also referred to as *asset management* ratios.

Profitability ratios relate *profit* to other financial parameters such as total assets or revenues. These ratios tend to be percentage ratios (i.e., numerator smaller than denominator), whereas many of the activity (revenue) ratios tend to be "times" ratios (i.e., numerator greater than denominator). The liquidity ratios also are typically given as times ratios. Liquidity should be greater than one to insure that current assets will cover current liabilities.

The actual ratios of interest in each of these categories can be very stylized, and the important ratios vary from industry to-industry and in some cases from company to company. In the profitability ratios, for instance, some firms use profit before taxes, others after-tax profit, and others gross profit as the basis for calculation. The most commonly used ratios in each of four categories and the references noting these ratios as useful in financial analysis are given in Table 6.4.

## TABLE 6.3 RMA Ratios for Commercial Contractors

| Ratios | Under $1MM | $1MM & less than $10MM | $10MM & less than $50MM | $50MM & over | All sizes |
|---|---|---|---|---|---|
| Quick | 1.7 | 1.4 | 1.2 | 1.2 | 1.4 |
| | 1.2 | 1.2 | 1.1 | 1.0 | 1.2 |
| | 0.8 | 1.0 | 0.9 | 0.8 | 0.9 |
| Current | 2.0 | 1.6 | 1.5 | 1.4 | 1.6 |
| | 1.4 | 1.3 | 1.2 | 1.3 | 1.3 |
| | 0.9 | 1.2 | 1.1 | 1.1 | 1.1 |
| Fixed Assets/worth | 0.2 | 0.1 | 0.2 | 0.3 | 0.2 |
| | 0.4 | 0.3 | 0.3 | 0.4 | 0.3 |
| | 1.0 | 0.5 | 0.6 | 0.9 | 0.6 |
| Debt/worth | 0.7 | 1.1 | 1.4 | 1.5 | 1.1 |
| | 1.8 | 2.1 | 2.4 | 2.1 | 2.0 |
| | 4.5 | 3.5 | 3.8 | 3.7 | 3.7 |
| Revenues/receivables | 11.9 | 9.1 | 8.1 | 15.2 | 9.4 |
| | 6.3 | 6.6 | 6.6 | 6.7 | 6.5 |
| | 4.2 | 4.8 | 5.2 | 5.8 | 4.8 |
| Revenues/working capital | 5.7 | 10.0 | 11.6 | 16.0 | 9.3 |
| | 14.2 | 17.7 | 23.9 | 24.1 | 17.9 |
| | −79.0 | 34.5 | 50.7 | 55.1 | 43.2 |
| Revenues/worth | 3.3 | 6.2 | 6.8 | 7.7 | 5.7 |
| | 7.6 | 10.0 | 10.8 | 11.0 | 9.9 |
| | 17.0 | 15.6 | 16.8 | 17.4 | 16.2 |
| % Profit before taxes/worth | 36.4 | 36.9 | 42.3 | 42.4 | 38.2 |
| | 14.0 | 17.7 | 23.7 | 29.8 | 19.1 |
| | −12.1 | 3.4 | 10.1 | 17.8 | 3.4 |
| % Profit before taxes/total assets | 15.6 | 11.5 | 10.4 | 14.2 | 12.0 |
| | 2.8 | 5.3 | 7.3 | 10.7 | 5.5 |
| | −8.4 | 0.8 | 2.9 | 5.1 | 0.6 |
| Contract revenues | 60,583M | 1,027,313M | 1,304,410M | 3,923,819M | 6,316,125M |
| Total assets | 133,535M | 406,029M | 511,766M | 934,366M | 1,985,696M |

M = $thousand; MM = $million.

Source: RMA Annual Statement Studies, Robert Morris Associates.

**TABLE 6.4   Ratios Commonly Used in the Construction Industry**

| | | Adrian | Van Horne | Robert Morris Assoc. | NECA | Brigham |
|---|---|---|---|---|---|---|
| *Liquidity Ratios* | | | | | | |
| 1. Current ratio $= \dfrac{\text{current assets}}{\text{current liabilities}}$ | times | X | X | X | X | X |
| 2. Quick ratio $= \dfrac{\text{curent assets} - \text{inventories}}{\text{current liabilities}}$ | times | X | X | X | | X |
| *Leverage Ratios* | | | | | | |
| 3. Total debt to total assets $= \dfrac{\text{total debt}}{\text{total assets}}$ | % | | X | | | X |
| 4. Total debt to net worth $= \dfrac{\text{total debt}}{\text{net worth}}$ | times | X | X | X | X | |
| 5. Times interest earned $= \dfrac{\text{profit}}{\text{long-term debt interest}}$ | times | | | | | X |
| *Activity Ratios* | | | | | | |
| 6. Turnover of total assets $= \dfrac{\text{revenues}}{\text{total assets}}$ | times | | X | | | X |
| 7. Revenues to receivables $= \dfrac{\text{revenues}}{\text{receivables}}$ | times | X | X | X | | X |
| 8. Quality of inventory $= \dfrac{\text{revenues}}{\text{inventory}}$ | times | X | | | | X |
| 9. Revenues to working capital $= \dfrac{\text{revenues}}{\text{WC}}$ | times | | X | X | X | |
| 10. Revenues to worth $= \dfrac{\text{revenues}}{\text{net worth}}$ | times | X | X | X | X | |
| 11. Revenues to fixed assets $= \dfrac{\text{revenues}}{\text{fixed assets}}$ | times | | | | | X |
| 12. Fixed assets to net worth $= \dfrac{\text{fixed assets}}{\text{net worth}}$ | % | | X | X | X | |

| | | Adrian | Van Horne | Robert Morris Assoc. | NECA | Brigham |
|---|---|---|---|---|---|---|
| **Profitability Ratios** | | | | | | |
| 13. Return on revenues $= \dfrac{\text{profit}}{\text{revenues}}$ | % | X | | | | |
| 14. Profit to total assets $= \dfrac{\text{profit}}{\text{total assets}}$ | % | | X | X | | X |
| 15. Profit to worth $= \dfrac{\text{profit}}{\text{net worth}}$ | % | X | X | X | X | X |
| 16. Profits to net working capital $= \dfrac{\text{profit}}{\text{WC}}$ | % | X | | | X | |

*Note:* The sources are as given in the reference section at the end of the text.

The calculation of some of these ratios using the Fudd Associates, Inc. (completed-contract method) statements of Figure 5.5 and 5.6 are presented in the following sections to demonstrate their nature and purpose.

## 6.6   LIQUIDITY RATIOS

Liquidity is the ability of the corporation to meet its current obligations or current liabilities. It must be noted here that billings in excess of costs on uncompleted contracts, although shown in Figure 5.6 as a current liability of the company, is not an obligation. It is actually a deferred credit to income and is in fact often shown as such. Similarly, costs in excess of billings on uncompleted contracts are not considered current assets for ratio analysis.

Current liabilities are those due and payable within one year. Under normal circumstances a business uses its current assets to pay its current liabilities, since both these items are involved in an annual receivable payment cycle.

The *current ratio* (current assets to current liabilities) provides a general picture of the adequacy of the working capital* of a company and of the company's ability to

---

* working capital = current assets − current liabilities.

meet its day-by-day payment obligations. It also measures the margin of safety provided for paying current debts in the event of a reduction in the value of current assets.

From Fudd Associates's balance sheet shown in Figure 5.6, the current ratio is

$$19X2: \quad \frac{428,753}{49,882 + 31,254} = 5.3$$

$$19X3: \quad \frac{280,038}{41,126 + 23,528} = 4.3$$

In commercial enterprises, a current ratio of 2 is considered acceptable. In Table 6.3, Robert Morris Associates gives a range of 1.2 to 1.6 for contractors with volume between 1 and 10 million dollars annually. Such a low current ratio is impractical in construction contracting unless the firm under study makes intensive use of subcontractors.

Current assets include three basic items: cash, accounts receivable, and inventory. When the liquidity of a company is evaluated on the basis of its current ratio, it is assumed that the inventory that figures as part of the total current assets is liquid. However, this assumption is not always realistic in contracting. Many construction inventories have to be built into the project before they can be sold (billed). Consequently, construction inventories move more slowly than those of the normal commercial enterprises.

For this reason a better evaluation of the company's liquidity can be obtained by the so-called *quick ratio*. The quick ratio is computed by taking the quick assets, which are defined as cash and accounts receivable, and relating them to total current liabilities.

For Fudd Associates, the ratio is

$$19X2: \quad \frac{56,753 + 153,112}{49,882 + 31,254} = 2.6$$

$$19X3: \quad \frac{18,111 + 92,801}{41,126 + 23,528} = 1.7$$

As a general rule of thumb, a quick ratio of 1.0 is usually deemed adequate, an assumption not necessarily true in construction. To illustrate this point, the character of accounts receivable in construction must be examined. This balance sheet item is made up of (1) unbilled work, (2) work billed but not yet paid, and (3) retainage withheld. The retainage on contracts that will not be completed in the next financial year cannot be considered truly current. In actual fact, if the company wants to maintain its volume, a fixed sum will constantly be invested in retainage. This sum depends on the conditions of payment in the market in which the company is competing and will probably be on the order of from 5 to 10 percent of billings on contracts in process. A better calculation for the quick ratio in construction can therefore be computed by relating quick assets minus 5 percent of billings on contracts in process to current liabilities. In any event, the amount of retainage in the accounts receivable item should be accounted for in calculating the quick ratio by direct subtraction or by making a percentage allowance.

For Fudd Associates this figure is:

$$19X2: \quad \frac{56,753 + 153,112 - 0.05(1,792,402)}{49,126 + 23,254} = 1.5$$

$$19X3: \quad \frac{18,111 + 92,801 - 0.05(1,284,406)}{41,126 + 23,528} = 0.7$$

It seems reasonable to assume that a value of 1.0 for this ratio is adequate. Table 6.3 indicates that a range between 1.0 and 1.4 is acceptable. It must therefore be concluded that Fudd Associates is under stress to meet its short-term financial commitments during 19X3.

The liquidity of a construction company varies drastically depending on the percentage of annual volume that is subcontracted. A company that makes extensive use of subcontractors can be expected to have far higher liquidity than the contractor who does most of the work with his in-house force.

## 6.7   LEVERAGE RATIOS

Liquidity refers to the ability of a company to maintain a sound financial position over the short term, and solvency refers to its ability to maintain a sound financial position over a longer term.

If a company wants to expand faster than the rate at which earnings can be retained in the company, financing of the expansion must be accomplished by the use of outside capital. When this occurs, the company can look either for long-term debt, involving repayment and interest, or for additional shareholders. The use of the company's assets to secure debt financing allows the company to take a small amount of the company's value and multiply it temporarily into larger amounts of borrowed capital. Use of money to gain access to a large multiple of the amount placed at risk is referred to as *leveraging*. A home owner leverages his down payment to gain access to the entire purchase price of a home through debt (mortgage) financing.

Shareholders' equity, which involves the issuance of stock, is almost always totally permanent capital. New shareholders have an equal claim to the profits generated, so the profits of the current owners are diluted. Long-term debt, in contrast, can increase the earnings of the owners if the invested capital has a higher return than the interest paid on the loan. The price paid, however, is that outside agencies gain an interest in the company.

The *debt-to-net-worth* ratio measures the amount of long-term debt in relation to the amount of shareholders' equity. The ratio is calculated by expressing the long-term debt in relation to total permanent capital (Eq. 4, Table 6.4). For Fudd Associates in 19X3, this is

$$\text{Debt-equity ratio} = \frac{\text{long-term liabilities}}{\text{net worth}}$$

$$= \frac{80{,}000}{200{,}000 + 54{,}247}$$

$$= 0.315$$

This means that 31.5 cents out of every dollar of permanent capital is covering long-term debt. The values for debt-to-net-worth range from 1.1 to 3.5. If the debt-equity ratio indicates that the long-term debtors of the company have invested more money than the shareholders (i.e., the debt-to-worth ratio is greater than one), there may be cause for concern about the adequacy of the permanent capital available in the company.

If the company obtains a disproportionate amount of its capital in the form of long-term debt, it may become burdened with excessively large payments associated with that debt. Long-term loans usually have a fixed annual interest cost attached to the use of the capital. As a general rule, when a corporation is unable to pay this cost, the creditor has the right to demand not only any interest due, but also the original sum loaned. For this reason it is important for the company to be able to meet its fixed annual interest obligations.

The *times-interest-earned* ratio is used to qualify the company's ability to support its long-term debt. This ratio is calculated by expressing the operating profit of the company in terms of the annual interest expense associated with the company's long-term debt. This ratio for Fudd Associates (based on the income statement of Figure 5.5) is $28{,}180/7{,}213 = 3.9$.

From the above calculation it can be seen that the 19X3 operating profit is almost four times greater than the interest on long-term debt. Therefore, a significant change in the level of the company's profitability would have to occur before its ability to meet annual interest payments would be seriously jeopardized.

## 6.8 ACTIVITY RATIOS

The activity ratios focus attention on the turnover of various aspects of the firm's operations. As we have noted, these ratios relate various aspects of the balance sheet, such as receivables and inventory, to the annual revenue. They are "times" ratios. In order to calculate these ratios, the construction volume for the company is required. The annual volume can be calculated by adding, for all contracts, the difference between gross billings at the end of the current year and gross billings at the end of the previous year. This calculation is shown in Table 6.5.

Calculating the *average collection period* is the standard test prescribed for testing the quality of accounts receivable. The average collection period is given as the inverse of Eq. 7, Table 6.4. That is, the average collection period is receivables/revenues. For Fudd Associates this would be $92{,}801/2{,}088{,}009$, or 0.044. Calculating in days

**TABLE 6.5   Construction Volume, Fudd Associates, Inc. (Fiscal Year Ended November 19X3)**

| Contract | Gross Billings Dec. 19X3 | Gross Billings Dec. 19X2 | Difference |
|---|---|---|---|
| K | $867,987 | $818,682 | $   49,305 |
| L | 749,431 | 325,989 | 423,442 |
| M | 978,587 | 444,827 | 533,760 |
| N | 217,162 | | 217,162 |
| O | 714,811 | 202,904 | 511,907 |
| P | 352,433 | | 352,433 |
| | | Construction volume = | $2,088,009 |

the calculation period is 92,801(360)/2,088,009, or 16 days. Because of the complexity of this balance sheet item in construction, this test will not suffice. There is, in fact, no clear and precise way of testing this quality.

An indication of the quality can be given by calculating what we would expect the amount of accounts receivable to be. This amount, again, depends on conditions of payment. It seems fair, however, to assume that it should be equal to retention withheld plus one month's construction volume (items billed but not yet paid).

$$\text{Estimated accounts receivable} = 0.05(\text{billings on contracts in process})$$

$$+ \frac{1}{12}(\text{annual volume})$$

$$= 0.05(1,284,406) + \frac{1}{12}(2,088,009)$$

$$= 238,221$$

This would seem a reasonable amount for the company to have tied up in accounts receivable. In actual fact, Fudd has only $92,801 in accounts receivable. Consequently, it must be concluded that the company has some arrangement with its clients to pay sooner than is expected. This strengthens the assumption that Fudd is in short-term financial distress.

Conversely, if accounts receivable are much higher than expected, a possible reason could be that the company has a large amount of claims that are under dispute, and the quality of the accounts receivable will again be suspect.

The quick ratio recognizes the potential liquidity problems in inventories but does not indicate their quality. To test the liquidity of inventories, the *inventory-turnover* ratio is used. This ratio is calculated by taking the cost of inventories used and dividing it by the average inventory for the year. In construction, this ratio is of limited value. In general, contractors maintain a very limited inventory of material because of the unique nature of each project and the cost of holding and transporting materials. Therefore, this ratio is not very informative.

Asset turnover ratios also yield information regarding the return on investment. The object of using this set of ratios is to determine the effectiveness with which the company puts the assets at its disposal to use in generating business. The best numerator for these ratios is probably the volume of construction done during the previous year. As the denominator, the average of the assets that the company had at its disposal during the year is appropriate. The calculation for average assets is shown below.

**Average Assets for the Year 19X2–19X3 (Completed Contract Method), Fudd Associates Co., Inc.**

| Assets | 19X2 | 19X3 | Average |
|---|---|---|---|
| Current assets | $428,753 | $280,038 | $354,396 |
| Fixed assets (book value) | 109,370 | 223,211 | 166,290 |
| Other assets | 1,247 | 2,428 | 1,838 |
| | $539,370 | $505,677 | $522,524 |

The following turnover ratios for Fudd Associates can now be calculated:

$$\text{a.} \quad \frac{\text{Total annual volume}}{\text{Average current assets}} = \frac{2,088,009}{354,396} = 5.9$$

$$\text{b.} \quad \frac{\text{Total annual volume}}{\text{Average fixed assets}} = \frac{2,088,009}{166,290} = 12.5$$

$$\text{c.} \quad \frac{\text{Total annual volume}}{\text{Average total assets}} = \frac{2,088,009}{522,524} = 4.0$$

By comparing these ratios to those for previous years, the company can determine whether its investment in assets is still effective; and by comparing them to industry figures, the company can determine whether it is keeping up with competition.

## 6.9 RETURN ON INVESTMENT AND PROFITABILITY RATIOS

Return on investment is actually a composite of two factors. The first is the rate at which the business turns its investment over or the rate at which assets are used to generate sales. The second factor that influences return on investment is the profit made on this turnover.

Different combinations of the profit and turnover elements can be used to give the same return on investment. The company must thus decide whether it is going to make its return by "fast pennies or slow nickels." That is, the firm must decide whether it will be a low-volume, high-profit or a low-profit, high-volume firm. This decision is best based on an analysis of trends in its profit structure and by a comparison of its profit structure to those of its competitors.

The profit structure of Fudd Associates is given by the following profit-to-sales ratios:

a. $\dfrac{\text{Gross profit earned on contracts}}{\text{Completed-contract sales}} = \dfrac{230,849}{2,596,005} = 8.9\%$

b. $\dfrac{\text{Gross profit earned on operations}}{\text{Completed-contract sales}} = \dfrac{124,425}{2,596,005} = 4.8\%$

c. $\dfrac{\text{Selling profit}}{\text{Completed-contract sales}} = \dfrac{103,101}{2,596,005} = 4.0\%$

d. $\dfrac{\text{Operating profit}}{\text{Completed-contract sales}} = \dfrac{28,180}{2,596,005} = 1.1\%$

Each of these ratios reflects on some aspect of the operations of the company. Ratio (a) reflects on the pricing policy and job cost control of the company, ratio (b) on its supervisory personnel, ratio (c) on its sales effort, and ratio (d) on the efficiency of its administration.

The all-important ratio is the return that the shareholders receive on their investment. In this ratio the numerator is the net income for the period and the denominator is the owners' net worth. Net worth can be calculated (1) on the basis of shareholders' equity at the end of the accounting period, or (2) on the basis of the average shareholders' equity at the beginning and end of the accounting period. Using the first approach, we calculate the return on investment for Fudd Associates as:

$$\dfrac{\text{Net income}}{\text{Net worth as of December 19X3}} = \dfrac{20,974}{254,247} = 8.3\%$$

A second approach for determining return on investment is to consider the investment as the total permanent capital of the company. Thus the long-term loans are also included as part of the investment. This calculation for Fudd Associates is as shown in the table.

**Return on Total Investment, Fudd Associates, Inc.**

| | | |
|---|---:|---:|
| Net income after tax | | $ 20,974 |
| Add back interest on long-term debt | $7,213 | |
| Tax deduction on interest expense at 50% | 3,607 | |
| Interest expense after tax | | 3,606 |
| Net income after tax, adjusted for interest | | 24,580 |
| Long-term debt | | 80,000 |
| Capital stock | | 200,000 |
| Retained earning | | 54,247 |
| Total capital | | $334,247 |

$$\dfrac{\text{Net income after tax, adjusted for interest}}{\text{Total capital}} = \dfrac{\$24,580}{\$334,247} = 7.4\%$$

It can be seen that the net income figure used in the ratio is adjusted for the interest paid on the long-term loan. Because long-term debt is used as part of the denominator, the income earned by this source of finance has to be added back to give the net income figure.

This formula for calculating the rate of return is usually used in comparing the company to other companies. The resulting figure reflects the return on the total capital of the company without regard to its source. It thus allows comparison to other companies irrespective of the makeup of those companies' permanent capital. For this reason it is a more accurate indication of the adequacy of managment's efforts to obtain an adequate return on investment.

A word of caution is necessary when financial statements prepared by the completed-contract method are used for ratio analysis. With this method of statement preparation, significant profit manipulation is possible with selection of a favorable closing date. Income statements must be inspected for such manipulation.

## 6.10   RATIO ANALYSIS OF
## APEX CONSTRUCTION COMPANY

Sufficient data are available in Table 6.1 and Figure 6.1 to perform a limited analysis of ratios for Apex Construction Company. A review of the firm's current ratio reveals excellent stability between the current assets and current liabilities. The stability gives the impression that the ratio of 1.4 will continue and the Apex Construction Co. will have the ability to meet it's day by day payment obligations.

| Year | Current Ratio (Current Assets/Current Liabilities) |
|------|----------------------------------------------------|
| 19X3 | 1.20 |
| 19X4 | 1.40 |
| 19X5 | 1.40 |
| 19X6 | 1.40 |
| 19X7 | 1.40 |

A better evaluation of the company's liquidity could be found by computing the quick ratio. The current ratio includes inventory as a current asset whereas the quick ratio does not include inventory as current asset.

Upon evaluation of the firm's total debt-to-total-asset ratio and the firm's total debt-to-net worth ratio, it appears that the firm's solvency is adequate. That is, it appears that Apex Construction Company has the ability to maintain a sound financial position over the longer term.

| Year | Total Debt-to-Total Assets (Long-Term Liabilities/Total Assets) |
|------|-----------------------------------------------------------------|
| 19X3 | 0.16 |
| 19X4 | 0.25 |
| 19X5 | 0.26 |
| 19X6 | 0.23 |
| 19X7 | 0.20 |

| Year | Total Debt-to-Net Worth (Long-Term Liabilities/Net Worth) |
|------|----------------------------------------------------------|
| 19X3 | 0.97 |
| 19X4 | 1.64 |
| 19X5 | 1.75 |
| 19X6 | 1.35 |
| 19X7 | 0.88 |

The trend in the above ratios, shows a decrease in the latter three years, indicating that less capital is covering long term debt. This is a positive indication.

In summary, Apex Construction Company has the capability of meeting both its long term and short term financial obligations. With this ability, the company will be able to obtain the necessary bonding and loans required to expand the firm's volume and therefore its earnings.

## REVIEW QUESTIONS AND EXERCISES

**6.1**   Name and discuss four principal tools of financial analysis.

**6.2**   How are financial documents analyzed using the common-size approach.

**6.3**   Give two examples of an activity ratio. What do these ratios reflect?

**6.4**   Using a common-size statement approach analyze these operations of the Jumbo Construction Company across a five-year period based on the data given on page 113.

**6.5**   Given the balance sheet on page 114, calculate the following ratios:

  (a) Current ratio
  (b) Quick ratio
  (c) Debt-equity ratio

**6.6**   How does trend analysis supplement the basic financial ratio calculations and their interpretation?

**6.7**   Given a current ratio of 1.7, to which percentage of book value would you have to liquidate current assets to be able to pay off current liabilities?

**6.8**   Given that the current ratio (i.e., current assets/current liabilities) for a company over the years is as shown, what can you say about the operation of this company?

| *19X1* | *19X2* | *19X2* | *19X3* | *19X4* |
|--------|--------|--------|--------|--------|
| 1.50 | 1.51 | 1.42 | 1.20 | 0.95 |

**6.9**   Given the financial statement shown on page 115 calculate the:

  (a) Current ratio
  (b) Quick ratio

## Jumbo Construction Company Five-Year Consolidated Summary of Operations

| | 19X5 | 19X4 | 19X3 | 19X2 | 19X1 |
|---|---|---|---|---|---|
| Revenues | $832,420,000 | $391,236,000 | $182,684,000 | $214,973,000 | $462,822,000 |
| Interest and dividend income | 2,526,000 | 3,802,000 | 2,627,000 | 1,304,000 | 1,005,000 |
| Cost of contracts | 775,552,000 | 347,425,000 | 148,846,000 | 183,862,000 | 434,120,000 |
| G & A expenses | 36,129,000 | 35,100,000 | 29,207,000 | 26,676,000 | 23,220,000 |
| Interest expense | 2,199,000 | 1,089,000 | — | — | — |
| Income before taxes | 21,066,000 | 11,424,000 | 7,258,000 | 5,739,000 | 6,487,000 |
| Provision for income taxes | 10,700,000 | 5,700,000 | 3,600,000 | 2,800,000 | 3,500,000 |
| Net income | 12,666,000 | 5,724,000 | 3,658,000 | 2,939,000 | 2,987,000 |
| Net income per share | 5.51 | 2.51 | 1.62 | 1.31 | 1.34 |
| Dividends per share | .70 | .45 | .35 | .33 | — |
| Average number shares | 2,298,000 | 2,281,000 | 2,252,000 | 2,236,000 | 2,230,000 |

**Balance Sheet, The Pecan Construction Company, Inc. Macon, Georgia (31 December 19X4)**

| *Assets* | | | *Liabilities* | | |
|---|---:|---:|---|---:|---:|
| **(a) Current assets** | | | **(f) Current liabilities** | | |
| Cash on hand and on deposit | $ 389,927.04 | | Accounts payable | | $ 306,820.29 |
| Notes receivable, current | 16,629.39 | | Due subcontractors | | 713,991.66 |
| Accounts receivable, including retainage of $265,686.39 | 1,222,346.26 | | Accrued expenses and taxes | | 50,559.69 |
| | | | Equipment contracts, current | | 2,838.60 |
| Deposits and miscellaneous receivables | 15,867.80 | | Provision for income taxes | | 97,816.66 |
| Inventory—construction material | 26,530.14 | | Total | | 1,171,826.90 |
| Prepaid expenses | 8,490.68 | | **(g) Deferred credits** | | |
| Total | 1,679,791.31 | | Income billed on jobs in progress at 31 December 19X4 | 2,728,331.36 | |
| **(b) Notes receivable, non-current** | 12,777.97 | | Costs incurred to 31 December 19X4 on uncompleted jobs | 2,718,738.01 | 9,593.35 |
| **(c) Property** | | | Total current liabilities | | 1,181,420.25 |
| Buildings | 5,244.50 | | Equipment contracts, non-current | | 7,477.72 |
| Construction equipment | 188,289.80 | | **(h)** Total liabilities | | 1,188,897.97 |
| Motor vehicles | 37,576.04 | | **Net worth** | | |
| Office furniture and equipment | 13,596.18 | | **(i)** Common stock, 4,610 shares | 461,000.00 | |
| Total | 244,706.52 | | Retained earnings | 184,655.32 | |
| **(d) Less accumulated depreciation** | 102,722.51 | | **(j)** Total net worth | | 645,655.32 |
| Net property | 141,984.01 | | | | |
| **(e) Total assets** | 1,834,553.29 | | **(k)** Total liabilities and net worth | | 1,834,553.29 |

*Source:* Clough, *Construction Contracting*, 3rd ed., copyright © 1975 by John Wiley and Sons, Inc. Reprinted by permission of John Wiley and Sons, Inc.

## Balance Sheet, Peachtree Construction (31 December 19X6)

| | |
|---|---:|
| Assets | |
| Current assets | |
| Cash | $243,146 |
| Accounts receivable | |
| Trade accounts | 201,573 |
| Retainage | 42,147 |
| Total accounts receivable | 243,720 |
| Material inventory | 1,873 |
| Work in process (costs and estimated earnings in excess of billings) | 76,142 |
| Prepaid expenses | 6,148 |
| Other current assets | 782 |
| Total current assets | 571,811 |
| Fixed assets | |
| Machinery and equipment | 542,173 |
| Cars and trucks | 49,214 |
| Furniture and fixtures | 5,812 |
| Total depreciable assets | 597,199 |
| Less accumulated depreciation | 291,314 |
| Net fixed assets | 305,885 |
| Total assets | 877,696 |
| Liabilities | |
| Current liabilities | |
| Accounts payable | 236,144 |
| Retainage payable | 21,312 |
| Notes payable | 84,211 |
| Accrued liabilities | 15,827 |
| Work in process (billings in excess of costs and estimated earnings) | 5,163 |
| Other current liabilities | 1,421 |
| Total current liabilities | 364,078 |
| Long-term liabilities | 138,481 |
| Total liabilities | 502,559 |
| Net worth | |
| Capital stock | 15,000 |
| Retained earnings | 360,137 |
| Total net worth | 375,137 |
| Total liabilities and net worth | $877,696 |

**6.10** Based on the income statement for Peachtree Construction shown below, what is the ratio of operating profit to completed contract sales. Is this ratio acceptable? Explain.

**6.11** Using the data given in Tables P2.1 and P2.2, calculate the following ratios for Pegasus International, Inc., for operations during 19X5:

(a) Quick ratio
(b) Debt ratio
(c) Return on net worth

**6.12** A five-year summary of key financial data for a construction firm is shown in the following figure. Evaluate and analyze the performance of this company during the period. What actions, if any, should be taken to improve the company's position?

**6.13** Explain why a firm's current ratio may differ from another in the same industry.

**6.14** Illustrate how a decrease in working capital can be accompanied by an increase in the current ratio.

## Income Statement, Peachtree Construction (Year ended 31 December 19X5)

|  | $ | % of Sales |
|---|---|---|
| Net sales—total revenue costs | 2,143,761 | 100.00 |
| Materials | 1,013,913 | 47.30 |
| Labor (includes all payroll taxes and union fringes) | 548,271 | 25.57 |
| Subcontracts | 201,798 | 9.41 |
| Other direct costs | 51,450 | 2.40 |
| Total direct costs | 1,815,432 | 84.68 |
| Gross profit | 328,329 | 15.32 |
| Operating expenses |  |  |
| Variable overhead |  |  |
| Auto and truck | 12,089 | .57 |
| Communications | 5,575 | .26 |
| Interest | 1,275 | .05 |
| Miscellaneous | 213 | .01 |
| Office supplies | 1,943 | .09 |
| Travel and entertainment | 871 | .04 |
| Total variable overhead | 21,966 | 1.02 |
| Fixed overhead |  |  |
| Contributions | 675 | .03 |
| Depreciation | 8,238 | .38 |
| Dues and subscriptions | 514 | .02 |
| Insurance | 12,475 | .58 |
| Legal and audit | 2,315 | .10 |
| Licenses and taxes | 175 | .01 |
| Payroll taxes (office only) | 51,790 | 2.42 |
| Rent | 2,506 | .12 |
| Repairs and maintenance | 375 | .02 |
| Salaries—office | 23,418 | 1.10 |
| Salaries—officers | 38,410 | 1.79 |
| Total fixed overhead | 140,891 | 6.57 |
| Total overhead | 162,857 | 7.59 |
| Net profit before taxes | 165,472 | 7.73 |
| State and federal taxes | 77,076 | 3.60 |
| Net profit | 88,396 | 4.13 |

Name __Clairmont Constuctors__

Year End __19X5__

Location __Anywhere, U.S.A.__

Legal Structure __Corporation__

| (000's) | 19X1 | 19X2 | 19X3 | 19X4 | 19X5 |
|---|---|---|---|---|---|
| Sales | 2,460 | 2,560 | 2,830 | 2,835 | 2,850 |
| % Sales growth | | 4 | 10.5 | 1 | 1 |
| Labor | 800 | 820 | 780 | 810 | 900 |
| Material | 925 | 900 | 1,010 | 1,010 | 1,040 |
| Cubcontr. | 500 | 550 | 720 | 765 | 680 |
| Gross margin | 235 | 290 | 320 | 250 | 230 |
| Gross margin % | 9.5 | 11.3 | 11.3 | 8.8 | 8.0 |
| G & A | 133 | 172 | 202 | 218 | 220 |
| G & A % sales | 5.4 | 6.7 | 7.1 | 7.7 | 7.7 |
| Income before tax | 102 | 118 | 118 | 32 | 10 |
| Income before tax% | 4.1 | 4.6 | 4.2 | 1.1 | 0.4 |
| Admin. salary | 24 | 27 | 32 | 38 | 39 |
| Assets | 800 | 812 | 830 | 850 | 830 |
| Sales to assets | 3.07 | 3.15 | 3.4 | 3.33 | 3.43 |
| Equity | 350 | 287 | 285 | 280 | 205 |
| Sales to equity | 7.03 | 8.92 | 9.93 | 10.12 | 13.90 |
| Current Assets | 520 | 532 | 475.8 | _510 | 511.5 |
| Investments | 100 | 110 | 120 | 110 | 120 |
| Investments | 100 | 110 | 120 | 110 | 120 |
| Property and equipment | 280 | 280 | 354.2 | 360 | 338.5 |
| Accu. depreciation | 100 | 110 | 120 | 130 | 140 |
| Current liabilities | 325 | 380 | 390 | 425 | 465 |
| Long term liabilities | 125 | 145 | 155 | 145 | 160 |
| Current ratio | 1.6 | 1.4 | 1.22 | 1.2 | 1.1 |
| Debt to equity | 1.28 | 1.83 | 1.91 | 2.04 | 3.04 |
| Income to equity % | 29.1 | 41.1 | 41.4 | 11.4 | 4.8 |
| Income to assets % | 12.7 | 14.5 | 14.2 | 3.8 | 1.2 |
| Sales to current assets | 4.73 | 4.81 | 5.94 | 5.56 | 5.57 |

*Source:* J. J. Adrian, Construction Accounting (Reston, Va.: Reston, 1978).

# Chapter 7

# COST CONTROL CONCEPTS

## 7.1 FINANCIAL VERSUS COST ACCOUNTING

The purpose and function of the financial accounting system has been discussed in some detail in the preceding chapters. In the following chapters the subject of cost planning and cost control is introduced and considered in detail. The topics of cost planning and cost control viewed jointly constitute cost management or, as it can be called, cost engineering. Among the ancillary topics within cost planning that must be thoroughly understood are estimate and budget development. For each project the contractor develops an estimate of costs, which is the basis of his price quotation to the client. When this estimate of costs is distributed across the life of the project by plotting cost amounts during each period of the project against time, the cost estimate is transformed into a budget.

At the center of the cost control and cost engineering activity is the job cost system, which is shown conceptually in its relation to the general ledger in Figure 7.1. This is a subsystem of the cost accounting system. The overall cost accounting system collects information on all costs to include, for instance, the general and administrative costs, selling costs, and other costs that are not directly project-related. The objective of the job cost system is to capture project expenses as they occur and allocate them to the physical and cost generating subelements of the project being constructed. Most costs associated with a project are tied to the placement of a physical subsection of the work, such as the casting of a floor slab or the installation of a bay of glass. Other costs occur in relation to a nonphysical entity, such as charges for the performance bond or premiums for builder's risk insurance. Such nonphysical costs must be captured by the job cost system.

The job cost system is purely a management information system in that it gathers information designed to aid the manager in controlling the project. As such, the job cost system is to the manager what the cockpit dials and instruments are to the pilot of an airplane. It is a monitoring system designed to provide timely feedback to the manager regarding actual performance on a given project vis-à-vis a project performance goal, such as a budget and/or cost estimate. A secondary function of the job cost system is the collection of data to be used as a basis for estimating future jobs. The

**FIGURE 7.1  General Accounting Flows**

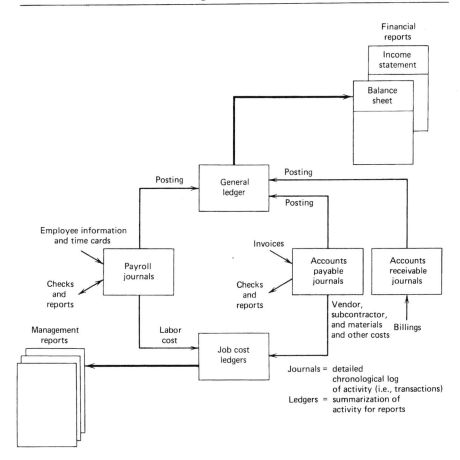

system collects historical data on actual costs of labor, equipment, and materials, which provide reference information in pricing out future projects. Therefore, the cost accounting system serves two major functions: (1) cost monitoring and control versus fixed target values and goals, and (2) collection of data for estimating future projects.

This system is therefore a management information system, and its extent and level of precision is strictly a management decision. The job cost system can be as complex or as cursory as management wants it to be. It can be maintained manually in pencil in a spiral notebook or electronically with advanced data-based concepts on large mainframe computers. If management desires it can dispense with the system entirely and maintain only a financial expense accounting similar to the two systems (BASIC systems 1 and 2) described at the end of Chapter 4. The fact that this is strictly a management system and is maintained at the discretion of company management is in contrast to the financial management system, which by law must be maintained for the purposes of taxation. The financial accounting system is also required for the preparation of reports to stockholders and for submission to prospective lenders.

Since the size and complexity of the job cost system is a management decision, the design of the system should be based on a trade-off between the cost of maintaining the system and the needs of management for current information. As noted above, many small contractors decide that a job cost system is a luxury they cannot afford. As a contractor grows, however, he usually realizes that a project can quickly get out of control without timely information. Let us again use an airplane analogy: When flying a light aircraft, the pilot will be satisfied with a minimal number of instruments. He may be able to fly visually without referring to the instruments and by simply relying on experience. As he moves into larger aircraft and more complicated situations, this practice is no longer adequate, and he must have instruments to monitor the performance of various aspects of the flight. In a large passenger jet aircraft, the complexity and number of systems is so great that several professionals must be in the cockpit to control the flight of the plane.

A large number of professionals and many sophisticated systems are normally required when one is building large and complex projects. In construction, management typically decides on a project-by-project basis how complex the monitoring systems will be. On small projects, rudimentary systems are used. On large and complicated projects, more sophisticated systems are selected. The job cost and cost control systems are tailored to the project.

The selection of the proper level of detail for the job cost system is usually very important to the success of a project. Just as the pilot will have difficulties if key instruments are not available to him, so the manager will encounter immediate difficulties with an inadequate cost control system. This chapter discusses the basic operation of a job cost system and looks at a manual approach to maintaining job cost information in coordination with general ledger account maintenance.

## 7.2   THE COST CONTROL CYCLE

The preparation of the cost plan is the first step in the cost control cycle, shown in Figure 7.2. Once the project commences, real-world conditions affect and cause deviations from the original plan, much as an aircraft experiences headwinds and other factors that influence its progress. The project status, which reflects real-world deviations from the ideal plan, must be monitored and reported to the cost engineer. Reports must be designed to detect cost deviations in a timely way. Decisions can then be made to correct such variations and keep the project within acceptable cost variance limits.

The map of the financial terrain across which the project will develop is the chart of cost accounts. Therefore, a contractor must have a chart of accounts that is sensitive enough to provide early warning of cost deviation; it should allow development of the cost plan in sufficient detail so that it is useful without being too complex. (The subject of cost account design is discussed further in Chapter 8.) The cost plan is based on

**FIGURE 7.2  Cost-Control Management Cycle**

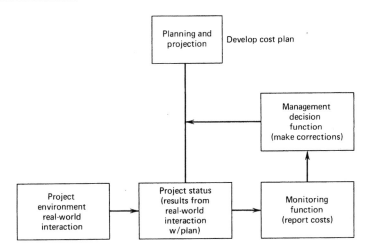

the cost estimate (developed to submit bid) and its scaling to the time frame allowed for the project.

Once the cost plan has been developed, actual cost data must be collected to reflect the project status. Several types of data must be collected including labor cost (i.e., payroll), equipment cost, material costs, and project overhead. Based on the collection of these data, reports that are responsive to the needs of management are developed. Because these reports must assist the cost engineer in detecting deviations from the original cost plan, the raw cost data collected must be arranged (i.e., distributed to the proper cost accounts within the chart of accounts) in useful formats. With proper reporting, the cost manager is in a position to take actions to bring the project back into line with the cost profile or budget.

## 7.3  FUNCTIONING OF THE JOB COST SYSTEM

The objective of the job cost system is to provide the manager with information regarding actual project cost performance, as contrasted with the established budget. The system monitors project expenses and distributes them to "cost centers" associated with the project. The cost centers are cost-generating subelements of the overall project and may be physical portions of the project, such as interior walls and doors, roofing, and the like, or nonphysical items, such as bonds and insurances associated with the project. Since the function of the job cost system is the collection and presentation of cost information, it receives its information from the payroll and accounts payable journals. That is, the job cost system is a receiver of information from other accounting activities. In a similar manner, the general ledger is a receiver of information from the payroll and accounts payable journals. This relationship is shown schematically in Figure 7.1. In effect, the same source document generates flows both to the appropriate project general ledger account (e.g., Expense or Construction in Progress) and to a job cost cost-center account. This parallel flow of information has been referred to previously as double debiting.* In the case of the general ledger account the data are aggregated by project for presentation in the balance sheet or income statement depending upon the method of accounting used. If the completed-contract method is used, the information is reflected in the account Construction in Progress (for the appropriate project). If the percentage-of-completion method is used, the information is reflected in the expense account for the appropriate project.

The parallel flow to the job cost system is distributed to a much more detailed set of accounts. The cost structure of the individual project is defined by those accounts from the chart of cost accounts that have been selected as active and relevant to the cost flows on the project under consideration. The complete chart of cost accounts acts as a menu from which appropriate cost center accounts are selected as required for a particular job. Therefore, the cost information that is aggregated to a general ledger account is distributed to a selected cost account in the job cost system. Again, the detail of the job cost accounts defined will vary from job to job and company to company. The job may be subdivided into as few as four or five cost centers or into thousands of line items, depending upon the complexity of the job and the detail of the chart of cost accounts. This job of allocation of cost to the various cost accounts becomes more complex as the size and nature of the job (as well as the cost system) become more complex. Since the purpose of this procedure is to compare actual costs with projected costs, the definition of cost accounts for a given project is closely related to the breakdown of the job into cost centers for the purpose of estimating. If the cost-generating components of the job are called "estimating" accounts, the cost accounts are defined so as to be compatible with the estimating accounts. In most cases there is a one-to-one relationship. That is, if an estimating element called "slab on grade— 6 in. and under" has been used for quantity and pricing development in the estimate, a cost account of the same description is used for collecting cost data during construction. This expedites comparison of actual costs incurred with costs projected as part of the estimate.

In summary, data flowing to the job cost system are sorted and allocated to the appropriate cost accounts on an ongoing basis throughout the life of the project. At the same time, parallel flows for the same data aggregate information in the appropriate general ledger accounts. The accounts in the job cost system can be thought of as subsidiary ledgers to the general ledger's accounts reflected directly in the balance sheet and the income statement. As noted previously, depending upon the desires of management, these subsidiary "cost center" ledgers may be maintained in balance with their corresponding (i.e., parent) general ledger accounts or may be allowed to go out of balance, depending upon policies regarding the operation of the job cost system.

---

* A debit is involved because an increase in balance occurs. Both the asset account, Construction in Progress, and the expense accounts increase their balances when debited.

## 7.4   A MANUAL JOB COST SYSTEM

In order to describe in greater detail the mechanics of data flow from source documents to both the general ledger and the job cost system, we now describe a manual system. The information described is based on the "Write It Once" system, which is a trademark product of the Burroughs Corporation. The Write It Once concept was originally documented in detail in publications of the Charles R. Hadley Company.* The system allows the bookkeeper to journalize and post information simultaneously by the use of a standard pegboard, standard forms, and carbon transfer paper. Entries from source documents are entered into a composite payroll and expense journal, the appropriate accounts payable ledger sheet, and the job cost ledger all in one operation. A standard sheet, the Record of Invoices and Work in Progress, is used to journalize entries. The columns of this sheet are arranged so that the appropriate ledger sheets for a given entry can be mounted over the journal sheet (by using the pegboard feature), a method that expedites simultaneous journalizing and posting. The standard sheet utilized is shown in Figure 7.3, and the overlaying of the forms is shown in Figure 7.4.

Each of the columns on the sheet has been defined to provide organization of the journalizing/posting operation while allowing for maximum flexibility in dealing with special situations. Entries based on source documents are made to the sheet and require transcription of data to certain of the columns depending upon the type of transaction being processed. The bookkeeper organizes the source documents (invoices, payroll summaries, and so on) so that they can be recorded in a sequential fashion. The function of each column on the sheet as well as the information required is discussed below. Columns 1 to 8 pertain to accounts payable (i.e., general ledger) ledger entries. Columns 11 to 19 pertain to job cost ledger entries.

*Column 1:* The date of the transaction is recorded. This would be the date on which the expense is billed (invoice date) or recognized (payroll date) when the accrual basis is used.

*Column 2:* This item identifies the source document or action that generated the entry. It may be an invoice number, a payroll journal page number, or similar control reference.

*Column 3:* All expenses result in a credit to the accounts payable column. An offsetting debit entry will be made to column 9 if the expense results in general inventory to the firm (e.g., purchase of lumber for stock, etc.). If the expense is directly related to a project, it will result in a debit to one of the work-in-progress columns (i.e., 15, 16, 17, or 19). An invoice that is nonproject or inventory related can be debited directly to the general ledger account in column 20. For instance, purchase of a fixed asset such as a typewriter or a piece of equipment could be debited through column 20 to the appropriate general ledger account.

---

* See Hadley Service Bulletin No. 194, *Contractor's Job Cost System,* April 1956.

**FIGURE 7.3 Record of Invoices and Work in Progress**

NO. 16 — RECORD OF INVOICES AND WORK IN PROGRESS — MONTH OF AUGUST 19 83 — Gen Ledger

| Date (1) | Reference (2) | Accounts Payable Debits (3) | Accounts Payable Credits (3) | Accrued Payroll (Credit) (4) | Overhead Absorbed (Credit) (5) | Other Credits Acct (6) | Other Credits Amount (6) | Prior Balance (Memo) (7) | Balance (Memo) (8) | Inventory DR (9) | Inventory CR (10) | Line | Date (11) | Job Number (12) | Vendor or Description (13) | Labor (14) | WIP Material (15) | WIP Sub-Contracts (16) | WIP Equip Charges (17) | WIP Overhead (18) | Other Costs Acct (19) | Other Costs Amount (19) | Gen Ledger Acct (20) | Gen Ledger Amount (20) |
|---|---|---|---|---|---|---|---|---|---|---|---|---|---|---|---|---|---|---|---|---|---|---|---|---|
| 8 Aug 83 | E26-2.. | | 500.00 | | | | | 375.00 | 875.00 | 580.00 | | 1 | | | Williams Brothers | | | | | | | | | |
| 10 Aug | 1625-8 | | 136.80 | | | | | 145.16 | 281.96 | | | 2 | 11 Aug 83 | 101 | Ace Hardware | | 136.80 | | | | | | | |
| 11 Aug | 2140 | | | | | | | | | | 450.00 | 3 | 11 Aug | 102 | Transfer to Job From Inventory | | 450.00 | | | | | | | |
| 12 Aug | 200-865 | | 100.00 | | | | | 418.25 | 518.25 | | | 4 | 12 Aug | 102 | Exxon, Inc | | | | 50.00 | | | | | |
| | | | | | | | | | | | | 5 | 12 Aug | 103 | Exxon, Inc. | | | | 50.00 | | | | | |
| 15 Aug | P-118 | | | 628.15 | | | | | | | | 6 | 15 Aug | 101 | Direct Labor - 8/15 | 628.15 | | | | | | | | |
| 18 Aug | 83-206 | | 637.50 | | | | | 158.20 | 795.70 | | | 7 | 18 Aug | 103 | Herrin Plumbing | | | 637.50 | | | | | | |
| 19 Aug | | | 150.00 | | | | | -0- | 150.00 | | | 8 | 19 Aug | 102 | Permit - City of Atlanta | | | | | | 750 | 150.00 | | |
| 22 Aug | 2144 | | | | | | | | | 325.00 | | 9 | 22 Aug | 103 | Transfer to Job 101 | | (325.00) | | | | | | | |
| 22 Aug | 2144 | | | | | | | | | | 325.00 | 10 | 22 Aug | 101 | Transfer From Job 103 | | 325.00 | | | | | | | |
| 23 Aug | Sched. | | | | 765.00 | | | | | | | 11 | 23 Aug | 101 | Overhead | | | | | 765.00 | | | | |
| 25 Aug | | | 198.00 | | | | | 216.00 | 414.00 | | | 12 | | | Georgia Power | | | | | | | | 751 | 198.00 |
| | | | 1722.36 | 628.15 | 765.00 | | | 1312.61 | 3034.91 | 825.00 | 775.00 | 13 | | | | 628.15 | 586.80 | 637.50 | 100.00 | 765.00 | | 150.00 | | 198.00 |
| | | | | | | | | | | | | 14 | | | | | | | | | | | | |
| | | | | | | | | | | | | 15 | | | | | | | | | | | | |

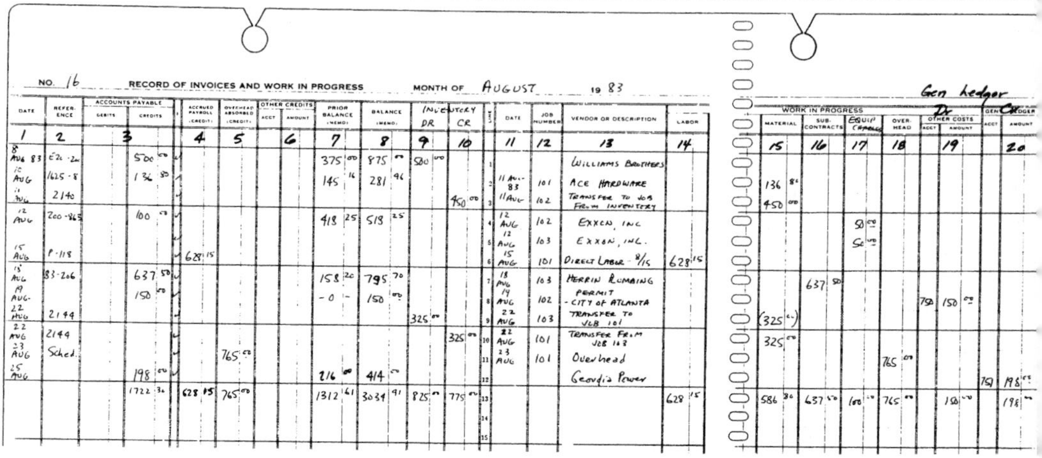

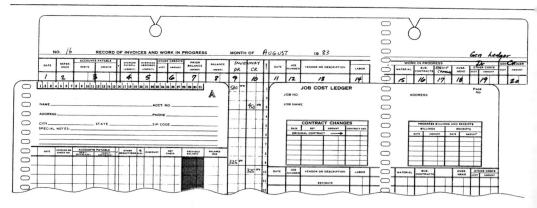

If a vendor issues a credit memo (reducing the billed amount or required payment), this would cause a debit (reduction in balance) to the account payable. This is recorded as a debit in column 3. Offsetting credits reflecting the reduction in billing would be made to the appropriate job, inventory, or other account.

*Column 4:* Payroll expenses are recorded in this column as a credit (increase in balance) entry. An offsetting debit entry in column 14 is made to reflect labor cost on the appropriate project.

**FIGURE 7.4 Accounts-Payable Ledger (*left*) and Job-Cost Ledger (*right*) Superimposed on the Record of Invoices and Work in Progress**

*Column 5:* This column is reserved for crediting overhead costs (salaries, site support charges, telephone, etc.) that are distributed to the project. An offsetting debit is made to the job overhead account in column 18.

*Column 6:* This is a user-defined column available for special-purpose use. Its use depends on how optional-use columns 9, 10, and 17 are used. Often debit corrections to the job must be made. This column provides a location for an offsetting credit if the other credit columns are not appropriate.

*Column 7:* This column is provided to record the original (prior to this entry) balance shown on the account payable being credited. It is designed to provide a control in the updating of the balance.

*Column 8:* The new account balance is calculated by adding the original balance (recorded in column 7) to the account of the transaction. That is, the amount of the invoice being processed is added to the original balance to get the new balance.

*Columns 9 and 10:* These are special-use columns that can be used for inventory control. Some companies purchase materials for nonjob inventory. Such materials are then later allocated to a particular job. Purchase of these materials to inventory can be credited directly in column 9. Sometimes materials are transferred from one job to another. One way to accomplish this is to go through a general ledger account. The transfer can be accomplished with these columns by crediting the materials account on the project (column 15) losing the materials (e.g., form material—Project 101) and debiting the company inventory (asset) account. The credit to the job in column 15 is shown by circling the entry. A second entry would then credit (reduce) the company materials inventory account and debit the materials to the job receiving them (e.g., form material—Project 103) in column 15.

Materials purchased to inventory or returned to stock (excess to the project) are debited directly in column 9. This is done by crediting the account payable or project expense account (circled entry in column 15) and debiting column 9.

Many contractors do not maintain a company-level inventory per se and charge all materials to the job. Therefore, columns 9 and 10 could be used for other entries. If a contractor charges a portion of his equipment fleet costs to a job account by a debit entry, the offsetting credit entry can be made here under the title "equipment charges" or "equipment rental (other income)." One of the two columns can be used for inventory, the other for equipment. In such cases, the credit entries are circled to distinguish them from debit entries.

*Column 11:* This and all columns to the right pertain to job cost ledger accounts (with the exception of column 20). The date on which the transaction is posted to the job cost ledger is recorded in this column.

*Column 12:* The job number is recorded so that journal entries are traceable to a particular project (unless they are not job related). This leads to a repetitive posting of the job number to the job cost ledger. This is justified as a check to verify that the proper job entry is being made to the proper job ledger.

*Column 13:* The vendor's name is recorded. If the transaction does not relate to a vendor (e.g., a payroll transaction), a short description of the nature of the transaction is shown here.

*Column 14:* Labor charges (payroll) on the job are recorded here. This is the debit that offsets the credit entry in column 4. This entry will normally be made weekly, based on payroll expenditures.

*Column 15:* Costs of materials charged to the job are posted here. This entry may relate to materials purchased from a vendor or taken from a company-level materials inventory.

*Column 16:* Subcontractor charges against the job are recorded here. This is the gross invoiced amount, regardless of whether a retainage is held against the subcontractor. The amounts retained will be reflected in the "balance" of the subcontractor's (account payable) ledger.

*Column 17:* This column is available for use as required. It may be used to debit equipment charges to the job in cases where equipment costs are prorated to various projects (i.e., absorbed). The offsetting credit would be recorded in column 9 or 10, as described above.

*Column 18:* This column allows for the charging of overhead to the job. Typically it is a prorate of home-office (G&A) overhead. This entry is offset by a credit entry in column 5 and represents absorption of nonjob specific overhead, such as home office rental and staff and support costs. This distribution is normally made on a monthly basis.

*Column 19:* Job-related charges other than those specified in columns 14 through 18 are debited (entered) here. These may be special items such as building permit fees, legal fees, and similar charges. Offsetting credit entries would be made in column 6.

*Column 20:* Items chargeable directly to a general ledger account rather than to a job or inventory would be recorded here. For instance, purchase of a typewriter could be debited directly to a fixed asset account using this column.

It can be seen that the *record-of-invoices and work-in-progress* sheet is organized so that the journalizing activity in effect causes posting of:

1. The accounts payable general ledger
2. The vendor accounts payable subledger
3. The work-in-progress general ledger
4. The job cost subledger for the appropriate job and cost center

The left side of this sheet (columns 1 to 8) acts both as a journal and as an accounts payable general ledger. The right side of the sheet (columns 11 to 19) acts as a journal and a work-in-progress general ledger [see Figure 7.5(a)]. When the individual vendor's accounts payable subledger is placed over the left side of the sheet, a single entry results in a journalizing and a "double crediting" action, crediting the accounts payable general ledger (by imprint through the carbon paper) and crediting the vendor's

**FIGURE 7.5(a)  Layout of Record-of-Invoices and Work-in-Progrss Sheet**

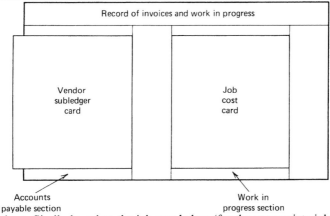

**FIGURE 7.5(b)  Positioning of Vendor Accounts Payable Subledger and Job-Cost Subledger on Record of Invoices and Work in Progress**

subledger sheet. Similarly, when the job cost ledger (for the appropriate job) is placed over the right side of the sheet, a "double debiting" account occurs, causing the general ledger account "Work in Progress" to be debited (again by imprint) and the job cost subledger to be debited [see Figure 7.5(b)].

The effect of the double crediting and double debiting action is to accomplish the posting flows shown in Figure 7.1 with a single entry. The standard vendor subledger card and job cost card are shown in Figures 7.6 and 7.7.

## 7.5  SOME TYPICAL ENTRIES TO THE JOB COST SYSTEM

In order to demonstrate the processing of various types of transactions, a number of typical entries are shown in Figure 7.3. In the case of all expense transactions, verification must be made of the amount and the fact that the services and/or goods involved

# FIGURE 7.6  Accounts-Payable Ledger Form

| | 1 | 2 | 3 | 4 | 5 | 6 | 7 | 8 | 9 | 10 | 11 | 12 | 13 | 14 | 15 | 16 | 17 | 18 | 19 | 20 | 21 | 22 | 23 | 24 | 25 | 26 | 27 | 28 | 29 | 30 | 31 | |

Ⓐ

NAME _____ ACCT. NO. _____

ADDRESS _____ PHONE _____

CITY _____ STATE _____ ZIP CODE _____
SPECIAL NOTES: _____
_____
_____

| | DATE | INVOICE OR CHECK NO. | ACCOUNTS PAYABLE | | ✓ | OTHER DEDUCTIONS | % DISC | DISCOUNT | NET CHECK | PREVIOUS BALANCE | BALANCE DUE |
|---|---|---|---|---|---|---|---|---|---|---|---|
| | | | DEBIT (DECREASE) | CREDIT (INCREASE) | | | | | | | |
| | | | | | | | | | | | |
| | | | | | | | | | | | |
| | | | | | | | | | | | |
| | | | | | | | | | | | |
| | | | | | | | | | | | |
| | | | | | | | | | | | |
| | | | | | | | | | | | |
| | | | | | | | | | | | |
| | | | | | | | | | | | |
| | | | | | | | | | | | |
| | | | | | | | | | | | |
| | | | | | | | | | | | |
| | | | | | | | | | | | |
| | | | | | | | | | | | |
| | | | | | | | | | | | |
| | | | | | | | | | | | |
| | | | | | | | | | | | |
| | | | | | | | | | | | |
| | | | | | | | | | | | |
| | | | | | | | | | | | |
| | | | | | | | | | | | |
| | | | | | | | | | | | |
| | | | | | | | | | | | |
| | | | | | | | | | | | |
| | | | | | | | | | | | |

# FIGURE 7.7   Job-Cost Ledger Form

| | | JOB COST LEDGER | | | | ADDRESS | | | PAGE No. | |
|---|---|---|---|---|---|---|---|---|---|---|

JOB NO.

JOB NAME

| CONTRACT CHANGES | | | |
|---|---|---|---|
| DATE | REF. | AMOUNT | CONTRACT AMT. |
| ORIGINAL CONTRACT ——▶ | | | |
| | | | |
| | | | |
| | | | |
| | | | |
| | | | |
| | | | |

| PROGRESS BILLINGS AND RECEIPTS | | | |
|---|---|---|---|
| BILLINGS | | RECEIPTS | |
| DATE | AMOUNT | DATE | AMOUNT |
| | | | |
| | | | |
| | | | |
| | | | |
| | | | |

| DATE | JOB NUMBER | VENDOR OR DESCRIPTION | LABOR | | MATERIAL | SUB-CONTRACTS | | OVER-HEAD | OTHER COSTS | |
|---|---|---|---|---|---|---|---|---|---|---|
| | | | | | | | | | ACCT | AMOUNT |
| | | ESTIMATE | | | | | | | | |

129

have been received, in order to avoid paying the wrong amount or paying for something not received. At the time of entry, the invoices or other source documents are arranged in order (usually in sequence based on the jobs to which they pertain). Approval to enter the data to the system is normally required. A total amount for the entries to be made is calculated; it is used later as a control check. The entries are described as follows:

1.  The first entry is to record Williams Brothers' invoice ($500) for lumber that will be placed in inventory. This requires use of the Williams Brothers accounts payable subledger.

2.  An invoice from Ace Hardware for materials to be charged to Job 101 is recorded. This requires the Ace Hardware subledger card and the Job 101 job cost ledger card.

3.  This transaction has to do with removing materials from company inventory and moving them to a particular job. The materials are charged to Job 102, so the inventory account is credited and the job ledger for Job 102 is debited (that is, a record of this expense is added to Job 102). The ledger card for Job 102 is required. No account payable subledger is required, since this transaction does not involve the vendor directly.

4.  This is a billing from Exxon that involves distribution for two different projects. The billing is for $100 in fuel costs; half is charged to Job 102 and half is charged to Job 103. A single entry is made on the left side of the Record of Invoices and Work in Progress to reflect the account payable to Exxon. This requires the Exxon account payable subledger card. Two entries are required on the work-in-progress side. The first entry is a debit to Job 102 in the amount of $50. This requires the presence of the Job 102 ledger card. The second is a debit to Job 103, again with the Job 103 ledger card. In this distributed entry, one line on the accounts-payable side generates two entries on the work-in-progress side.

5.  This is a payroll entry for direct labor generated by summary information from the payroll system. A credit is generated to accrued payroll and a debit to work in progress and the job cost ledger of Job 101. This requires the job cost subledger card. The vendor account payable subledger card is not used, since no vendor is involved.

6.  A subcontractor billing from Herrin Plumbing is entered in the amount of $637.50. This is for work on Job 103. This requires the accounts payable subledger card for Herrin Plumbing and the job cost ledger for Job 103.

7.  The cost of a building permit from the City of Atlanta is charged to Job 102. This is recorded in column 19 with an account reference. An offsetting credit to accounts payable is made in column 3. No vendor subledger is involved. The job cost ledger for Job 102 is required.

8.  Materials are transferred from Job 103 to Job 101. This is accomplished by moving them from Job 103 to inventory and then from inventory to Job 101. Two transactions are needed. The first credits the Job 103 job cost ledger in the amount of $325 for the cost of the material. An offsetting debit in the same amount is made to company inventory. A second entry credits (reduces) inventory

and debits Job 101 in the amount of transfer ($325). This set of transactions requires the job cost subledger for Job 103 and then the job cost subledger card for Job 101.

9. A distribution of G&A overhead occurs at the end of the month. Column 5, overhead absorbed, is credited and column 18, job overhead, is debited. The job cost subledger is required.

10. A general charge for heating of the main office, $198, is credited to the vendor account, Georgia Power, and debited to the appropriate general ledger G&A expense accounts.

These transactions illustrate the flexibility of the "write-it-once" system. Further, they illustrate how both general ledger and job cost accounts are maintained and updated simultaneously. Many different manual systems are available. The intent of this discussion is primarily to convey a better understanding of how costs flow to the job cost subledgers so that management is able to monitor the buildup of expense versus initial estimate projections.

It should be clear now that the level of detail achieved in segregating costs on a given job (into cost centers or subelements) is a function of how many job cost subledger cards are associated with each job. If only one job cost ledger is used, only an accumulation of costs by job is achieved. As more subdivisions are defined, more subledgers per job are required. As the hierarchy of cost breakdown increases, a set of job cost cards (or cost accounts) on each job, as indicated in Figure 4.10, will develop. Again, the complexity of the subledgers defined is a management decision. The structure and number of cost accounts and the values of multilevel systems are discussed in the next chapter.

## REVIEW QUESTIONS

**7.1** What is the relationship between the general ledger and the job cost systems? Is it necessary to keep them balanced? Explain.

**7.2** Are cost accounting figures found in annual reports of a publicly held construction company? Why?

**7.3** What are the two essential components of a cost-control management system cycle?

**7.4** What are the two major reasons for maintaining a job cost system?

**7.5** What is a cost account and what is it used for? What is the relationship between cost accounts and estimating accounts?

**7.6** Using the transactions for CPM Construction Company given in Problem 5.8, make appropriate entries to a record of invoices and work-in-progress sheet similar to that shown in Figure 7.3.

# Chapter 8

# COST ACCOUNT STRUCTURES

## 8.1 COST CONTROL AS A MANAGEMENT TOOL

The early detection of actual or potential cost overruns in field construction activities is vital to management. It provides the opportunity to initiate remedial action and increases the chance of eliminating such overruns or minimizing their impact. Since cost overruns increase project costs and diminish profits, it is easy to see why both project management and upper-level management must become sensitive to the costs of all project activities.

An important byproduct of an effective cost reporting system is the information that it can generate for management on the general cost performance of field construction activities. This information can be brought to bear on problems of great interest to project management. The determination of current project status, effectiveness of work progress, and preparation of progress payment requests require data generated by both project planning and cost control reporting systems. Project cost control data are important not only to project management in decision-making processes, but also to the company's estimating and planning departments because these data provide feedback information essential for effective estimates and bids on new projects. Thus a project control system should both serve current project management efforts and provide the field performance data base for estimating future projects.

A conceptual diagram of the relationship of the financial accounting and cost accounting systems is shown in Figure 8.1. Cost management focuses on monitoring and controlling the cash flowing into and through the project cost accounts. In effect, the project expense accounts are the cost accounts that are to be managed. Estimating and cost control are closely linked, and one builds on the other. Both are refined as field data and actual field experience become integrated.

In order to bid on a job, an estimate, or cost plan, is prepared. This estimate is prepared by referring to data on past projects of a similar nature and references containing data not available in company files. The cost centers represented by line items in the job cost system are utilized to break the project down into estimating accounts. There is normally a *one-to-one* relationship between the estimating accounts and the cost accounts. The costs developed during the estimating phase are transferred

**FIGURE 8.1 Cost Accounting System Uses Field Reports to Detect Variances and Retains Data for Future Estimates**

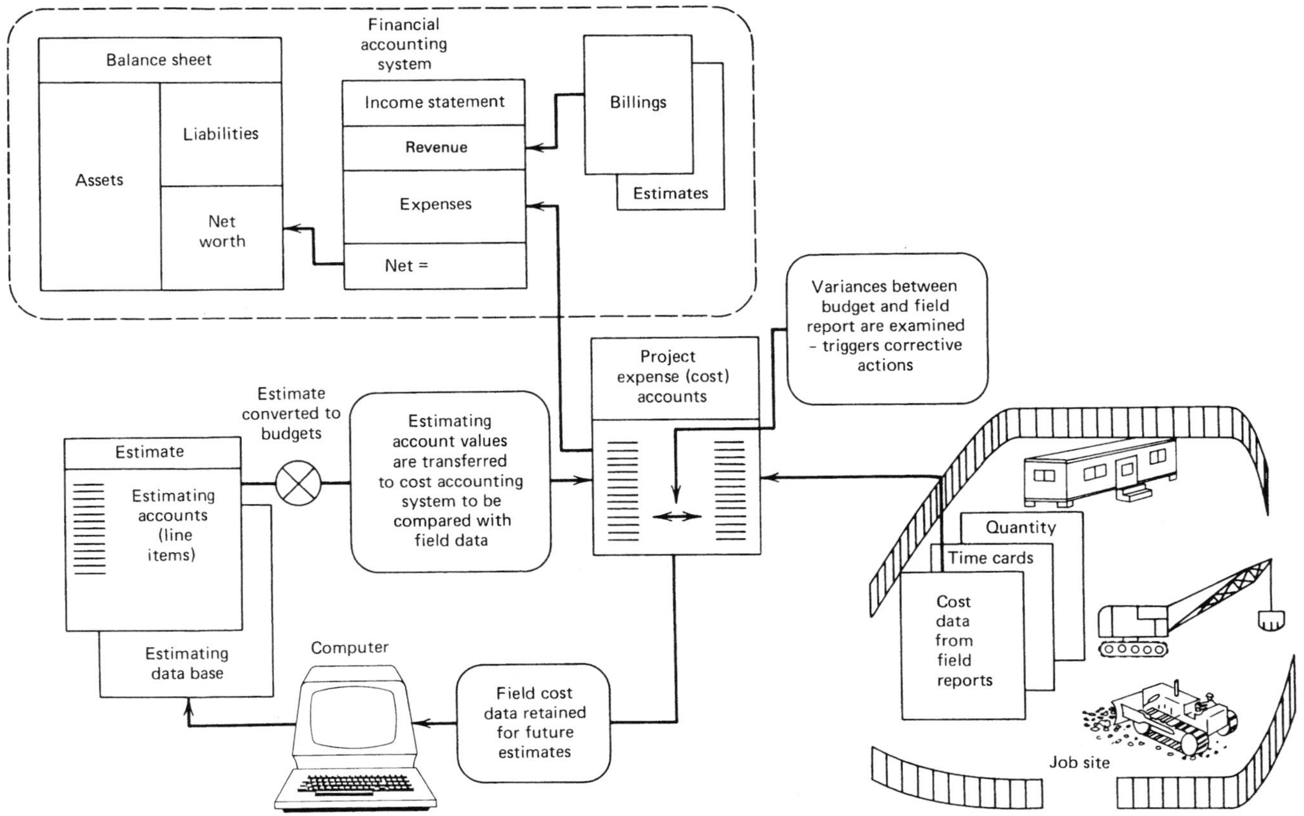

Variances between budget and field report are examined - triggers corrective actions

Quantity

Time cards

Cost data from field reports

Job site

Financial accounting system

Billings

Estimates

Income statement

Revenue

Expenses

Net =

Balance sheet

Liabilities

Net worth

Assets

Project expense (cost) accounts

Estimating account values are transferred to cost accounting system to be compared with field data

Field cost data retained for future estimates

Estimate converted to budgets

Computer

Estimate

Estimating accounts (line items)

Estimating data base

to the job cost system to act as target values against which field costs and performance can be measured. As construction proceeds, field data are collected, distributed, and recorded in the appropriate job cost accounts. At this point, the cost manager is in a position to compare actual field data with the expectations reflected in the estimate. Variances are monitored between expected costs, as projected by the estimate and budget (time-scaled estimate), and the field production rates and charges. By examining the variances, the manager can detect which accounts are seriously deviating from planned progress and can take corrective action.

As information is collected to perform the control function, it also is available to be stored and maintained for use in estimating new projects. Typically, cost data are stored on the computer so that they can be retrieved and utilized for estimates of future projects. The cost accounting system thus performs the double function of allowing control of the job in progress by variance analysis as well as providing data in an organized fashion for future projects.

The form and design of a cost control system depends upon the needs of the contractor and the effort that must be expended to implement and maintain the system. As with all management systems, the effort and time expended in operating the system must provide benefits that exceed the cost of the system. To be effective, the cost control system must produce the right amount of information at the right time. Too much information too late is of little help in management decision processes. Similarly, too little information may be ineffective, if significant cost overrun problems go undetected until a final project accounting. The extent to which project management supports a project control system depends upon management attitude, the financial risks inherent in the form of contract being used, and the size of the profit margin. In lump sum and unit price contracts, the contractor binds himself under the terms of the contract to complete the project within certain costs. In other types of contracts, such as cost-plus and construction management contracts, the contractor may not be in the same cost-constrained situation.

The design, implementation, and maintenance of a project cost control system can be considered a multistep process. The five steps shown schematically in Figure 8.2 form the basis for establishing and maintaining a cost control system. The following questions regarding each step in the implementation of the cost control system must be addressed.

1. *Chart of cost accounts.* What will be the basis adopted for developing estimated project expenditures, and how will this basis be related to the firm's general accounts and accounting functions? What will be the level of detail adopted in defining the project cost accounts, and how will they interface with other general ledger accounts?

2. *Project cost plan.* How will the cost accounts be utilized to allow comparisons between the project estimate and cost plan with actual costs as recorded in the field? How will the project budget estimate be related to the construction plan and schedule in the formation of a project cost control framework?

3. *Cost data collection.* How will cost data be collected and integrated into the cost reporting system?

**FIGURE 8.2   Steps in Cost Control**

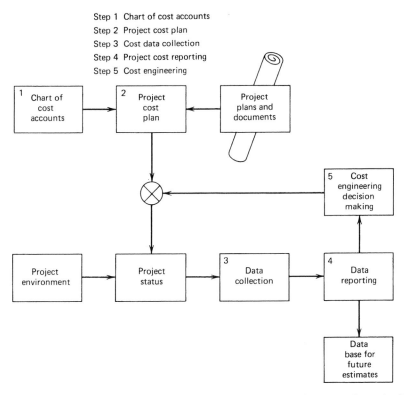

Step 1  Chart of cost accounts
Step 2  Project cost plan
Step 3  Cost data collection
Step 4  Project cost reporting
Step 5  Cost engineering

4.  *Project cost reporting*. What project cost reports are relevant and required by project management in its cost management of the project?

5.  *Cost engineering*. What cost engineering procedures should project management implement in its efforts to minimize costs?

These are basic questions that management must address in setting up the cost control system. The structure of cost accounts will be discussed in detail in this chapter. The other questions will be considered in subsequent chapters.

## 8.2.   THE CHART OF COST ACCOUNTS

The chart of cost accounts defines the level of detail at which job cost information will be maintained. As noted in Chapter 7, it may be very rudimentary or very detailed depending on the complexity of the project and the needs of management. The more complex the project, the finer the level of detail must be. The numerical or alphanumeric designations of individual line items defined for control purposes are called *cost codes*.

Therefore, the chart of cost accounts is defined in terms of a cost code or numbering system. This system is normally developed to reflect various levels of detail. A hierarchical coding system is used to present levels of information, since summarization of cost information at various levels is desirable. The multilevel nature of cost accounting was introduced in Chapter 4 (Sections 4.9 and 4.10). The definition of a cost coding system that is flexible enough to reflect the desired levels of job cost information and is responsive to the needs of management is critical to the overall design of the cost control system.

It should be noted that the chart of cost accounts is a part or subelement of the overall chart of accounts introduced in Section 4.1 and referred to in Section 5.3. The cost accounts can be thought of as a subsection of the overall chart of accounts. Depending upon the desires of management and the purpose of the job cost system, these accounts may be fully integrated into the general ledger system and maintained in balance in conformance with general bookkeeping principles. In such cases, the individual job cost accounts can be thought of as subsidiary ledgers within the general ledger system. This is the approach described in Section 4.10. If a large number of accounts is defined, a cumbersome posting procedure requiring multilevel debiting may result. If the subsidiary ledger system is used, one approach to reducing the complexity of the posting activity is to reduce the number of summary levels. In some situations cost account subsidiary ledgers are consolidated directly to the project expense or construction in progress account (depending on the method of accounting used) as with the manual system described in Chapter 7. This means that the multilevel system of Section 4.10 (see Figure 8.3) is reduced to a two-level system with double rather than multilevel debiting.

At the upper level are the project expense ledgers shown in Figure 8.3 as level 2. The lower level would combine levels 3, 4, 5, and 6 into a single set of items detailed at level 6. For instance, an entry for truck driver's pay in regard to activity 2 on line item 3 is made as a single debit to the composite account (that is, a single subledger account card combining the attributes of levels 3 through 6). A parallel debit is made to the project level (level 2) work-in-progress account for the project in question. This approach does away with the debiting of the intervening level accounts at levels 5 through 3. Since the multiple level subsidary ledgers on the intermediate levels are dispensed with, the number of accounts and the number of entries generated is greatly reduced. Due to the stylizing or tailoring of cost account structures to each job, some accounts will not be active (utilized) on a given project. That is, they will not be necessary and therefore will not be defined. Only the required or relevant subsidiary ledgers will be established.

Certain contractors find it desirable to uncouple the job cost system from the balancing requirements of a general ledger system. That is, the job cost system is set up as a management information collection system separate and distinct from the general ledger and, therefore, not subject to the balancing requirements of general accounting principles. The individual cost collection items are still referred to as cost "accounts," and the chart of cost accounts is still considered as a subelement of the overall chart of accounts. The job cost system, however, simply receives data from the payroll and accounts payable as shown in Figure 7.1 and no attempt is made to maintain it in

# FIGURE 8.3 A Comprehensive Multilevel Cost Accounting System

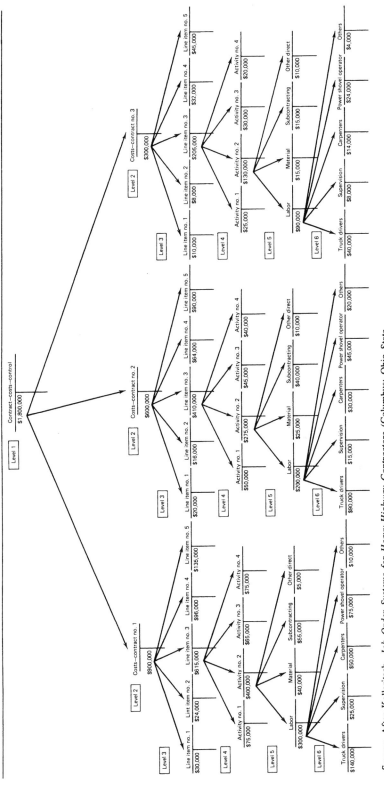

*Source:* After Kollaritsch, *Job Order Systems for Heavy-Highway Contractors* (Columbus: Ohio State University, college of Administrative Science, 1974), p. 20.

balance (from an accounting point of view) with the general ledger. Both the general ledger and the job cost systems receive data from the journals (payroll and accounts payable). However, one is a financial record and is balanced in accordance with the requirements of the Internal Revenue Service and accounting practice. The job cost system is subject only to the requirements of management, and may therefore be balanced or unbalanced as desired.

## 8.3.   COST ACCOUNT STRUCTURE

The first step in establishing a cost control system for a construction job is the definition of project-level cost centers. The primary function of the cost account section of a chart of accounts is to divide the total project into significant units of work, each consisting of a given type of work that can be measured in the field (see Figure 8.4). Once job cost accounts are established, each account is then assigned an identifying code known as a cost code. Once segregated by associated cost centers, all the elements of expense (direct labor, indirect labor, materials, supplies, equipment costs, and the like) constituting a work unit can be properly recorded by cost code.

The design, structure, and development of a cost coding system and its associated set of expense accounts have a significant impact on the cost management of a company or project. As mentioned previously, the job cost accounting system is essentially an accounting information system. Therefore, management is free to establish its own chart of accounts in any way that helps it in reaching specific financial and cost control objectives, whether these objectives are related to general company performance, to the control of a specific project, or to specific contract requirements.

A variety of cost coding systems exists in practice, and standard charts of accounts are published by organizations such as the American Home Builders Association, American Road Builders Association, Associated General Contractors, and the Construction Specifications Institute. In many industries, cost codes have a company-wide accounting focus emphasizing expense generation based on a departmental breakdown of the firm. In some construction firms, cost systems have a structured sequence corresponding to the order of appearance of the various trades or types of construction processes typical of the company's construction activity. In most construction companies, rather detailed project cost accounts such as those shown in Figure 8.4 are used. This method recognizes the fact that construction work is project oriented and that to achieve the cost management goal of maximizing profit, projects must be accounted for individually. One project may be a winner while another is losing money. Such situations may be masked in the accounting system unless job cost accounts are maintained on a project-by-project basis. Therefore, both billings (revenue) and cost (work in progress) accounts are typically maintained for each project. The actual account descriptions or designations vary in accordance with the type of construction and the technologies and placement processes peculiar to that construction. Building contractors, for instance, are very interested in accounts that describe the cost aspects of forming and casting in place structural concrete as used in building frames. Heavy construction contractors, on the other hand, are interested in earthwork-related accounts

# FIGURE 8.4 List of Typical Project Expense (Cost) Accounts

| MASTER LIST OF PROJECT COST ACCOUNTS |
| :---: |
| *Subaccounts of General Ledger Account 80.000* |
| *PROJECT EXPENSE* |

| Project Work Accounts 100–699 | | Project Overhead Accounts 700–999 | |
| --- | --- | --- | --- |
| 100 | Clearing and grubbing | 700 | Project administration |
| 101 | Demolition | .01 | project manager |
| 102 | Underpinning | .02 | office engineer |
| 103 | Earth excavation | 701 | Construction supervision |
| 104 | Rock excavation | .01 | superintendent |
| 105 | Backfill | .02 | carpenter foreman |
| 115 | Wood structural piles | .03 | concrete foreman |
| 116 | Steel structural piles | 702 | Project office |
| 117 | Concrete structural piles | .01 | move in and move out |
| 121 | Steel sheet piling | .02 | furniture |
| 240 | Concrete, poured | .03 | supplies |
| .01 | footings | 703 | Timekeeping and |
| .05 | grade beams | | security |
| .07 | slab on grade | .01 | timekeeper |
| .08 | beams | .02 | watchmen |
| .10 | slab on forms | .03 | guards |
| .11 | columns | 705 | Utilities and services |
| .12 | walls | .01 | water |
| .16 | stairs | .02 | gas |
| .20 | expansion joint | .03 | electricity |
| .40 | screeds | .04 | telephone |
| .50 | float finish | 710 | Storage facilities |
| .51 | trowel finish | 711 | Temporary fences |
| .60 | rubbing | 712 | Temporary bulkheads |
| .90 | curing | 715 | Storage area rental |
| 245 | Precast concrete | 717 | Job sign |
| 260 | Concrete forms | 720 | Drinking water |
| .01 | footings | 721 | Sanitary facilities |
| .05 | grade beams | 722 | First-aid facilities |
| .07 | slab on grade | 725 | Temporary lighting |
| .08 | beams | 726 | Temporary stairs |
| .10 | slab | 730 | Load tests |
| .11 | columns | 740 | Small tools |
| .12 | walls | 750 | Permits and fees |
| 270 | Reinforcing steel | 755 | Concrete tests |
| .01 | footings | 756 | Compaction tests |
| .12 | walls | 760 | Photographs |
| 280 | Structural steel | 761 | Surveys |
| 350 | Masonry | 765 | Cutting and patching |
| .01 | 8-in. block | 770 | Winter operation |
| .02 | 12-in. block | 780 | Drayage |
| .06 | common brick | 785 | Parking |
| .20 | face brick | 790 | Protection of adjoining |
| .60 | glazed tile | | property |
| | | 795 | Drawings |

*Continued*

**FIGURE 8.4   Continued**

| MASTER LIST OF PROJECT COST ACCOUNTS<br>*Subaccounts of General Ledger Account 80.000*<br>PROJECT EXPENSE | | | |
|---|---|---|---|
| *Project Work Accounts*<br>*100–699* | | *Project Overhead Accounts*<br>*700–999* | |
| 400 | Carpentry | 796 | Engineering |
| 440 | Millwork | 800 | Worker transportation |
| 500 | Miscellaneous metals | 805 | Worker housing |
| .01 | metal door frames | 810 | Worker feeding |
| .20 | window sash | 880 | General clean-up |
| .50 | toilet partitions | 950 | Equipment |
| 560 | Finish hardware | .01 | move in |
| 620 | Paving | .02 | set up |
| 680 | Allowances | .03 | dismantling |
| 685 | Fencing | .04 | move out |

such as grading, ditching, clearing and grubbing, and machine excavation. Standard cost accounts published by the American Road Builders Association emphasize these accounts, while the Uniform Construction Index (UCI), published by the Construction Specifications Institute, emphasizes building-oriented accounts. A breakdown of the major classifications within the UCI cost account system is shown in Table 8.1. A portion of the second level of detail for classifications 0 to 3 is shown in Figure 8.5.

## 8.4   PROJECT COST CODE STRUCTURE

The UCI code identifies three levels of detail. At the highest level the major work classification as given in Table 8.1 is defined. Also at this level major subdivisions within the work category are established. For instance, 3 or 300 level accounts pertain to concrete while 3.1 or 310 accounts are accounts specifically dealing with concrete

**TABLE 8.1   Classification of Accounts: Major Divisions in Uniform Construction Index**

| Cost Centers | |
|---|---|
| 0 Conditions of the contract | 9 Finishes |
| 1 General requirements | 10 Specialties |
| 2 Site work | 11 Equipment |
| 3 Concrete | 12 Furnishings |
| 4 Masonry | 13 Special construction |
| 5 Metals | 14 Conveying system |
| 6 Carpentry | 15 Mechanical |
| 7 Moisture prevention | 16 Electrical |
| 8 Doors, windows, and<br>  glass | |

# Figure 8.5 Detailed Codes for Classification within Uniform Construction Index

*0 Conditions of the contract*
0000–0099. unassigned

*1 General Requirements*
0.100. Alternates of Project Scope
0.101–0109. unassigned
0110. Schedules and Reports
0111–0119. unassigned
0120. Samples and Shop Drawings
0121–0129. unassigned
0130. Temporary Facilities
0131–0139. unassigned
0140. Cleaning Up
0141–0149. unassigned
0150. Project closeout
0151–0159. unassigned
0160. Allowances
0161–0169. unassigned

*2 Site Work*
0200. Alternates
0210–0209. unassigned
0120. Clearing of Site
0211. Declination
0212. Structures moving
0213. Clearing and grubbing
0214–0219. unassigned
0220. Earthwork
0221. Site grading
0222. Excavating and backfilling
0223. Dewatering
0224. Subdrainage
0225. Soil poisoning
0226. Soil compaction control
0227. Soil stabilization
0228–0229. unassigned

0230. Piling
0231–0234. unassigned
0235. Caissons
0236–0239. unassigned
0240. Shoring and bracing
0241. Sheeting
0242. Underpinning
0243–0249. unassigned
0250. Site drainage
0251–0254. unassigned
0255. Site utilities
0256–0259. unassigned
0260. Roads and Walks
0261. Paving
0262. Curbs and gutters
0263. Walks
0264. Road and parking
Appurtenances
0265–0269. unassigned
0270. Site Improvements
0271. Fences
0272. Playing fields
0273. Fountains
0274. Irrigation systems
0275. Yard improvements
0276–0279. unassigned
0280. Lawns and Planting
0281. Soil Preparation
0282. Lawns
0283. Ground covers and other plants
0284. Trees and shrubs
0285–0289. unassigned
0290. Railroad Work
0291–0294. unassigned
0295. Marine Work
0296. Boat Facilities
0297. Protective Marine Structures
0298. Dredging
0299. unassigned

*3 Concrete*
0300. Alternates
0301–0309. unassigned
0310. Concrete Formwork
0311–0319. unassigned
0320. Concrete Reinforcement
0321–0329. unassigned
0330. Cast-in-Place Concrete
0331. Heavyweight aggregate concrete
0332. Lightweight aggregate concrete
0333. Post-tensioned concrete
0334. Nailable concrete
0335. Specially finished concrete
0336. Specially placed concrete
0337–0339. unassigned
0340. Precast Concrete
0341. Precast concrete panel
0342. Precast structural concrete
0343. Precast prestressed concrete
0344–0349. unassigned
0350. Cementitious Decks
0351. Poured gypsum deck
0352. Insulating concrete roof decks
0353. Cementitious unit decking
0354–0399. unassigned

forming. In a similar manner, 3.2 or 320 accounts are reserved for cost activity associated with concrete reinforcement.

At the next level down, a designation of the physical component or subelement of the construction is established. This is done by adding two digits to the work classification two-digit code. For instance, the two-digit code for columns is 25. Therefore, the code 3.1–25 indicates an account dealing with concrete forming costs for columns.

At the third and lowest level, digits specifying a more precise definition of the physical subelement are used. For instance, an account code of 3.1-25-650 can indicate that this account records costs for forming concrete columns that are rectangular in cross section and dimensioned 24 in. × 24 in. (See Figure 8.6.) At this level the refinement of definition is very great, and the account can be made very sensitive to the peculiarities of the construction technology to be used. For instance, the costs of forming a rectangular column are different from those of forming a circular column. Further refinement could differentiate between forming columns that are 8 feet or less in height from those which are greater than 8 feet. At this level, the cost engineer and construction manager have a great deal of flexibility in reflecting unique aspects of the placement technology that lead to cost fluctuations and thus must be considered in defining cost centers. Clearly, if extreme variation or change of detail is required, more than three digits may be required (i.e., more than 999 variations of a physical subelement can obtain).

Large and complex projects in industrial and energy-related construction may require cost codes that reflect additional information, such as the project designation, the year

## FIGURE 8.6   UCI Cost (Line Item) Structure

| 3.1 | **FORMWORK** | CREW | DAILY OUTPUT | UNIT | BARE COSTS | | | TOTAL INCL. O&P |
|---|---|---|---|---|---|---|---|---|
| | | | | | MAT. | INST. | TOTAL | |
| 25 | FORMS IN PLACE, COLUMNS | | | | | | | |
| 650 | 24″ × 24″ plywood columns, 1 use | C-1 | 190 | S.F.C.A. | 1.34 | 2.54 | 3.88 | 5.05 |

Line number determination:

Major UCI subdivision = 03.1
(2 digits plus decimal
point plus last digit)

Major classification within
UCI subdivision = 25
(2 digits)

Item line number = 650
(3 digits)

Complete line number
= 3.1—25—650

*Source:* This information is copyrighted by Robert Snow Means Co., Inc. It is reproduced from *Buidling Construction Cost Data 1981,* p. 1, with permission.

**FIGURE 8.7   Classification of Accounts: Typical Data Structure for a Computerized Cost Code**

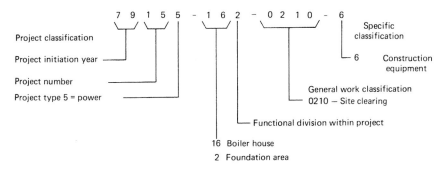

in which the project was started, and the type of project. Long and complex codes in excess of 10 digits can result. An example of such a code is shown in Figure 8.7. This code, consisting of 13 digits, specifically defines the following items:

1.  Year in which project was started (1979)
2.  Project control number (15)
3.  Project type (5 for power station)
4.  Area code (16 for boiler house)
5.  Functional division (2, indicating foundation area)
6.  General work classification (0210, indicating site clearing)
7.  Distribution code (6, indicating construction equipment)

The distribution code establishes what type of resource is being costed to the work process (i.e., clearing), the physical subelement (i.e., foundations) in what area of which project. Typical distribution codes might be as follows:

1   Labor
2   Permanent materials
3   Temporary materials
4   Installed equipment
5   Expendibles
6   Construction equipment
7   Supply
8   Subcontract
9   Indirect

Clearly, a high concentration of information can be achieved by proper design of the cost code. Such codes are also ideally suited for data retrieval, sorting, and assembly of reports on the basis of selected parameters (e.g., all construction equipment costs for concrete forming on project 10 started in 19X2). The desire to cram too much

**Figure 8.8  Industrial Project Cost Account Structure**

|  | Site Subfoundation and Excavation 10.00 | Concrete 20.00 | Superstructure 30.00 | Finishes and Specialties 40.00 | Equipment 50.00 | Piping 60.00 | Heating, Ventilation, and Air Conditioning 70.00 | Electrical 80.00 | Overhead and Distributables 90.00 |
|---|---|---|---|---|---|---|---|---|---|
| Yard — 01.00 | 11.00 Site preparation and improvement | 21.00 Outside concrete work | 31.00 Miscellaneous site structures | 41.00 Outside specialties | 51.00 Yard equipment | 61.00 Outside piping | 71.00 | 81.00 Outside electrical work | 91.00 Tests and samples |
| Building — 02.00 | 12.00 Excavation for building foundations | 22.00 Building foundation concrete work | 32.00 Structural frame and miscellaneous metals | 42.00 | 52.00 Commercial equipment | 62.00 Underfloor building piping | 72.00 | 82.00 | 92.00 Supervision and expense |
| Building — 03.00 | 13.00 Excavating for slab on grade, pits, trenches | 23.00 Slab on grade, pit, trench concrete work | 33.00 Moisture protection | 43.00 Doors, windows, glass | 53.00 Institutional equipment | 63.00 Building service piping | 73.00 Building comfort HVAC | 83.00 Building service electrical | 93.00 Temporary facilities and utilities |
| Building — 04.00 | 14.00 | 24.00 Superstructure concrete work | 34.00 Masonry work | 44.00 Building finishes | 54.00 Residential equipment | 64.00 | 74.00 | 84.00 | 94.00 Engineering, legal expense, contingency, etc. |

| Process[a] | 15.00 | 25.00 | 35.00 | 45.00 | 55.00 | 65.00 | 75.00 | 85.00 | 95.00 |
|---|---|---|---|---|---|---|---|---|---|
| 05.00 | | Special construction concrete work | Carpentry and millwork | Building specialties | Elevators, lifts, and hoists | Special building piping systems | | Communication systems | Construction equipment and small tools |
| 06.00 | Excavation for process equipment foundations | Process equipment foundation substructure concrete work | | Process painting | Process equipment general | Process piping | Process cooling and heating systems | Process electrical equipment | Consumable supplies and stock accounts |
| 07.00 | | Process equipment foundation superstructure concrete work | Process equipment structural steel supports | Process insulation | Process equipment general | Process piping | Process air and ventilation systems | Process electrical raceways | Taxes, insurance, permits, licenses, bonds, overtime, and fee |
| 08.00 | Excavation for special process equipment foundations | Special process equipment foundation concrete work | Process steel structures | Process refractories | Process equipment general | Process piping | Dust collection systems | Process electrical wire, cable, bus | Extras |
| 09.00 | | | | | Process equipment specific | Pneumatic instrumentation | | Electronic instrumentation | Backcharges |

[a]Excluding 90.00 Overhead Series.

145

information into cost codes, however, can make them so large and unwieldy (not to mention confusing to upper-level management) that even main frame computers have problems dealing with them.

A matrix showing the cost account structure used by a large contractor for industrial construction projects is shown in Figure 8.8. This four-digit scheme corresponds to the *functional division code* shown in Figure 8.7 and specifies the work type (10.00 to 90.00) as well as locational information (01.00 to 09.00). It must be emphasized that cost account coding is based on the needs of the individual company (whether client or contractor) maintaining the cost control. Therefore, even though standards such as the UCI code exist, the cost coding systems inevitably vary from contractor to contractor, client to client, and agency to agency. (Some government agencies require submission of cost data within a given cost code format.) Codes become stylized to the needs of the user. They evolve from an initial structure to uniquely tailored systems of cost/expense centers reflecting the cost control needs of management. In some contracting companies, the number of cost accounts per project may be as low as 10 to 15. On complex nuclear or fossil fuel power plants, 10 to 15 thousand cost accounts may be maintained.

## 8.5   EFFECT OF CONTRACT TYPE ON NUMBER OF JOB COST ACCOUNTS

The results of a study of building contractors operating in the southeastern U.S. are shown in Figure 8.9.* Information is organized by categorizing 30 contractors as to the form of contracts used in their work with clients.

1.   Group A contractors work on competitively bid fixed-price contracts more than 75 percent of the time.
2.   Group B contractors work between 50 and 75 percent of the time on fixed-price contracts, although the remainder of the work is done on a negotiated cost-plus-fee basis.
3.   Group D contractors are primarily cost-plus-fee contractors, working in this format more than 75 percent of the time.

Although the sample is small, it is clear that the contractors working on fixed-price contracts are much more interested in detailed cost control (as reflected by the higher number of cost codes per project). This is to be expected, since their cost plan is the basis for a fixed quotation for the work and overrun costs will cut into profit. The cost "pilot" in this case must submit a much more precise "flight plan" and cannot allow large variations from this plan without losing money. The cost-plus-fee contractors (Group D), on the other hand, appear not to maintain the same level of detail as the

* Thomas W. Gibb, Jr., "Building Construction in Southeastern United States" (Atlanta: School of Civil Engineering, Georgia Institute of Technology, 1975).

**FIGURE 8.9   Estimating Practices for 30 Building Contractors**

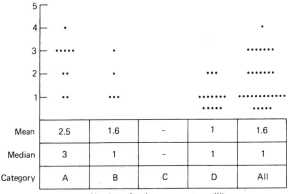

Number of estimators per ten million-
dollar annual volume

| | | | | | |
|---|---|---|---|---|---|
| Mean | 2.5 | 1.6 | – | 1 | 1.6 |
| Median | 3 | 1 | – | 1 | 1 |
| Category | A | B | C | D | All |

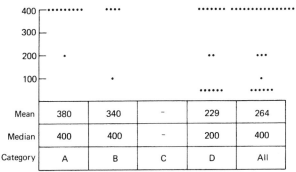

Typical number of coded line items
on two-million-dollar project

| | | | | | |
|---|---|---|---|---|---|
| Mean | 380 | 340 | – | 229 | 264 |
| Median | 400 | 400 | – | 200 | 400 |
| Category | A | B | C | D | All |

fixed-price group; their fee cannot be impacted (in most cases) by overrunning a fixed cost level. If costs vary from original plan, the fee cannot decrease and, in the case of the seldom-used cost-plus-a-percentage-fee arrangement, can actually expand (see Section 4.6 in *Construction Management* by Halpin and Woodhead). The study indicates a variation in the number of cost accounts maintained per project based on the type of contract under which most work is performed.

The structure of the cost codes and the definition of project accounts have a far-reaching effect, since these accounts or line items define the cost centers that are used both for estimating and for cost control. During the development of the estimate, the individual line items in the project section of the chart of accounts are used to break the job down so that cost projections can be developed. Values of estimated cost for each of the line items are calculated by using standard estimating methods. These values become the target or cost profile against which actual costs will be compared. At the time of estimate preparation, the line items can be thought of as *estimating accounts*.

After work commences, actual cost data are collected and compared on a line-item-by-line-item basis to the estimated costs. During this phase of cost management, the line items into which the project has been subdivided for cost control purposes (at the time of estimate) are referred to as *job cost accounts*.

## 8.6   SELECTION OF COST SYSTEM CODES

The contractor can take several approaches to selection of a cost code structure for a particular project.

1. Some trade organizations recommend the adoption of a standard set of codes (such as those shown in Figure 8.4 or the UCI code) that are used for all projects.
2. A second approach is to use a standard set of codes that can be modified to incorporate unique features of a particular project. This allows a certain amount of customizing within the basic framework of the standard coding system.
3. Certain firms may decide to develop customized sets of codes for each individual project. That is, each project has its own unique set of cost codes. This approach may be mandatory in large firms doing different types of construction (i.e., heavy, building, and industrial).

The Associated General Contractors and others recommend that a contractor adopt a standardized chart of accounts on a company-wide basis as opposed to developing a new set of accounts for each new job. Maintaining a standardized system of estimating and cost control can result in the following benefits to the contractor:*

1. The company has a checklist in order to keep estimating errors to a minimum.
2. The system allows the contractor to allocate correctly the distribution of field labor costs. (Maintaining a standard form accustoms personnel to proper allocation, thus minimizing mistakes.)
3. A standard chart of accounts facilitates the use of computers in the contractor's organization.

In addition, use of standard job cost codes simplifies the collection and maintenance of historical data, which can be used for estimating on future jobs. Field data collection is enhanced since field personnel become very familiar with each account, and allocation of cost and quantity to individual accounts becomes more precise and consistent.

All contractors do not use standardized cost coding systems, although suggested standards have been prepared by trade associations and contractor groups. As noted previously, two of the best-known coding systems are the *Uniform Construction Index,*

---

* The Associated General Contractors of America, *Cost Control and CPM in Construction* (Washington, D.C.: AGC, 1968), p. 32.

sponsored by the Associated General Contractors (AGC)* and the *Cost Accounting Manual for Highway Contractors, a System for Cost Control,*† sponsored by the American Road Builders Association.

The advantages of using a standardized system on a company-wide basis are the greatest when the company's jobs are very similar or small. If this system is applied to jobs of a large, complex nature or to jobs different from those for which the system was originally developed, several disadvantages can exist:‡

1. Many items in the index become so small as to be negligible on some jobs. When this happens, each item is still shown separately, wasting time on small amounts of work without giving useful information.
2. There is the likelihood that several work units could be omitted from the standard cost code currently in use. In the case of a decimal cost coding system, checking for the possibility of this omission requires considerable time.
3. With a standard system (that is, standardized numbers for all items, not just major designations), the numbering system must be changed for any item so large that it needs to be broken apart into subelements.
4. A decimal system is subject to more errors in reporting, as it requires more digits in each number and, most importantly, requires a period (a hard mark to recognize when written under field conditions).

## 8.7  CUSTOMIZING STANDARD COST CODES

It can be argued that each job should have its own unique cost code as opposed to having a company-wide standard cost system. The uniform number system can tend to trick estimators into using historical cost data from the account without reviewing its past history. Second, each job is unique with its individual peculiarities. In standard systems the cost data from various jobs are typically stored in accordance with the company's standard cost format. Job peculiarities become merged in the average unit costs that the firm uses for estimating.

Suppose, for instance, that a standard account for casting *concrete slab on grade* is being used for estimating a new project. Six sets of cost data for this account (from previous jobs) are available. All of these jobs were slabs cast in the open for warehouse projects. Access was simple, and the concrete was placed directly with a crane and bucket. The job being estimated, however, requires slab concrete to be placed in an

---

* The Associated General Contractors of America, *Uniform System for Building Construction Specifications, Data Filing and Cost Accounting* (Washington, D.C.: AGC, 1966).

† Dan S. Brock, *Cost Accounting Manual for Highway Contractors* (Washington, D.C.: American Road Builders Association, 1971).

‡ King Royer, *The Construction Manager* (Englewood Cliffs, N.J.: Prentice-Hall, 1974), p. 40.

underground vault through an opening in a structural frame and wall system that has been previously constructed. Access will be difficult, and transport of concrete to the placement location will require special equipment. The average cost for this account on the previous jobs is $100 per cubic yard. Clearly, this value will be low for the job being estimated, since it is based on easy access and no special material transport. If the estimator is not sensitive to these differences, he may take the standard value and "plug it in." Using such low average values (unconverted for site conditions) will lead to the contractor's bidding low and being awarded projects that will ultimately prove to be "unprofitable."

Another difficulty occurs after the system has been in use for some time. New foremen, superintendents, estimators, and related personnel involved in field data collection do not understand the nomenclature of the cost code, and the reported quantities become confused.

Several of these problems associated with using a company-wide standardized cost system can be corrected by using modifications to the standard codes. First, if a contracting firm adopts a standardized cost coding system that has been prepared by a trade association, there is much less likelihood that work units pertaining to bid items on an unusual project would not be included in the cost code. Second, the problems of differing levels of significance of cost items on various projects can be resolved in part by using the approach adopted in the American Road Builders Association standardized system.

Cost accounts as used in this system are normally identified by an eight-digit alphanumeric code. The significance of code characters is illustrated as follows:

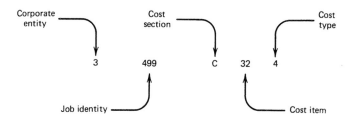

Corporate entity—ABC subsidiary
Job identity — Job 499
Cost section — Mobilization, site preparation and grading
Cost item — Excavation — earth
Cost type — equipment

If the amount of earth excavation were not significant on this particular project, cost item 32 would be sufficient to allocate all costs associated with earth excavation. As such, cost item 32 represents the total costs that would be allocated to earth excavation without regard to the method of accomplishment.

On the other hand, if this cost item were a significant portion of the project and required subdivision to achieve proper control, the levels of detail could be increased.

The cost code illustrated below represents the total equipment costs associated with earth excavation when a wheel scraper is used. Thus the system permits the level of detail to be increased as required by the nature of the project.

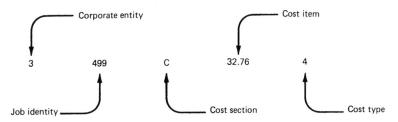

In addition, the opportunity for confusion among cost items is less when data recorded by this system are used in estimating future work. Cost item 32, which typically represents the total cost of earth excavation, would be present in job cost records only when the total amount of earth excavation on a project was not significant. In the case where this component represents a substantial portion of the project cost, a four-digit cost item (e.g., 32.XX) could be used to insure proper segregation of data.

The cost accounting section of the *Uniform Construction Index* does not provide for such detail with regard to each cost item, since it was felt that classification of material below the section level would make the cost analysis format too inflexible. However, it is possible for contractors to make individual modifications to this system to allow for variations in levels of significance of cost items.

## 8.8 NONSTANDARD CODES FOR EACH PROJECT

A third technique for eliminating the weaknesses of standard cost codes is to compile a list of applicable cost accounts for each new project from the firm's standard chart of cost accounts. In the process of making this selection, cost control personnel can make any necessary allowances for the level of detail in individual cost items and for the peculiarities of the project they deem necessary. Next, a written description of each of these cost items is prepared and reviewed with the on-site personnel responsible for cost control.

If this procedure proves too confusing to the foreman or other on-site personnel, cost engineers or timekeepers who understand the chart of accounts must be available to assist each foreman in preparation of the cost data source documents (time sheets and quantity bills). This and the preparation of a written description of the individual cost items can be made part of the regular duties of the cost engineer. Admittedly, this technique of insuring understanding of the cost codes requires some additional expense, yet in terms of total cost this is the most economical alternative.

To illustrate, a contractor might be engaged in two projects: One is primarily a site preparation, whereas the second has to do with building construction. Site-development-oriented accounts are not well adapted to controlling building construction, and vice versa. Therefore, working from a master set of expense/cost accounts, the cost engineer stylizes accounts for each job. For the site preparation project the following accounts are defined:

*Site Preparation—Job 101*

| Code | Description | Units |
|------|-------------|-------|
| 100 | Set up tool warehouse | Job |
| 110 | Clearing and grubbing | Acre |
| 115 | Rough grading | Yd² |
| 120 | Fine grading | Yd² |
| 200 | Drainage excavation | Yd³ |
| 201 | Culvert installation | Linear feet |

For the building project, a different set of accounts as well as a different coding system is used.

*Transformer Building—Job 102*

| Code | Description | Units |
|------|-------------|-------|
| 101M | Machine excavation | Yd³ |
| 101H | Hand excavation | Yd³ |
| 150F | Foundation concrete | Yd³ |
| 150S | Structural concrete | Yd³ |
| 201 | Masonry | Blocks |
| 210 | Inside work | Job |
| 220 | Roofing | Squares |
| 250 | Electrical & plumbing | Job |
| 300 | Walkways & landscaping | Job |

By developing unique accounts for each of these projects, the cost engineer has been able to incorporate only those items that require cost control and has greatly condensed the effort required for estimating and cost control to just the accounts defined. It is just this uniqueness, however, which means that greater care will be required in developing the estimate (since historical data from standard accounts may not be directly available) and in field data collection. Field personnel will not be familiar with these specially defined accounts, so close supervision to insure consistency in quantity measurement and cost reporting must be maintained.

## 8.9   COST ACCOUNTS FOR INTEGRATED PROJECT MANAGEMENT

In large and complex projects, it is advantageous to break the project into common building blocks for control both of cost and of time. In the past 10 years, the concept of a common unit within the project that integrates both scheduling and cost control has led to the development of the *work breakdown* approach. The basic common

FIGURE 8.10  Project Control Matrix

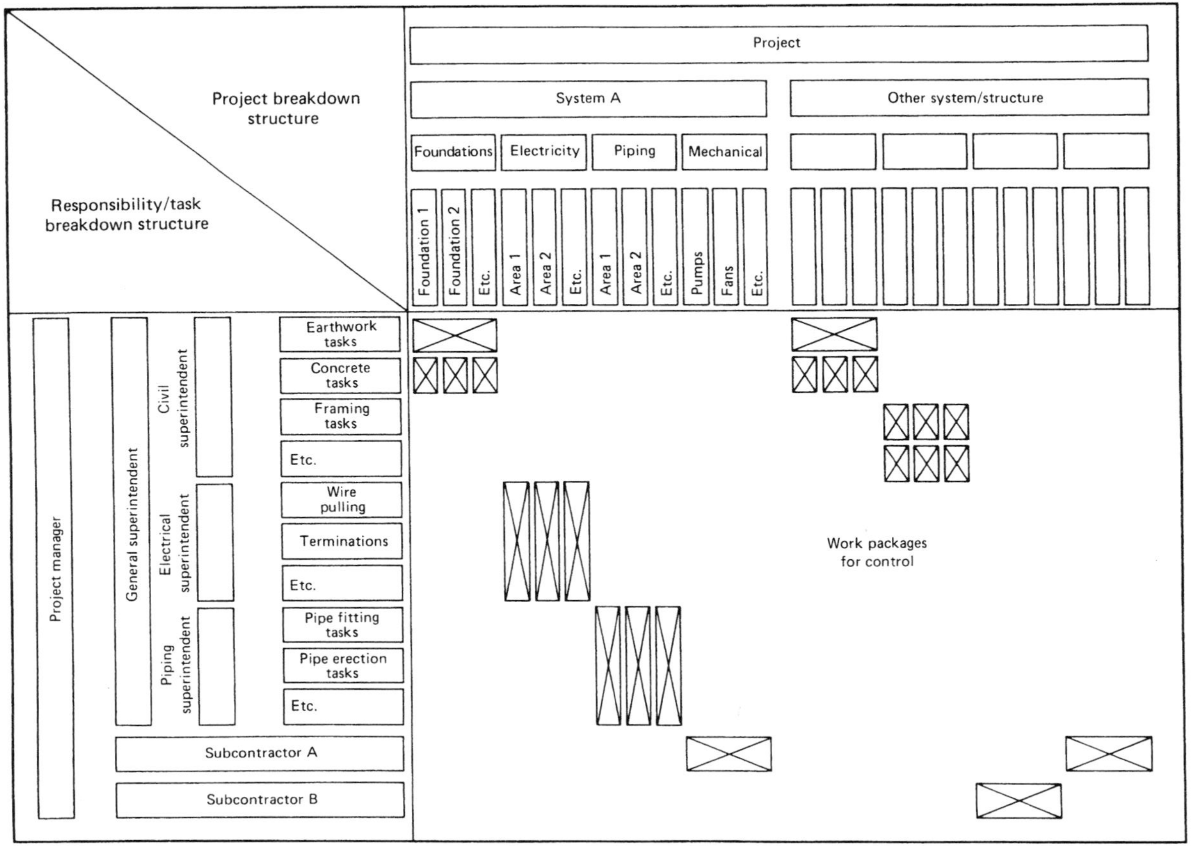

denominator in this scheme is the *work package,* which is a subelement of the project on which both cost and time data are collected for project status reporting. The collection of time and cost data based on work packages has led to the term "integrated project management." That is, the status reporting function has been integrated at the level of the work package. The set of work packages in a project constitutes its *work breakdown structure* (WBS).

The work breakdown structure and work packages for control of a project can be defined by developing a matrix similar to the one shown in Figure 8.10. The columns of this matrix are defined by breaking the project down into physical subcomponents. Thus we have a hierarchy of levels that begins with the project as a whole and, at the lowest level, subdivides the project into physical end items such as foundations and areas. As shown in the figure, the project is subdivided into systems. The individual systems are further divided into disciplines (e.g., civil, mechanical, electrical). The lowest level of the hierarchy indicates physical end items (foundation 1, etc.).

The rows of the matrix are defined by technology and responsibility. At the lowest

**FIGURE 8.11   Three-Dimensional Visualization of Work-Package-Oriented Cost Accounts**

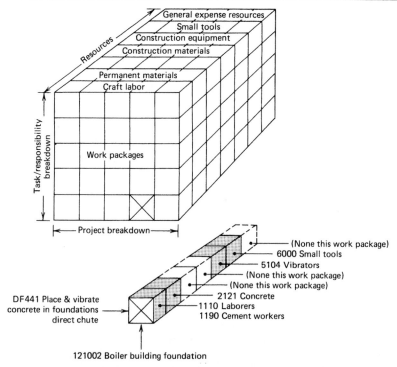

*Source:* James M. Neil, "A System for Integrated Project Management *Proceedings of the Conference on Current Practice in Cost Estimating and Cost Control* (Austin, Texas, April 1983).

**FIGURE 8.12    Basic Cost-Code Structure**

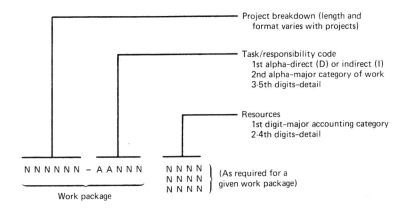

level of this hierarchy, the responsibilities are shown in terms of tasks, such as concrete, framing, and earthwork. These tasks imply various craft specialties and technologies. Typical work packages then are defined as concrete tasks on foundation 1 and earthwork on foundations 1 and 2.

This approach can be expanded to a three-dimensional matrix by considering the resources to be used on each work package (see Figure 8.11). Using this three-dimensional breakdown, we can develop definition in terms of physical subelement, task, and responsibility, as well as resources commitment. A cost code structure to reflect this matrix structure is given in Figure 8.12. This 15-digit code defines units for collecting information in terms of work package and resource type. Resource usage in terms of monetary units, quantities, man-hours, and equipment-hours for a foundation in the boiler building would be collected under work package code 121002. If this work relates to placement and vibration of concrete by using a direct chute, the code is expanded to include the alphanumeric code DF441. The resource code for the concrete is 2121. Therefore, the complete code for concrete in the boiler building foundations placed by using a chute would be 121002-DF441-2121. This code allows collection of cost data at a very fine level. Scheduling of this work is also referenced to the work package code as shown in Figure 8.13. The schedule activities are shown in this figure as subtasks related to the work package.

## 8.10    SUMMARY

Collection of cost data on a project allows the manager to determine project status and monitor progress to date. The number of cost accounts utilized varies from company to company and project to project. The type of contract may influence the level of detail of cost data collection and the number of accounts. Standard account structures are commonly used by firms constructing projects of a similar type and construction such as contractors specializing in the construction of school buildings. The short-

**FIGURE 8.13  Project-Control Matrix with Scheduling of Subtasks**

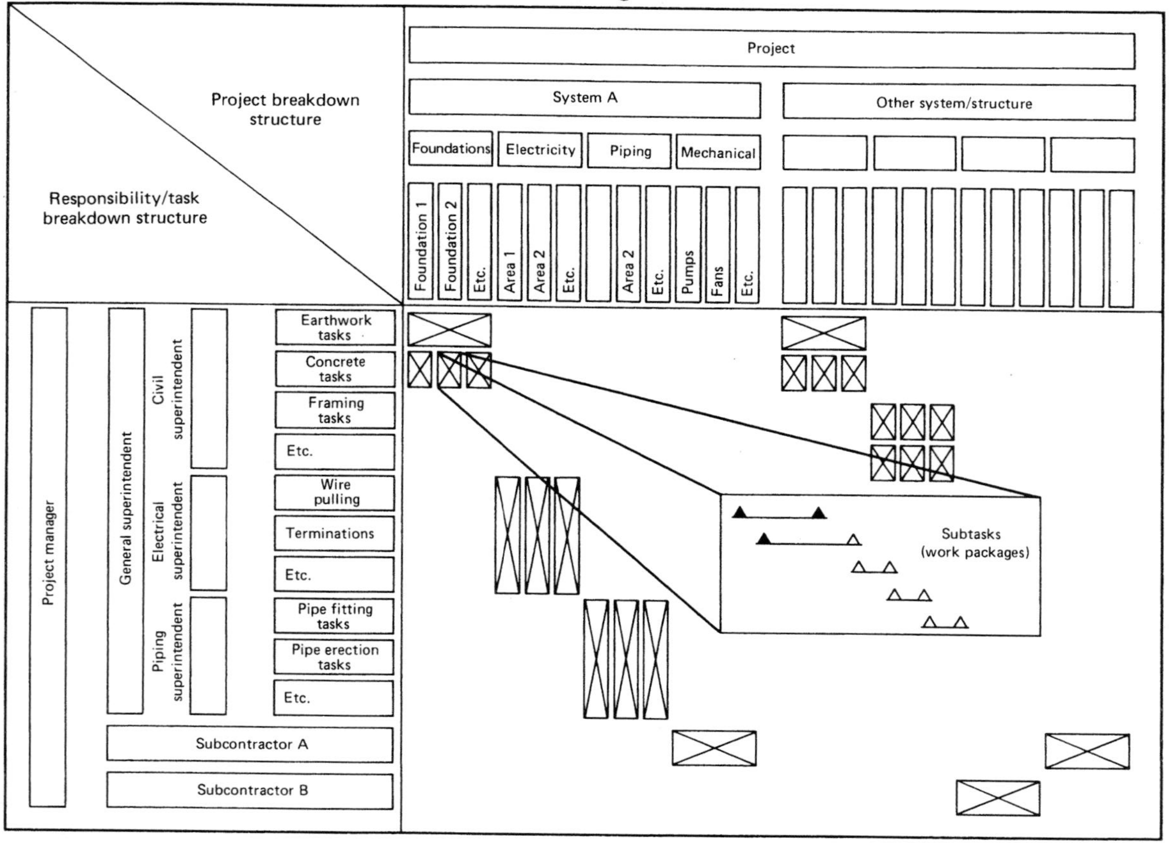

comings inherent in standardized accounts can be overcome to some extent by stylizing one standard account structure to deal with the peculiarities of a particular job. Some contractors prefer to develop individual codes for each project. The work package approach designed to integrate time and cost data collection has become increasingly popular over the past decade. This approach to structuring cost accounts is based on a subdivision of individual projects into work packages by using a work breakdown structure.

## REVIEW QUESTIONS

**8.1** As a construction project manager, what general categories of information would you want to have on a cost-control report to properly evaluate what you think is a developing overrun on an operation, "place foundation concrete," that is now underway and has at least five weeks to go before it is completed?

**8.2** What are the major functions of a project-coding system?

**8.3** List advantages and disadvantages of the UCI-coding system.

**8.4** Assume you are the cost engineer on a new $12 million commercial building project. Starting with your company's standard cost code, explain how you would develop a project cost code for this job. Be sure the differences in purpose and content between these two types of cost codes are clear in your explanation. Specify any additional information that may be needed to draw up the project cost code.

**8.5** Develop a cost-code system that gives information regarding:
   (a) When project started
   (b) Project number
   (c) Physical area on project where cost accrued
   (d) Division in Uniform Construction Index
   (e) Subdivision
   (f) Resource classification (labor, equipment)

# Chapter 9

# BID PREPARATION AND COST RECOVERY

## 9.1   COST FACTORS IN BID DEVELOPMENT

In order to prepare a bid, the estimator must have a good knowledge of the costs that will be incurred and must be recovered. Four major price development categories must be considered in preparing the bid, as follows:

1. Direct costs related to placing construction
2. Subcontractor cost and repayment
3. Job indirect costs (required to mobilize)
4. Markup

Some of the cost recovery subelements that must be addressed in categories 1 through 4 are shown in Figure 9.1. Items in these categories simply relate to costs that will be incurred by the contractor in constructing the project. These costs must be covered, or the contractor will be providing the construction to the client at less than cost. Profit and recovery of a charge for the management service provided by the contractor are included in the markup.

## 9.2   DIRECT COST RECOVERY

Direct costs are those resource costs required to physically place the elements of construction in the project. The costs involved in placing a unit of construction (e.g., cubic yard of concrete, square of roofing, etc.) are

1. Labor costs
2. Material costs

**FIGURE 9.1  Cost Factors in Bid Development**

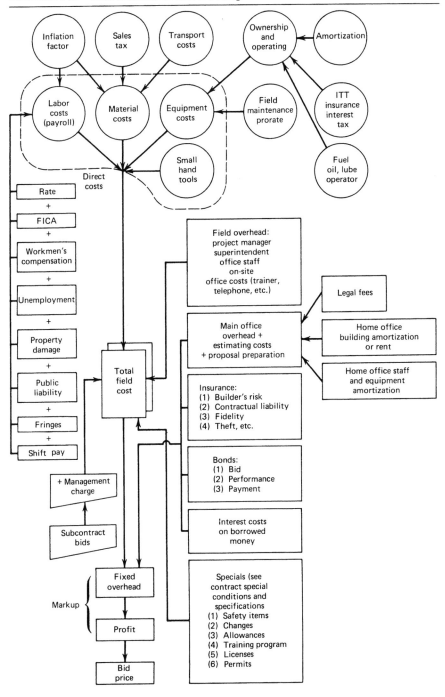

3.   Equipment costs
4.   Small hand tools

These costs are generally readily discernible and are easily assigned to a particular cost account. They are incurred as a physical piece of the construction is placed and are associated directly with that physical end item (e.g., footer, panel of glass, section of ducting). The physical elements of the construction are typically related directly to a cost account.

For example, all forming costs for rectangular columns 24 × 24 in. that are less than 8 feet in height are aggregated in an account defined in the job cost system. The costs may be incurred in forming such a column in the parking garage, on the third floor of the main structure, or wherever. Regardless of where the physical column actually stands in the construction, the costs involved in constructing it are aggregated in an appropriate account. In order to construct the form a crew, materials, equipment, and hand tools are required. The crew cost is known, since it consists of the pay rate of the men involved as well as burdens that must be paid related to their salaries. These rates can be developed on an hourly basis, so the dollars per hour rate required for the labor crew is readily developed. The cost factors contributing to the labor cost are shown in Figure 9.1.

The material cost usually is determined by using a basic quantity parameter such as square foot contact area* for forming, or pieces of concrete masonry units in block wall construction, and multiplying by a unit price. The unit price used is normally a combination of quotations received from materials suppliers and experience with the technology used to place the material. For instance, a contractor may decide to have concrete supplied from a transit mix vendor. In such a case, he can simply receive a quotation for concrete received FOB (free on board) at the construction site. Similarly, a steel vendor will quote the price of reinforcing steel delivered to the construction site. These quotations are simply factored directly into the bid; they cover cost, insurance, transport to site, and tax. The production costs are combined in the vendor quotation.

If the contractor decides to set up his own batch plant and manufacture his own concrete on site, he must include factors for cost of cement, cost of aggregate, water costs, amortization of batch plant, and transit trucks to pouring location. As shown in Figure 9.1, the cost of production or procurement of the material plus transportation to the site and taxes as applicable must be included in the bid. As with labor costs, the potential for cost escalation or inflation during the project should also be considered.

Equipment costs are usually associated with the particular estimating/cost account for which the equipment is to be used. There are, however, exceptions, and some contractors prefer to charge certain pieces of equipment to the project as an entity rather than charging to individual cost accounts. This practice leads to certain accounts that reflect only equipment charges (i.e., no material). This is common in heavy and highway work that is equipment intensive. Notwithstanding, whether the equipment is charged against an account describing a physical end item (e.g., slab on grade, steel

---

* Square foot contact area (SFCA) is the area of form in contact with the concrete at time of placement.

erection) or as a piece or fleet of equipment available as a resource pool to the project, certain costs must be recovered. The ownership costs are normally concerned with recovery of

1. Amortization (i.e., charge to allow replacement of the equipment)
2. Insurance of equipment
3. Interest on debt financing of equipment
4. Taxes on equipment
5. Storage and transportation to site

These are the fixed costs associated with owning the equipment and are incurred whether or not the equipment is operational. The costs incurred while the equipment is operational include:

1. Fuel
2. Oil
3. Lubrication
4. Preventative maintenance and small repair
5. Tire replacement (where applicable)
6. Operator's salary

These costs are discussed in detail in Chapters 9 and 10 of *Construction Management* by Halpin and Woodhead and in the *Contractors Equipment Ownership Expense Manual*, published by the Associated General Contractors. Some contractors prefer to select a charge rate for equipment that recovers these costs by referring to a recognized charging standard such as that used by equipment rental firms. Reference rates are published on an annual basis by the *Associated Equipment Distributors* (AED) and certain publishing houses. The contractor may opt to recover the equipment costs noted above by charging some percentage of the AED rate or other applicable rate. Therefore, if the AED rate for an entrenching machine is $110 per hour, the contractor may charge 60 percent of that rate, or $66 per hour. The actual percent factor selected would be based on the age of the equipment and the nature of the job.

Hand tools such as hammers, shovels, and other items controlled by the tool room (and having negligible depreciable value) are normally considered to be consumed in the process of completing the construction. They are normally handled as direct costs allocated to particular cost accounts.

## 9.3  DIRECT LABOR COSTS*

The large number of contributions and burdens associated with the wage of a worker makes the determination of a worker's cost to the contractor a complex calculation. The contractor must know how much cost to put in the bid to cover the salary and

* Material in Sections 9.3 and 9.4 is taken with acknowledgements from Halpin and Woodhead, *Construction Management* (New York: Wiley, 1980), Chapter 14.

associated contributions for all of the workers. Assuming that the number of carpenters, ironworkers, operating engineers, and other craft workers required is known and that the hours for each can be estimated, the average hourly cost of each craft can be multiplied by the required craft hours to arrive at the total labor cost for a particular cost center. The hourly average cost of a worker to the contractor consists of the following components:

1. Direct wages
2. Fringe benefits
3. Social security contributions (FICA)
4. Unemployment insurance
5. Workmen's compensation insurance
6. Public liability and property damage insurance
7. Subsistence pay
8. Shift pay differentials

The direct wages and fringe benefits can be determined by referring to a summary of wage rates such as the one shown in Figure 9.2.

All workers must pay social security on a portion of their salary. For every dollar the worker pays, the employer must pay a matching dollar. The worker pays a fixed percentage on every dollar earned up to a cutoff level. After the annual income has exceeded the cutoff level, the worker (and the worker's employer) need pay no more. An indication of the increasing level of contribution since the Federal Insurance Contributions Act (FICA) was enacted in 1937 is shown in Figure 9.3. The FICA contribution in 1984 was required on the first $37,800 of annual income at the rate of 6.7 percent. Therefore, a person making $37,800 or more in annual income would contribute $2532.60, and the person's employer or employers would contribute a like amount.

Unemployment insurance contributions are required of all employers. Each state sets a percentage rate that must be paid by the employer. The premiums are escrowed on a monthly or quarterly basis and sent periodically to the state unemployment agency. The amount to be paid is based on certified payrolls submitted by the employer at the time of paying this contribution. The fund established by these contributions is used to pay benefits to workers who are temporarily out of work through no fault of their own.

The states also require employers to maintain workmen's compensation insurance for all workers in their employ. This insurance reimburses the worker for injuries incurred in the course of employment. Labor agreements also specifically state this requirement, which recognizes the employer's responsibility to provide a safe working environment and the employer's obligation to provide support to disabled workers. Without this insurance, workers injured in the course of their work activity could become financially dependent on the state. The rates paid for workmen's compensation are a function of the risk associated with the work activity. The contribution for a pressman in a printing plant is different from that of a worker erecting steel on a high-rise building. A typical listing of construction specialties and the corresponding rates

# FIGURE 9.2  Labor Organizations and Wage Rates

## LABOR ORGANIZATIONS AND WAGE RATES

| CRAFT AND BUSINESS REPRESENTATIVE | WAGE RATE PER HOUR | FOREMAN | OVER TIME RATE | W—WELFARE P—PENSION A—APPRENTICE V—VACATION | TRAVEL PAY SUBSIS- TENCE | AUTOMATIC WAGE INCREASES | AUTOMATIC FRINGE INCREASES | EXPIRA- TION DATE |
|---|---|---|---|---|---|---|---|---|
| Asbestos Workers Local No. 18 Robert J. Scott, BR 946 North Highland Indianapolis, Indiana 46202 317-638-4234 | $7.20 | | Double | W—20c P—20c A—6c V—60c Deduct | $11 per day | | | 5-31-71 |
| Boilermakers Local No. 60 George Williams, BR 400 North Jefferson Peoria, Illinois 61603 309-673-9131 | $7.15 | 50c—F $1.00—GF | Double | W—40c P—65c A—01c | | 30-M—$6-D 60-M—$8-D | $1.00— 9-1-71 | 8-31-72 |
| Boilermakers Local No. 363 Anthony Moceri B.M. 19 S. 97th St. Belleville, Illinois 62223 618-397-7779 | $7.35 | 50c—F $1.00—GF | Double | W—40c P—65c A—01c | | 30-M—$5-D 60-M—$7-D | | 9-2-71 |
| Carpenters Local No. 44 Gene Stirewalt, BR 212 W. Hill St. Champaign, Illinois 61820 217-356-5463 | $6.29 | 12% | Double | W—17½c P—30c A—05c IAF—02c ISC—½c | | 35c—10-15-70 40c—4-15-71 35c—10-15-71 | P—10c 4-15-71 | 4-15-72 |
| Carpenters Local No. 347 Lee V. Foreman, BR P.O. Box 774 Mattoon, Ill. 61938 | $5.93 | 50c | Double | W—25c P—15c A—02c | | 80c—6-1-71 | P—05c 6-1-71 | 6-1-72 |
| Cement Finishers Local No. 143 Francis E. Ducey, BR 212½ South First St. Champaign, Illinois 61820 Office 217-356-9313 Home 217-485-3515 | $6.62½ | 50c 15% GF | Double | W—17½c | | 30c—1-24-71 | | 7-24-71 |
| Electricians Local No. 601 Jack Hensler, BR 212 South First St. Champaign, Illinois 61820 217-352-1741 | $6.35 | 10% 20%-GF | Double | W—20c A—7⁄10% | | 45c—11-1-70 70c—5-1-71 30c—11-1-71 | | 4-30-72 |
| Electricians Local No. 146 Larry Lawler, BM 2955 N. Woodford Decatur, Illinois 62526 217-877-4604 | $6.60 | 10%—F 15%—GF | Double | W—20c V—15c Deduct A—1⁄10% P—1% | | 60c—2-21-71 | | 2-20-71 |
| Electricians Local No. 489 William Dittamore, BM 106 S. 19th St. Mattoon, Illinois 61938 | $6.30 | 10%—F 20%—GF | Double | W—20c P—1% A—$25 | | 45c—1-1-71 35c—7-1-71 45c—1-1-72 | | 8-31-72 |

*Source:* D. W. Halpin and R. W. Woodhead, Construction Management, copyright © 1980 by John Wiley and Sons, Inc. Reprinted by permission of John Wiley and Sons, Inc.

**FIGURE 9.3 Levels of Social Security contributions**

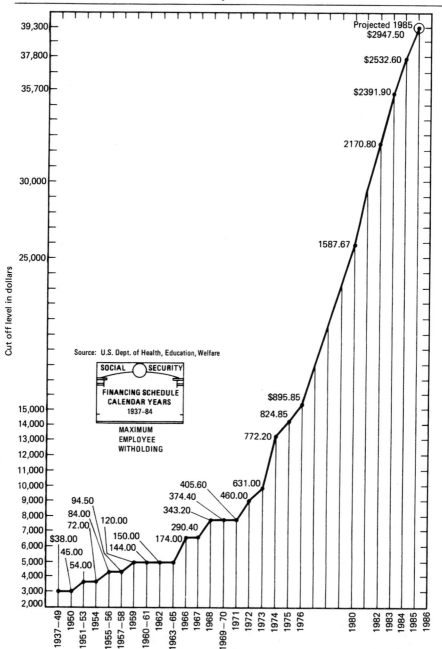

*Source:* D. W. Halpin and R. W. Woodhead, Construction Management, copyright © 1980 by John Wiley and Sons, Inc. Reprinted by permission of John Wiley and Sons, Inc.

is given in Table 9.1. Similar summaries are printed in the Quarterly Cost Roundup issues of the *Engineering News Record*. The rates are quoted in dollars of premium per $100 of payroll. The rate for iron or steel erection, for example, is $12.82 per one hundred dollars of payroll paid to ironworkers and structural steel erectors.

The premium pay for public liability and property damage (PL and PD) insurance is also tied to the craft risk level and is given in Table 9.1. When a construction project is underway, accidents occurring as a result of the work can injure persons in the area or cause damage to property in the vicinity. If a bag of cement falls from an upper story of a project and injures persons on the sidewalk below, these persons will normally seek a settlement to cover their injuries. The public liability arising out of this situation is the responsibility of the owner of the project. Owners, however, normally pass the requirements to insure against such liability to the contractor in the form of a clause in the general conditions of the construction contract. The general conditions direct the contractor to have sufficient insurance to cover such public liability claims. Similarly, if the bag of cement falls and breaks the windshield on a car parked near the construction site, the owner of the car will seek to be reimbursed for this damage. This is a property damage situation that the owner of the construction project becomes liable to pay. Property damage insurance carried by the contractor (for the owner) covers this kind of liability. Insurance carriers normally quote rates for PL and PD insurance on the same basis as for workmen's compensation insurance. Therefore, to provide PL and PD insurance, the contractor must pay $1.50 for PL and $0.94 for each $100 of steel erector salary paid on the job. These rates vary over time and geographical area and can be reduced by maintaining a safe record of operation. The total amount of premium is based on a certified payroll submitted to the insurance carrier.

Subsistence is paid to workers who must work outside of the normal area of the union local. As a result, they incur additional cost because of their remoteness from home and the need to commute long distances or perhaps live away from home. If an elevator constructor in Atlanta, Georgia must work in Macon for two weeks, he will be outside of the normal area of his local and will receive subsistence pay to defray his additional expenses.

Shift differentials are paid to workers in recognition that it may be less convenient to work during one part of the day than during another. Typical provisions in a sheet metal worker's contract are given in Figure 9.4.* In this example, the differential results in an add-on to the basic wage rate. Shift differential can also be specified by indicating that a worker will be paid for more hours than he works. A typical provision from a California ironworkers contract is as follows:

For shift work the following standards apply

1. If two shifts are in effect, each shift works $7\frac{1}{2}$ hours for 8 hours of pay.
2. If three shifts are in effect, each shift works 7 hours for 8 hours of pay.

---

* Agreement between Atlantic Building Systems, Inc. and Local Union 93, Sheet Metal Worker's International Association.

**TABLE 9.1  Damage Insurance for Certain Common Construction Classifications (California rates on 1 October 1968)[a]**

| Classification | Workman's Compensation | | PL and PD[b] | | | |
| --- | --- | --- | --- | --- | --- | --- |
| | Code Number | Rate | Code Number | Bodily Injury | Property Damage | Total |
| Bridge building—metal | 5040 | $16.35 | 3452 | $1.50 | 0.94 | 18.79 |
| Bridge and trestle construction—wood | 6209 | 17.70 | k6209 | 0.40 | 0.26 | 18.36 |
| Caisson work—not pneumatic | 6252 | 10.55 | 3438 | 0.39 | xcu[d] | |
| Canal construction | 6361 | 4.24 | 6229 | 0.29 | 0.55xu | |
| Carpentry—not otherwise covered | 5403 | 8.35 | 3457 | 0.40 | 0.26 | 9.01 |
| Carpentry—shop only | 2883 | 3.81 | 2464 | 0.10 | 0.05 | 3.96 |
| Carpentry—new, private residences | 5645 | 4.70 | 5645 | 0.29 | 0.18 | 5.17 |
| Churches, clergy, professional assistants | 8840 | 0.52 | | — | — | |
| Churches, all other employees | 9015 | 3.23 | | — | — | |
| Concrete construction—bridge and culverts with clearance over 10 ft | 5222 | 9.62 | 5213 | 0.38 | 0.27 | 10.27 |
| Concrete construction—including forms, etc. | 5213 | 7.09 | 5213 | 0.38 | 0.27 | 7.74 |
| Concrete building construction—tilt-up method | 5214 | 2.69 | | | | |
| Concrete or cement work—sidewalks, floors, etc. | 5200 | 2.69 | | | | |
| Contractors—executive supervisors | 5606 | 1.54 | 3759 | 0.21 | 0.11 | 1.86 |
| Dam construction—concrete | 5207 | 7.07 | 5213 | 0.38 | 0.27 | 7.74 |
| Dam construction—not otherwise covered | 6011 | 5.21 | 6019 | e | e | |
| Electric wiring—within buildings | 5190 | 2.74 | 5190 | 0.17 | 0.21 | 3.12 |
| Engineers, consulting | 8601 | 0.78 | 3759 | 0.21 | 0.17 | 1.16 |
| Excavation—rock, no tunneling[e] | 1605 | 8.03 | 3470 | 0.93 | 1.00xcu | |
| Iron or steel erection | 5059 | 12.82 | 3452 | 1.50 | 0.94 | 18.79 |
| Iron or steel erection—not otherwise covered | 5057 | 11.88 | 5057 | 0.99 | 0.80 | 13.16 |
| Iron, steel, brass, aluminum erection—nonstructural, within building | 5102 | 4.19 | 3442 | 0.37 | 0.36 | 4.92 |
| Iron or steel—steel shop | 3030 | 8.36 | 3431 | 0.21 | 0.20 | 8.77 |
| Logging | 2702 | 13.57 | 2702 | 0.18 | 0.18 | 13.93 |
| Painting | 5474 | 5.11 | 3429 | 0.13 | 0.44 | 5.68 |
| Painting—steel structures or bridges | 5040 | 16.35 | 3452 | 1.50 | 0.94 | 18.70 |

| | | | | |
|---|---|---|---|---|
| Pile driving—building foundation only | 6003 | 13.81 | 0.93 | 1.00xcu |
| Pile driving—including timber wharf building | 6003 | 13.81 | 0.24 | 0.83cu |
| Pile driving—sonic method | 6003 | 13.81 | — | — |
| Plumbing | 5183 | 2.97 | 0.24 | 0.59u |
| Railroad construction | 6701 | 9.75 | 0.36 | 0.28x |
| Railroad construction—laying track and contractor maintenance | 7855 | 6.28 | 0.36 | 0.28x |
| Reinforcing steel installation—placing for concrete construction | 5225 | 7.09 | — | — |
| Sewer construction | 6306 | 7.58 | 0.89 | 1.00xcu |
| Street and road construction—paving, etc. | 5506 | 5.33 | 1.00 | 0.55xcu |
| Street and road construction—grading | 5507 | 5.20 | 1.40 | 0.91xcu |
| Tunneling[f]—all work to completion, including lining | 6251 | 17.47 | 0.39 | *(5)xcu |
| Tunneling[f]—pneumatic, all work to completion | 6260 | 29.89 | 0.39 | *(5)xcu |
| Water mains or connections—construction | 6319 | 4.71 | 0.89 | 1.00xcu |
| Excavation—general, not otherwise covered[e] | 6217 | $ 3.48 | $0.93 | 1.00xcu |

[a]Premium rates are to be applied to the base of $100 payroll. In California premium can be computed on the straight-time portion (i.e., no overtime at the rate shown is less than a given minimum, the contractor must pay this minimum. In some cases where the contractor's accident record is good, sizeable refunds or dividends may be paid back to him after the insurance period is over.

[b]Limits of Coverage:

*Public liability.* Maximum coverage under these rates—$5000 per person or $10,000 per accident. Cost of higher coverage $10,000/$20,000, 1.26 × basic rate; $25,000/$50,000, 1.47 × basic rate; $50,000/$100,000, 1.59 × basic rate; $300,000/$300,000, 1.78 × basic rate.

*Property damage.* Maximum coverage under these rates—$5,000 per accident and $25,000 per policy. Cost of higher coverage $25,000/$100,000, 1.23 × basic rate; $50,000/$100,000, 1.30 × basic rate.

*Workmen's compensation.* The employer may be required to pay to an injured employee an additional 50 percent of the compensation award ($7,500 maximum + costs up to $250) where accident was caused by employer's serious and willful misconduct.

*Territory covered by quoted rules.* Territory No. 01 (Alameda, Contra Costa, San Francisco, Santa Clara) as of August 21, 1968.

[c]Foremen and superintendents in charge of erection or construction work, watchmen, timekeepers, or cleaners shall be assigned to the government classification.

[d]Symbol meaning:

x—explosion hazard     c—collapse hazard     u—underground hazard

The presence of "xcu" hazards results in increaess in property damge rates commensurate with the hazardous condition.

[e]Schedule rating—type of merit rating by which basic manual rates are modified to fit the physical conditions of the individual plant in accordance with the industrial compensation rating schedule.

[f]Tunneling subject to basic pneumoconiosis surcharge.

## FIGURE 9.4   Shift work provision.

A shift differential premium of ten (10) cents per hour will be paid for all time worked on the afternoon or second shift, and a shift differential of fifteen (15) cents per hour will be paid for all time worked on the night or third shift as follows:

(1) *First Shift.* The day or first shift will include all Employees who commence work between 6 a.m. and 2 p.m. and who quit work at or before 6 p.m. of the same calendar day. No shift differential shall be paid for time worked on the day or first shift.

(2) *Second Shift.* The afternoon or second shift shall include all Employees who commence work at or after 2 p.m. and who quit work at or before 12 midnight of the same calendar day. A shift differential premium of ten (10) cents per hour shall be paid for all time worked on the afternoon or second shift.

(3) *Third Shift.* The night or third shift shall include all Employees who commence work at or after 10 p.m. and who quit work at or before 8 a.m. of the next following calendar day. A shift differential premium of fifteen (15) cents per hour shall be paid for all time worked on the night or third shift.

(4) *Cross Shift.* Where an Employee starts work during one shift, as above defined, and quits work during another shift, as above defined, said Employee shall not be paid any shift differential premium for time worked, if any, between the hours of 7 a.m. and 3 p.m.; but shall be paid a shift differential of ten (10) cents per hour for time worked, if any, between the hours of 3 p.m. and 11 p.m., and a shift differential premium of fifteen (15) cents per hour for all time worked, if any, between the hours of 11 p.m. and 7 a.m.

*Source:* D. W. Halpin and R. W. Woodhead, *Construction Management,* copyright © 1980 by John Wiley and Sons, Inc. Reprinted by permission of John Wiley and Sons, Inc.

This means that if a three-shift project is being worked the ironworker will receive overtime for all time worked over seven hours. In addition, he will be paid eight hours' pay for seven hours' work.

## 9.4   AVERAGE HOURLY COST CALCULATION

A summary of data required to calculate the hourly cost to the contractor of various crafts in a given work area is given in Table 9.2. Assume that the hourly cost of a brick mason working on a project in this area is required. The job involves constructing a condominium project, and the brick mason works ten-hour shifts six days a week. The last two hours of the weekdays (first five) are considered to be overtime. The entire sixth day is worked on an overtime basis. Overtime is paid at time and a half. The contractor carries $300,000/$300,000 public liability and $50,000/$100,000 property damage insurance. The calculation of the hourly amount that must be recovered by the contractor to maintain his brick mason in the field is shown in Figure 9.5.

It is necessary to determine which hours are paid at premium rate and which hours

**FIGURE 9.5.   Sample Wage Calculation**

|  | Hours Worked | Straight-Time Hours (ST) | Premium Time (PT) |
|---|---|---|---|
| Monday–Friday | $5 \times 8 = 40$ | $5 \times 8 = 40$ | |
| | $5 \times 2 = 10$ | $5 \times 2 = 10$ | $\frac{1}{2} \times 5 \times 2 = 5$ |
| Saturday | $1 \times 8 = 8$ | $1 \times 8 = 8$ | $\frac{1}{2} \times 8 = 4$ |
| | $1 \times 2 = 2$ | $1 \times 2 = 2$ | $\frac{1}{2} \times 2 = 1$ |
| | 60 | 60 | 10 |

Base rate = \$9.35/hour
ST 60 hours @ \$9.35/hour = \$561.00
PT 10 hours @ \$9.35/hour = $\underline{\quad 93.50}$
  Gross pay          = \$654.50

| Fringes (based on ST hours only) | | |
|---|---|---|
| Pension | = | \$ 0.50/hour |
| Health and welfare | = | 0.55/hour |
| Vacation | = | 0.65/hour  (deferred wage = $60 \times 0.65 = 39.00$) |
| Miscellaneous promotion | = | $\underline{0.08/\text{hour}}$ |
| | | \$ $\overline{1.78} \times 60$  = \$106.80 |

Insurances (per \$100 payroll)
WC = \$3.55                         WC, PL, and PD $= \$ 4.58 \times \left( \dfrac{\$561.00}{\$100.00} \right)$

PL = $(0.38) \times (1.78)$   = \$0.68           = 25.69
PD = $(0.27) \times (1.3)$    = $\underline{\$0.35}$
                              \$4.58

FICA        = $(0.067) \times (\$654.50 + 39.00) = (0.67) \times (\$693.50) = \$46.46$
Unemployment = $(0.05) \times (\$654.50 + 39.00) = (0.05) \times (\$693.50) = \$34.68$

Total cost = base + fringes + WC, PL, PD + FICA + Unemployment = \$849.57

Average hourly cost (to contractor) = $\dfrac{\$868.13}{60}$ = \$14.47

are paid at the straight time rate. It is important to differentiate between the two, since certain burdens paid by the contractor are paid only on straight hours, while others are paid on the entire pay amount received by the worker. Fringe benefits paid to the union as well as workmen's compensation are based on the straight-time hours only. Social security and unemployment compensation, on the other hand, are paid on the worker's entire wage.

In this example, 60 hours are worked. The rate in calculating premium pay is that the basic hour worked, whether on normal time or overtime, is considered to be straight time. The portion of the overtime rate greater than one is considered to be paid at premium pay. Therefore, a single hour worked at a rate of $1\frac{1}{2}$ will result in one hour being paid as a straight-time hour and one $\frac{1}{2}$ hour being paid as a premium hour. If the effective rate is double time, one straight-time hour will be paid and one premium hour will be paid for a single hour worked. In the example, the last two hours of the weekdays and the sixth day are overtime. This results in 20 hours of overtime, yielding

**TABLE 9.2  Building Craft Wage and Insurance Rates**

| Locals | Wages | Pension | Health and Welfare | Vacation | Apprentice Training | Miscellaneous | Workmen's Compensation[a] | Public Liability | Property Damage[c] |
|---|---|---|---|---|---|---|---|---|---|
| Asbestos workers | $10.15 | $0.60 | $0.55 | | $0.10 | | $6.09 | $1.00 | $0.55 |
| Boilermakers | 10.25 | 0.75 | 1.05 | | 0.02 | | 6.46 | 0.37 | 0.36 |
| Bricklayers | 9.35 | 0.50 | 0.55 | $0.65 | | $0.08 promotion | 3.55 | 0.38 | 0.27 |
| Carpenters | 9.45 | 0.45 | 0.50 | | 0.02 | | 5.67 | 0.40 | 0.26 |
| Cement masons | 8.90 | 0.55 | 0.40 | | | 0.20 building | 2.51 | 0.41 | 0.29 |
| Electricians | 10.45 | 1.1% | 0.9% | 0.8% | 0.05% | | 2.19 | 0.17 | 0.21 |
| Operating engineers | 9.35 | 0.75 | 0.50 | | 0.07 | 0.10 administration | 5.61 | 0.93 | 1.00 |
| Ironworkers | 9.60 | 0.57 | 0.65 | | 0.07 | | 14.59 | 1.50 | 0.94 |
| Laborers | 6.25 | 0.33 | 0.20 | | | 0.05 education | 3.75 | 0.19 | 0.20 |
| Painters | 9.45 | 0.65 | 0.65 | | 250/yr | | 3.59 | 0.13 | 0.44 |
| Plasterers | 9.17 | 0.55 | 0.40 | | | 0.20 building / 0.10 promotion | 3.48 | 0.39 | 0.27 |
| Plumbers | 10.75 | 0.50 | 0.65 | | 0.11 | 0.06 promotion / 0.02 national | 2.80 | 0.29 | 0.59 |
| Sheet metal | 10.20 | 0.70 | 0.50 | | 0.04 | 0.09 ind.[d] | 3.57 | 0.21 | 0.20 |
| Unemployment: 5% | | | | | | | | | |
| Social security: 6.7% | | | | | | | | | |

Rates are applied per $100 of pay.

[a] Public liability. Maximum coverage under these rates: $5000/person, $10,000 per accident. For higher coverage: $10,000/20,000, 1.26 × basic rate; $25,000/50,000, 1.47 × basic rate; $50,000/100,000, 1.59 × basic rate; $300,000/300,000, 1.78 × basic rate;

[c] Maximum coverage under these rates: $5000 per person, $25,000 per accident. For higher coverage: $25,000/100,000, 1.23 × basic rate; $50,000/100,000, 1.30 × basic rate.

[d] Industry advancement fund.

10 premium hours and 20 hours of straight time. The total number of straight-time hours worked is 60. This results in a straight-time wage of $561 and a premium amount of $93.50.

As noted above, the calculation of fringe benefits is based on the straight-time wage only. For brick masons four types of fringes are paid by the contractor. Fringes for health and welfare, pension, and an amount for promotion of the craft are paid at the rates shown. A vacation payment in the amount of $0.65 per hour is also paid and must be considered separately. It constitutes a deferred wage and must be included in calculating the FICA and unemployment contributions. The total of fringe benefits results in the contractor's paying an additional $1.78 per straight-time hour.

Amounts to cover insurance premiums for workmen's compensation, public liability, and property damage must also be escrowed by the contractor for payment to the insurance carrier at a later time. The appropriate WC premium for bricklayers is $3.55 for $100 of straight-time payroll. To achieve the required coverages the basic PL rate must be multiplied by 1.78 and the PD rate is 1.3. The total premium amount is $4.58 per $100 of straight-time wage. In this case, $25.69 must be paid by the contractor for the 60 hours of bricklayer time.

The social burdens include the FICA and unemployment amounts paid by the contractor. This contribution is based on the total wage paid to include the deferred wage represented by the vacation fringe. The contractor pays $46.03 to the Social Security Administration as his matching amount to that paid by the worker. An amount of $34.35 is paid into the unemployment trust based on a 5 percent contribution rate (see Table 9.2). The total cost of the worker's 60 hours to the contractor is

|               |          |
|---------------|----------|
| Gross pay     | $654.50  |
| Fringes       | 89.00    |
| WC, PL, PD    | 25.69    |
| FICA          | 46.03    |
| Unemployment  | 34.35    |
|               | $849.57  |

Hourly rate = $849.57/60 = $14.16/hour

This is clearly considerably more than the basic pay rate of $9.35. Amounts for the additional "burdens" on basic pay must be recovered, or the contractor will grossly underbid the project.

## 9.5   A BIDDING GAME

A bidding game developed by Au, Bostleman, and Parti is very useful as a vehicle for becoming familiar with the problems facing a construction contractor in bidding projects over an extended time frame. This game is described in detail in the article "Construction Management Game—A Deterministic Model."* The format of the game

---

* Published in *Journal of the Construction Division,* American Society of Civil Engineers, July, 1969.

allows five teams or individuals acting as contractors in a given market to compete against one another. The teams (player/contractors) are confronted each period of play during the game with jobs that are available to be bid (i.e., the market). The number of jobs varies from period to period, depending upon what part of the year (e.g., spring, summer, etc.) is being simulated. The periods are one quarter (three months) in duration. The fluctuation of the market is established by the game controller and can appear as shown in Figure 9.6. The contracting teams are presented with jobs available to be bid and subcontractors (or trade contractors) who are prepared to bid on portions of the overall project. The game provides four subcontractor bids on each of the four subelements of each job. Each job is assumed to consist of four major phases as follows:

1. Site preparation
2. Foundation
3. Structural frame
4. Finishes

Four subcontractors submit bids on these four phases so that the player is provided with 16 specialty contract bids on each project. The player/contractor is in the position of a construction manager or prime contractor who has 100 percent of the work subcontracted out. The decisions required of each contracting team during each period are summarized in the flow chart of Figure 9.7.

First, the team or player must decide whether to bid or not bid each of the jobs available in the market. If the decision is to bid a particular project, the next consideration is which four subcontractors are to be selected to carry out the four phases of the work. On the face of it, it would seem that the lowest-bidding subcontractor in each of the four phases of the work should be selected. This may not always be the best decision.

**FIGURE 9.6  Typical Demand Curve**

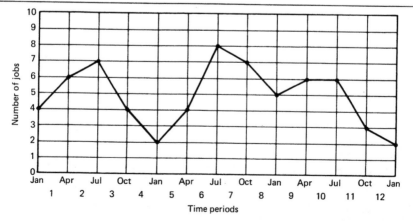

**FIGURE 9.7   Decision Flow Graph for Bidding Game**

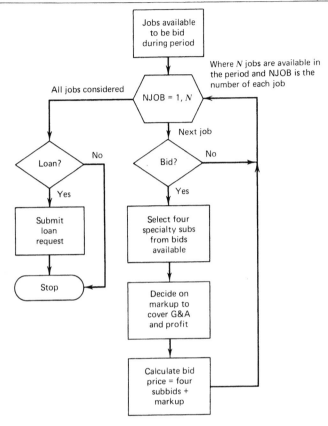

The game departs from reality to a certain degree in the way in which it considers subcontractors. In the real world, subcontractors typically submit fixed bids and must abide by them. In the game, the subcontractors may overrun or underrun the quotation given for player/contractor consideration. Each subcontractor bidding on each phase has a reliability factor based on his past performance. The reliability factors range from "A" for very reliable to "D" for highly unreliable. In addition, locational factors indicate whether the subcontractor can hold his stated price. Nine zones, as shown in Figure 9.8, are defined in the game, and the location of the job versus the location of the subcontractor may lead to excessive mobilization and demobilization costs. If the subcontractor is in the same zone as the job, the effect is minimal. If the subcontractor must mobilize across several zones, however, his price may be higher. The player must analyze the subcontractor prices, locations, and reliability and decide on one subcontractor for each of the four phases of the project.

Once the four subcontractors to be used have been selected, a basic price for the work can be calculated by adding the four quotations together. Figure 9.9 shows a

**FIGURE 9.9    Listing of Subcontractor Bids for a Typical Period**

| Job Number[a] (1) | Job Location (2) | Subcontractor (3) | Subcontractor Location (4) | Subcontract Bid (5) |
|---|---|---|---|---|
| 22 | 1 | 18 | 9 | $198,708.00 |
| 22 | 1 | 16 | 5 | 154,722.00 |
| 22 | 1 | 13 | 2 | 148,039.00 |
| 22 | 1 | 15 | 6 | 154,550.00 |
| 22 | 1 | 28 | 3 | 206,193.00 |
| 22 | 1 | 25 | 7 | 215,327.00 |
| 22 | 1 | 27 | 2 | 217,890.00 |
| 22 | 1 | 26 | 8 | 191,521.00 |
| 22 | 1 | 33 | 4 | 227,318.00 |
| 22 | 1 | 34 | 8 | 247,369.00 |
| 22 | 1 | 38 | 8 | 218,081.00 |
| 22 | 1 | 32 | 7 | 234,201.00 |
| 22 | 1 | 42 | 4 | 181,425.00 |
| 22 | 1 | 49 | 9 | 217,900.00 |
| 22 | 1 | 48 | 5 | 209,579.00 |
| 22 | 1 | 45 | 2 | 188,873.00 |
| 23 | 6 | 10 | 1 | 283,045.00 |
| 23 | 6 | 12 | 1 | 280,055.00 |
| 23 | 6 | 11 | 4 | 327,434.00 |
| 23 | 6 | 15 | 6 | 285,505.00 |
| 23 | 6 | 20 | 9 | 287,928.00 |
| 23 | 6 | 21 | 1 | 361,565.00 |
| 23 | 6 | 25 | 7 | 359,780.00 |
| 23 | 6 | 27 | 2 | 343,068.00 |
| 23 | 6 | 37 | 1 | 451,972.00 |
| 23 | 6 | 31 | 3 | 382,588.00 |
| 23 | 6 | 34 | 8 | 392,981.00 |
| 23 | 6 | 33 | 4 | 407,970.00 |
| 23 | 6 | 48 | 5 | 288,822.00 |
| 23 | 6 | 44 | 2 | 232,694.00 |
| 23 | 6 | 42 | 4 | 246,091.00 |
| 23 | 6 | 49 | 9 | 262,378.00 |

[a]Job number 22 and job number 23 are the only jobs in the period.

typical output with subcontractor bids for each of two jobs. If, for instance, it is decided to bid job 22 and the four subcontractors selected are 18, 25, 38 and 45, the base price for the work would be $820,989.

| Subcontractor | Bid |
|---|---|
| 18 | $198,708 |
| 25 | 215,327 |
| 38 | 218,081 |
| 45 | 188,873 |
| | $820,989 |

The next decision that must be made is how much of a markup should be added to this base figure to defray G & A costs and recover a certain profit.

Quite a number of considerations come into play here. The level of G & A costs incurred by the contractor can be best understood by studying the income statement in Figure 9.10. As can be seen, four major costs are deducted from the gross profit. These are

1. Home office operating cost (item E)
2. Information costs (item F)
3. Bidding costs (item G)
4. Interest costs (item H)

**FIGURE 9.10 Typical Income Statement from Bidding Game (Contractor 4 in Period 4)**

| *Statement of Earnings*[a] | | |
|---|---|---|
| *(1)* | *(2)* | *Amount* |
| A. Income from construction contracts | | $ 412,510.00 |
| | *Cost of Contracts* | |
| B. Subcontracts not supervision of subcontractors | | 347,891.00 |
| C. Field overhead | | 2,063.00 |
| | | $  62,556.00 |
| | *Administrative and General Expenses* | |
| E. Office operating cost | | $  10,421.00 |
| F. Information costs | | 650.00 |
| G. Bidding costs | | 2,333.00 |
| H. Interest on existing loans | | 900.00 |
| I. Earnings before federal income taxes | | $  48,352.00 |
| J. Federal income taxes | | 17,411.00 |
| K. Net earnings | | 30,941.00 |
| L. Retained earnings at beginning of period | | 208,422.00 |
| M. Liquid assets | | $ 239,363.00 |
| | *Loans* | |
| N. Existing loans | | $  60,000.00 |
| O. New loans (one year notes) | | 25,000.00 |
| P. Loans due this time period | | 10,000.00 |
| Q. Total cash on hand | | 314,363.00 |
| R. Retained earnings at end of period | | $ 239,363.00 |
| S. Percentage gain of loss up to end of period | | + 19.7 |

[a]General contractor number 4, time period 4.

Bidding costs are assessed at one-fourth of one percent of the bid amount. Therefore, the decision to bid implies commitment of this amount to get into the competition. For example, if a job is bid at the price of $5 million, the bidding costs will be $12,500. Therefore, the decision to bid has cost ramifications, and this bidding cost must be recovered in the markup.

The cost of home office overhead is assessed in the amount of 5 percent of the retained earnings reported in the income statement at the beginning of the period. Therefore, if the team or player has retained earnings of $140,000 at the beginning of the quarter, $7000 in expense will be incurred for home office support during the period. This flow of cash period by period throughout the play of the game must be estimated and recovered in the markup.

The game allows the player to request information concerning the reliability of the subcontractors and the number of jobs anticipated to be available in future periods. This provision is designed to act as an information source similar to the Dun and Bradstreet credit service and the Dodge Reporting Service for jobs available. The player may access these services periodically and pays a fixed fee for these data.

Interest costs must also be considered in markup. The contractor can borrow a total amount not to exceed two times his liquid assets. This provision allows him to expand his bonding capacity and, hence, his ability to bid large jobs or several small jobs.

All of these costs must be kept in mind when the markup is determined. The strategy of how much profit to add to the recovery of direct and G & A costs in the light of the amount of competition has been considered in detail in numerous articles and texts.* The interested reader should consult Appendix L in *Construction Contracting* by Clough. One approach based on the expected rate of return desired on the project is discussed in Appendix C. If it is decided to use a markup of 10 percent on Job 22, the bid price would be $903,087 ($820,989 + 10 percent markup).

Once a final bid price for each of the jobs to be bid has been determined by adding the markup to the base costs of the four subcontractors, the player can exercise the option of taking a loan at the bank. Money can be borrowed at the cost of $1\frac{1}{2}$ percent simple interest per quarter for a period of one year. The obvious question is why it would be necessary to borrow money. The answer to this has to do with the concept of bonding capacity as utilized in the game.

The bonding capacity of a contractor is the amount of work that the bonding company will allow the contractor for at any given time. The types of bonds and their purposes are discussed in detail in Sections 2.13 and 2.14 of *Construction Management* by Halpin and Woodhead. Bonding capacity limits the contractor's ability to pursue work. The amount of work that a contractor can bid in any period is the bonding capacity minus the amount of work remaining to complete. The bonding capacity for contractors in the bidding game is calculated as 40 times the individual contractor's cash on hand from the previous period of play. To illustrate, assume that the cash on hand from the previous period is $140,000, which implies a bonding capacity of $5.6 million. If the remaining amount of work to do on three jobs in progress is $3.6 million, the contractor cannot bid a job in excess of $2 million.

* See, for instance, William R. Park, *Construction Bidding for Profit* (New York: Wiley, 1979).

By taking a loan the player can expand the value of cash-on-hand and hence the bonding capacity. If, in the above example, the contractor/player borrows $100,000, the bonding capacity will expand in the next period of play by $4,000,000. The amount borrowed at any given time may not exceed two times the liquid assets reported on the most recent income statement. In this illustration, by borrowing $100,000 at a cost of $6000 for the year, the contractor has access to bidding $4 million more in work. This use of borrowing to expand bonding capacity does not strictly conform to the real construction situation. However, it gives the player practice in trying to meet expanding markets and emphasizes the need to plan ahead. If the contractor wins jobs at a low profit against heavy competition and exhausts the available bonding capacity, there may not be sufficient bonding capacity to enter a "fat" market with many jobs available for bid and minimal competition. Successful contractors must be able to anticipate when the market will expand and must be prepared with appropriate bonding capacity.

After all job bids have been submitted and loans requested, a report similar to that shown in Figure 9.11 indicates the following information for each player/contractor:

1. Low bidders and low bids on all jobs available.
2. Other bids submitted on jobs for which contractor submitted bids.
3. Award statements to low-bidding contractors with indication of amount to perform work and amount of profit/loss to be incurred on job, as well as number of periods of duration.
4. Income statements as of the period of play (see Figure 9.10).
5. Relative standings on each of the five player/contractors based on the retained earnings ratio.

The fact that the amount of profit and loss on each job is immediately known to the low bidder is not realistic, since this would only become apparent at the time of completing the job. Since this is a game, however, the level of realism can only be approximate.

**FIGURE 9.11  Typical Information Regarding Bidding**

| Job Number (1) | General Contractor Number (2) | Contract Amount (3) |
|---|---|---|
| 5 | All bids have been rejected by the owners; the project is postponed indefinitely. | — |
| 6 | 3 | $ 262,420.00 |
| 7 | No bids were received on this project. | — |
| 8 | 4 | 830,627.00 |
| 9 | 1 | 593,575.00 |
| 10 | 3 | 1,178,844.00 |

*Information for a Company Receiving an Award*
General contractor number 4
Time period 2
You have been awarded job number 8

The gaming environment as described here is supported by a computer program that prints game markets and results as shown in figures 9.9, 9.10, and 9.11. The game is designed to be played over a three-year period of simulated time. Over this gaming horizon some of the five player/contractors will emerge as winners, while others will be losers. The game, even though it cannot duplicate reality, does confront the players with the same kind of decisions and thought processes that company-level management must consider in deciding when and what to bid. From this point of view, playing the game is an excellent learning experience.

## REVIEW QUESTIONS AND EXERCISES

**9.1**  Compute the hourly cost to a contractor for a carpenter working in the Atlanta area. The operator works 10 hours per day on a 20 yd$^3$ shovel. The last two hours are considered overtime. The contractor carries $300,000/$300,000 public liability and $50,000/$100,000 property damage insurance. Use wage rates and other data given in Table 9.2. He works six days a week and overtime is paid at double time. On this job he receives shift pay and is paid 8 hours for the first 7$^1/_2$ hours he works.

**9.2**  Given the project data below, and the minimum attractive annual rate of return of 30% how much would you mark up the project based on cash flow? Lag factors for all costs incurred are zero. No home office overhead is considered. Income is received one period after expense incurred. Retainage = 10% throughout job. Use methods described in Appendix C.

## Table of Expenses

| Month | Mobilization Demobilization | Subcontractors | Materials | Payroll | Equipment | Field Overhead |
|---|---|---|---|---|---|---|
| 0 | $40,000 | $0 | $0 | $0 | $0 | $0 |
| 1 | 0 | 10,000 | 10,000 | 10,000 | 20,000 | 1,000 |
| 2 | 0 | 30,000 | 20,000 | 15,000 | 10,000 | 5,000 |
| 3 | 0 | 30,000 | 30,000 | 20,000 | 20,000 | 6,000 |

**9.3**  Compute the average hourly cost to a contractor for a cement mason. The job involves the construction of tilt-up panels for a warehouse. The mason works ten hours a day six days a week. The last two hours on weekdays are considered overtime and are paid at double time. All work on the sixth day (Saturday) is considered overtime, and is paid at time and a half. The contractor carries $300,000/$300,000 public liability and $50,000/$100,000 property damage insurance as required. Use wage and other data as given in Table 9.2.

# Chapter 10

# THE ESTIMATING PROCESS

## 10.1 ESTIMATING CONSTRUCTION COSTS

The key to a good job and successful cost control is the development of a good estimate as the basis for bid submittal. The estimate represents the cost "flight plan" that will be followed by the constructor and which will aid him in achieving profit. If the flight plan is unrealistic or contains basic errors, the contractor will lose money on the job. If the estimate is well thought out and correctly reflects the costs that will be encountered in the field, the chances of a profitable job are greatly increased.

Estimating is the process of looking into the future and trying to predict project costs and resource requirements. Studies indicate that one of the major reasons for the failure of construction contracting firms is incorrect and unrealistic estimating and bidding practices. If 20 estimators or contractors were furnished the same set of plans and specifications and told to prepare an estimate of cost and resources, it would be safe to assume there would not be more than two estimates in the entire twenty that had been prepared on the same basis or from the same units. Therefore, a consistent procedure or set of steps for preparing an estimate is needed to minimize errors and achieve reliable results.

## 10.2 TYPES OF ESTIMATES

Estimating methods vary in accordance with the level of design detail that is available to the estimator. Prior to the commencement of design, when only conceptual information is available, a comprehensive unit such as a square foot of floor space or a cubic foot of usable space is used to characterize the facility being constructed. The representative unit is multiplied by a price per unit to obtain a gross estimate ($\pm 10$ percent accuracy) of the facility cost. A table of square foot and cubic foot building costs as given in the publication *Building Construction Cost Data* published by R. S. Means is shown in Figure 10.1. Such information is available in standard references and can be used for preliminary cost projections based on minimal design data. This *conceptual estimate* is useful in the schematic or budgetary phase, when design details

# FIGURE 10.1  Costs Based on a Representative Unit

| 17.1 | S.F., C.F. and % of TOTAL COSTS | UNIT | UNIT COSTS | | | % OF TOTAL | | |
|---|---|---|---|---|---|---|---|---|
| | | | 1/4 | MEDIAN | 3/4 | 1/4 | MEDIAN | 3/4 |
| 01 950 | Total: Mechanical & Electrical | Apt. | 4,330 | 5,490 | 7,250 | | | |
| 02 | APARTMENTS, Mid Rise | S.F. | 26.75 | 32.25 | 48.15 | | | |
| 002 | Total project costs | C.F. | 2.37 | 3.42 | 4.09 | | | |
| 180 | Equipment | S.F. | .35 | .71 | 1.34 | 2.50% | 2.50% | 3.20% |
| 272 | Plumbing | | 2.22 | 2.41 | 2.91 | 8.40% | 8.90% | 9.50% |
| 277 | Heating, ventilating, air conditioning | | 1.09 | 1.13 | 1.40 | 4.10% | 4.30% | 10% |
| 290 | Electrical | | 1.80 | 2.19 | 4.13 | 7.10% | 7.90% | 11.40% |
| 310 | Total: Mechanical & Electrical | ▼ | 5.75 | 7.60 | 11.60 | 21% | 26.60% | 30% |
| 900 | Per apartment unit, total cost | Apt. | 21,170 | 31,970 | 45,430 | | | |
| 950 | Total: Mechanical & Electrical | " | 6,160 | 10,330 | 13,040 | | | |
| 03 | APARTMENTS, High Rise | S.F. | 37.90 | 43.60 | 50.70 | | | |
| 002 | Total project costs | C.F. | 3.70 | 4.52 | 4.96 | | | |
| 180 | Equipment | S.F. | .64 | 1.05 | 1.31 | 1.60% | 2.20% | 3% |
| 272 | Plumbing | | 3.10 | 4.08 | 4.92 | 8.20% | 8.90% | 10.30% |
| 277 | Heating, ventilating, air conditioning | | 1.27 | 3.01 | 4.48 | 3.70% | 6.30% | 9% |
| 290 | Electrical | | 2.83 | 3.46 | 4.21 | 6.70% | 7.50% | 8.70% |
| 310 | Total: Mechanical & Electrical | ▼ | 7.70 | 10.90 | 12.80 | 19.80% | 24.30% | 27.80% |
| 900 | Per apartment unit, total cost | Apt. | 31,890 | 39,030 | 48,960 | | | |
| 950 | Total: Mechanical & Electrical | " | 7,680 | 9,060 | 12,920 | | | |
| 04 | AUDITORIUMS | S.F. | 37.70 | 56.80 | 75 | | | |
| 002 | Total project costs | C.F. | 1.76 | 2.27 | 2.94 | | | |
| 272 | Plumbing | S.F. | 2.04 | 2.86 | 3.68 | 3.20% | 4.30% | 6% |
| 277 | Heating, ventilating, air conditioning | | 6.40 | 7 | 9.60 | 10.70% | 13.20% | 17.40% |
| 290 | Electrical | | 4.46 | 6.30 | 8.20 | 18.10% | 10.40% | 13.90% |
| 310 | Total: Mechanical & Electrical | | 11.30 | 16.30 | 20.40 | 22.50% | 27.60% | 32.20% |
| 05 | AUTOMOTIVE SALES | ▼ | 20.60 | 28.90 | 34.50 | | | |
| 002 | Total project costs | C.F. | 1.51 | 1.71 | 2.40 | | | |
| 272 | Plumbing | S.F. | 1.03 | 1.97 | 2.40 | 4.40% | 6.50% | 8% |
| 277 | Heating, ventilating, air conditioning | | 1.30 | 3.09 | 3.73 | 5.10% | 8% | 10.80% |
| 290 | Electrical | | 2.48 | 3.80 | 4.75 | 8.80% | 11% | 12.40% |
| 310 | Total: Mechanical & Electrical | | 6.35 | 9.10 | 11.10 | 22.60% | 28.20% | 32.30% |
| 06 | BANKS | ▼ | 54.20 | 69.20 | 92.60 | | | |
| 002 | Total project costs | C.F. | 3.78 | 5.10 | 6.80 | | | |
| 180 | Equipment | S.F. | 2.59 | 9.32 | 16.96 | 4.40% | 9.90% | 17.70% |
| 272 | Plumbing | | 2.09 | 2.96 | 4.25 | 3.30% | 4.40% | 5.80% |
| 277 | Heating, ventilating, air conditioning | | 3.64 | 5.40 | 7.20 | 5.80% | 8.20% | 10.70% |
| 290 | Electrical | | 4.88 | 6.90 | 9.60 | 8.30% | 10.30% | 12.40% |
| 310 | Total: Mechanical & Electrical | ▼ | 11.60 | 15.60 | 20.90 | 19.60% | 23.80% | 27.70% |
| 350 | See also division 11.1-9 | | | | | | | |
| 13 | CHURCHES | S.F. | 37.70 | 46.40 | 56.70 | | | |
| 002 | Total project costs | C.F. | 2.41 | 2.92 | 3.59 | | | |
| 180 | Equipment | S.F. | .43 | .91 | 1.65 | .90% | 1.80% | 3.60% |
| 272 | Plumbing | | 1.57 | 2.14 | 3.08 | 3.60% | 4.80% | 6.30% |
| 277 | Heating, ventilating, air conditioning | | 3.43 | 4.45 | 6.10 | 7.90% | 9.90% | 11.70% |
| 290 | Electrical | | 2.95 | 3.81 | 5.20 | 7% | 8.20% | 10% |
| 310 | Total: Mechanical & Electrical | | 8.80 | 10.60 | 13.10 | 20.40% | 23.60% | 27% |
| 350 | See also division 11.1-12 | | | | | | | |
| 15 | CLUBS, COUNTRY | S.F. | 36.20 | 46.20 | 56.90 | | | |
| 002 | Total project costs | C.F. | 3.48 | 3.77 | 5.10 | | | |
| 272 | Plumbing | S.F. | 3.26 | 3.99 | 5.70 | 7.20% | 9% | 11.20% |
| 277 | Heating, ventilating, air conditioning | | 2.50 | 4.64 | 5.90 | 8.40% | 9.40% | 10.50% |
| 290 | Electrical | | 2.49 | 4.69 | 6.15 | 7.70% | 9.40% | 10.60% |
| 310 | Total: Mechanical & Electrical | ▼ | 10.35 | 13.35 | 15.85 | 27.10% | 30.10% | 31.30% |

For expanded coverage of these items see Means' *Appraisal Manual 1981*

*Source:* This information is copyrighted by Robert Snow Means Co., Inc. It is reproduced from *Building Construction Cost Data 1981*, p.277, with permission.

180

**FIGURE 10.2 Project Proposal: Layout Sketch and Outline Specifications**

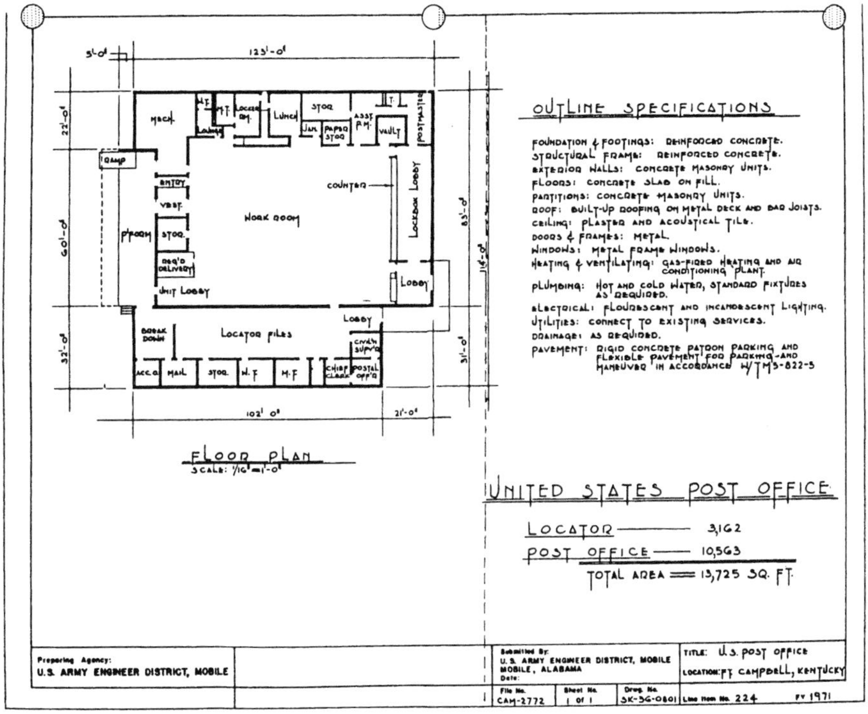

*Source:* D. W. Halpin and R. W. Woodhead, Construction Management, copyright © 1980 by John Wiley and Sons, Inc. Reprinted by permission of John Wiley and Sons, Inc.

are not available. The figures developed are of limited use for project control, and their use should be discontinued as soon as design data are available. The conceptual drawings for a small building project are shown in Figure 10.2. The conceptual estimate for this building is given in Figure 10.3.

As the level of design detail increases, the designer typically maintains estimates of cost to keep the client informed of the general level of costs to be expected. The production of the plans and specifications usually proceeds in two steps. The first step is called *preliminary design* and offers the owner a pause in which to review construction before detail design commences. A common time for this review to take place is at 40 percent completion of the total design. The preliminary design extends the concept documentation.* At this point in the design process, a *preliminary estimate* is prepared by the architect or architect/engineer to reflect expected costs based on more definitive data.

_____

* D. W. Halpin and R. W. Woodhead, *Construction Management* (New York: Wiley 1980), p. 25.

## FIGURE 10.3  Current Working Estimate for Budget Purposes

| TO: Chief of Engineers<br>Department of the Army<br>Washington 25, D.C. | FROM: Mobile District<br>Corps of Engineers<br>Mobile, Alabama 36601 | FISCAL YEAR<br>1971 | DATE PREPARED<br>14 Oct 69 |
|---|---|---|---|

| | | NAME AND ADDRESS OF A.E.<br>N.A. | |
|---|---|---|---|
| | | BASIS OF ESTIMATE<br>Code "A" Budget Sketch & 1391 | A.E. FEE<br>N.A. |

| NAME AND LOCATION OF INSTALLATION<br>Ft. Campbell, Kentucky | TYPE OF CONSTRUCTION<br>Permanent | STATUS OF DESIGN<br>Preliminary 0% complete  Final 0% complete |
|---|---|---|
| LINE ITEM NUMBER<br>224 | DESCRIPTION OF FACILITY<br>Post Office | FINAL DESIGN COMPLETION DATE<br>Not Authorized |

| DESCRIPTION | QUANTITY | UNIT | UNIT PRICE | TOTALS ($000) |
|---|---|---|---|---|
| 1. *Building* | | | | |
| General Construction | 13,725 | Sq ft | $21.12 | $289.9 |
| Plumbing | 13,725 | Sq ft | 1.21 | 16.6 |
| Heating and Ventilating | 13,725 | Sq ft | 1.34 | 18.4 |
| Air Conditioning (50-ton) | 13,725 | Sq ft | 3.81 | 52.3 |
| Electrical | 13,725 | Sq ft | 2.83 | 38.8 |
| Subtotal | 13,725 | Sq. ft | 30.31 | 416.0 |
| | | | | |
| 2. *Utilities* | | | | |
| a. *Electrical* | | | | |
| Transformers | 112.5 | kVA | 26.40 | 3.0 |
| Poles with X-arms, Pins, Insulation, etc. | 4 | Each | 356.40 | 1.4 |
| Dead Ends | 6 | Each | 39.56 | 0.2 |
| Down Guys and Anchors | 4 | Each | 89.10 | 0.4 |
| Fused Cutouts and L.A. | 6 | Each | 59.94 | 0.4 |
| #6 Bare Cu. Conductor | 2,400 | lin ft | .30 | 0.7 |
| #3/0 Neoprene Covered Service | 160 | lin ft | 1.48 | 0.2 |
| Parking Area Lights on Aluminum Pole | 7 | Each | 933.41 | 6.5 |
| 3C #8 DB 600-V Cable | 210 | lin ft | 1.70 | 0.4 |
| 2C #8 DB 600-V Cable | 740 | lin ft | 1.27 | 0.9 |
| 3-in. Duct Conc. Encased U.G. | 100 | lin ft | 4.75 | 0.5 |
| Subtotal | | | | 15.0 |
| | | | | |
| b. *Water* | | | | |
| 3-in. Water Line | 365 | lin ft | 4.30 | 1.6 |
| 3-in. Gate Valve and Box | 1 | Each | 118.80 | 0.1 |
| Fire Hydrants | 2 | Each | 534.60 | 1.1 |
| Connections to Existing Lines | 3 | Each | 273.24 | 0.8 |
| Subtotal | | | | 4.0 |
| | | | | |
| c. *Sewer* | | | | |
| 6-in. Sanitary Sewer | 215 | lin ft | 5.94 | 1.3 |
| 8-in. Sanitary Sewer | 375 | lin ft | 6.89 | 2.6 |
| Manhole | 2 | Each | 534.60 | 1.1 |
| Connection to Exist. Manhold | 1 | Each | 118.80 | 0.1 |
| Subtotal | | | | 5.0 |
| | | | | |
| d. *Gas* | | | | |
| 1 1/4 in. Gas Line | 1,000 | lin ft | 3.09 | 3.1 |
| 1 1/4 in. Plug Valve and Box | 1 | Each | 118.80 | 0.1 |
| Connect to Existing | 1 | Each | 237.60 | 0.2 |
| Street and Parking Area Crossing | 280 | lin ft | 1.54 | 0.4 |
| | | | | 4.0 |

Once the preliminary design has been approved by the owner, final or detail design is accomplished. The detail design phase culminates in the plans and specifications that are given to the constructor for bidding purposes. In addition to these detailed design documents, the architect/engineer produces a final *engineer's estimate* indicating the total job cost minus markup. This estimate should achieve approximately ± 3 percent accuracy, since the total design is now available. The owner's estimate is used

| DESCRIPTION | QUANTITY | UNIT | UNIT PRICE | TOTALS ($000) |
|---|---|---|---|---|
| 3. *Site Work* | | | | |
| Clearing and Grubbing | 2.4 | Acre | 495.01 | 1.2 |
| Borrow Excavation | 10,000 | cu yd | 3.46 | 34.6 |
| Remove B.T. Paving | 1,070 | sq yd | 1.44 | 1.5 |
| Subtotal | | | | 37.0 |
| 4. *Paving* | | | | |
| Paving—1½ A.C. and 8-in. Stab. | 3,950 | sq yd | 6.19 | 24.5 |
| Aggr. Base | 2,250 | lin ft | 4.70 | 10.6 |
| 6-in. P.C. Concrete Paving | 380 | sq yd | 9.70 | 3.7 |
| 3-in. Painted Parking Lines | 1,680 | lin ft | 0.27 | 0.5 |
| Concrete Sidewalk | 440 | sq yd | 8.10 | 3.6 |
| Subtotal | | | | 43.0 |
| 5. *Storm Drainage* | | | | |
| 15-in. Concrete Cl. II Pipe | 40 | lin ft | 8.91 | 0.4 |
| 15-in. Concrete Cl. III Pipe | 20 | lin ft | 10.22 | 0.2 |
| Reinf. Drainage Structure Concrete | 8 | cu yd | 207.90 | 1.7 |
| C.I. Grates and Frames | 1,900 | lb | 0.37 | 0.7 |
| Subtotal | | | | 3.0 |
| 6. *Landscaping* | | | | |
| Sprigging and Seeding | 1.6 | Acre | 945.00 | 1.5 |
| Landscaping | | Job | | 2.1 |
| | | | | 4.0 |
| 7. *Communications* | | | | |
| a. Telephone | | LS | 756.00 | 1.0 |
| b. Support (Within Building) | | | | |
| 100 Pr. DB Pic Cable | 600 | LF | 1.26 | 0.8 |
| 51 Pr DB Pic Cable | 550 | LF | 0.72 | 0.4 |
| Splicing Sleeves and Material | | LS | $480.00 | 0.5 |
| Labor | | LS | | 2.4 |
| | | | | 5.0 |
| Total estimated cost (excluding design, but including reserve for contingencies and supervision and administration (S&A) | | | | 536.0 |
| 1. Estimated contract cost | | | | 460.6 |
| 2. Reserve for Contingencies | 10 | % | | 46.0 |
| 3. Supervision and administration (S&A) | | | | 29.4 |
| Total estimated cost (excluding design, but including reserve for contingencies and supervision and administration | | | | 536.0 |
| 4. *Design* | | | | |
| District Expenses (Preliminary and Final) | | | | 35.0 |
| Subtotal | | | | $35.0 |

*Source:* D. W. Halpin and R. W. Woodhead, Construction Management, copyright © 1980 by John Wiley and Sons, Inc. Reprinted by permission of John Wiley and Sons, Inc.

(1) to ensure that the design produced is within the owner's financial resources to construct (i.e., that the architect/engineer has not designed a gold-plated project), and (2) to establish a reference point in evaluating the bids submitted by the competing contractors.

On the basis of the final drawings and specifications the contractor prepares his estimate of the job's cost to include a markup for profit. This is the *bid estimate*. Both the engineer's and bid estimates require a greater level of effort and a considerable number of estimator hours to prepare. A rough rule of thumb states that the preparation of a bid estimate by the contractor will cost one-fourth of one percent of the total bid price. From the contractor's point of view this cost must be recovered as overhead on jobs that are won. Therefore, a prorate based on the number of successful bids versus total bids must be included in each quotation to cover bid costs on unsuccessful bids.

In building construction, these four levels of estimates are the ones most commonly encountered. To recapitulate, the four types of estimates are

1. Conceptual estimate
2. Preliminary estimate
3. Engineer's estimate
4. Bid estimate

These four levels of precision reflect the fact that as the project proceeds from concept through preliminary design to final design and the bidding phase, the level of detail increases, allowing the development of a more accurate estimate. Estimating continues during the construction phase to establish whether the actual costs agree with the bid estimate. This type of "estimating" is what allows the contractor to project profit or loss on a job after it is in progress.

A listing of estimates commonly developed in conjunction with large and complex industrial projects (e.g., power plants, chemical process plants, and the like) is given in Figure 10.4. This list includes a magnitude level estimate that is similar in purpose to the conceptual estimate used in building construction. That is, it is used to reflect gross costs for planning and decision purposes before the preliminary and definitive design phases begin. The definitive estimate, as used on complex industrial projects, is a prefinal estimate developed just prior to the production of final drawings and specifications. The definitive estimate can be prepared when all components comprising the project scope definition have been quantitatively determined and priced by using actual anticipated material and labor costs. This estimate is normally prepared when the project scope is defined in terms of firm plot plans, mechanical and process flow diagrams, equipment and material specifications, and engineering and design layouts. The pricing bases are formal vendor quotations for all major items and current predictable market costs for all commodity accounts. The amount of variability inherent in each level of estimate is reflected by the contingency curves shown on the right in Figure 10.4. The variability is, of course, quite high at the magnitude level and decreases to the 3 to 5 percent range as bid level documents become available.

## 10.3   DETAILED ESTIMATE PREPARATION

The preparation of a detailed bid level estimate requires that the estimator break the project into its cost centers or cost subelements. That is, the project is broken down into subcomponents that will generate costs. It is these costs that the estimator must develop on the basis of the characteristic resources required. The word "resource" is

**FIGURE 10.4 Estimate Types**

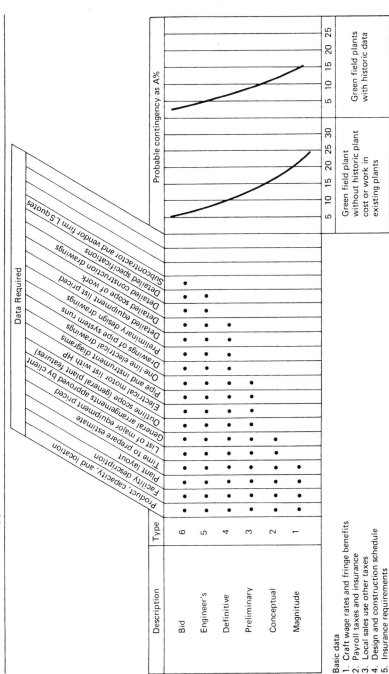

| Description | Type | Product, capacity, and location | Facility description | Plant layout | Time to prepare estimate | List of major equipment priced | General arrangements approved by client | Outline scope (general plant features) | Electrical motor list with HP | Pipe and instrument diagrams | One-line electrical drawings | Drawings of pipe system runs | Preliminary design drawings | Detailed equipment list priced | Detailed scope of work | Detailed construction drawings | Detailed specifications | Subcontractor and vendor firm L.S quotes |
|---|---|---|---|---|---|---|---|---|---|---|---|---|---|---|---|---|---|---|
| Bid | 6 | • | • | • | • | • | • | • | • | • | • | • | • | • | • | • | • | • |
| Engineer's | 5 | • | • | • | • | • | • | • | • | • | • | • | • | • | • | • | | |
| Definitive | 4 | • | • | • | • | • | • | • | • | • | • | • | • | | | | | |
| Preliminary | 3 | • | • | • | • | • | • | • | • | | | | | | | | | |
| Conceptual | 2 | • | • | • | | | | | | | | | | | | | | |
| Magnitude | 1 | • | • | | | | | | | | | | | | | | | |

Data Required

Probable contingency as Δ%

| Green field plant without historic plant cost or work in existing plants | Green field plants with historic data |
|---|---|
| 5  10  15  20  25  30 | 5  10  15  20  25 |

Basic data
1. Craft wage rates and fringe benefits
2. Payroll taxes and insurance
3. Local sales use other taxes
4. Design and construction schedule
5. Insurance requirements

*Source:* Frederick S. Merritt, ed., Building Construction handbook, 3rd ed. (New York: McGraw-Hill, 1975).

185

used here in the broad sense and applies to the man-hours, materials, subcontracts, equipment-hours, and dollars needed to accomplish the work or meet the requirements associated with each cost center. Typically in construction the cost center relates to some physical subcomponent of the project, such as foundation piles, excavation, steel erection, interior dry wall installation, and the like. Certain nonphysical components of the work generate costs, however, and these cost centers must also be considered. Many of the items listed as "indirects" in Figure 9.1 are typical of costs that are not directly connected with physical components or end items in the facility to be constructed. Such items do, however, generate cost that must be recovered. These costs include insurance and bonding premiums, fees for licenses and permits required by contract, expense for special items relating to safety and minority participation programs, and home office overheads projected as allocated to the job. These items are sometimes referred to as "general conditions" or "general requirements" items, although they may or may not be specifically referred to in the contract documents. Accounts relating to these items fall into the categories for "conditions of contract" and "general requirements" as listed in the major divisions of the Uniform Construction Index (Table 8.1). As estimators prepare bids, they have a general framework for cost recovery in mind, such as Figure 9.1. In addition, they have a knowledge of the technologies involved in building the project, which allow them to divide projects into individual pieces of work (physical subcomponents, systems, etc.). These work packages consume resources, generating costs that must be recovered from the client. Typically, the chart of cost accounts acts as a guide or checklist as the estimator reviews the plans and specifications to highlight what cost centers are present in the contract being estimated.

Although the process of estimating is part art, part science, the estimator generally follows certain steps in developing the estimate:

*1.* Break the project into cost centers.
*2.* Estimate the quantities required for cost centers that represent physical end items (e.g., cubic yards of earth, lineal feet of pipe, and so forth). For physical systems this procedure is commonly called "quantity take-off." For those cost centers that relate to nonphysical items, determine an appropriate parameter for cost calculation (e.g., the level of builder's risk insurance required by the contract, or the amounts of the required bonds).
*3.* Price out the quantities determined in step 2 using historical data, vendor quotations, supplier catalogues, and other pricing information. This pricing may be based on a price per unit (unit cost) basis or a lump sum (one job) basis. Price development for physical work items may require an analysis of the production rates to be achieved based on resource analysis. If this analysis is used, the estimator must:

    a. Assume work team composition to include number of workers (skilled and unskilled) and equipment required.
    b. On the basis of team composition, estimate an hourly production rate based on the technology being used.

   c.  Make an estimate of the efficiency to be achieved on this job, considering site conditions and other factors.
   d.  Calculate the effective unit price.

4.  Calculate the total price for each cost center by multiplying the required quantity by the unit price. This multiplication is commonly called an "extension," and this process is called "running the extensions."

The estimator usually summarizes the values for each cost center on a summary sheet, such as that shown in Figure 10.5.

## FIGURE 10.5   Typical Estimate Summary

Jefferson Starship Contractors, Inc.
ESTIMATE SUMMARY
Estimate No. ___6692___   By: ___DWH___      Date: _1 August 19X4_
Owner: ___NASA___             Project: _VA Building_

| Code | Description | MH | Labor | Material | Sub | Owner | Total |
|------|-------------|-----|-------|----------|------|--------|---------|
| 01 | Site improvements | | | | | | |
| 02 | Demolition | | | | | | |
| 03 | Earthwork | | | | | | |
| 04 | Concrete | | | | | | |
| 05 | Structural Steel | 1,653 | 18,768 | 15,133 | | | 33,901 |
| 06 | Piling | | | | | | |
| 07 | Brick & masonry | | | | | | |
| 08 | Buildings | | | | | | |
| 09 | Major equipment | 2,248 | 26,059 | 1,794 | | | 27,853 |
| 10 | Piping | 2,953 | 34,518 | 57,417 | 1,500 | 34,541 | 127,976 |
| 11 | Instrumentation | | | | 33,000 | | 33,000 |
| 12 | Electrical | | | | 126,542 | | 126,542 |
| 13 | Painting | | | | 14,034 | | 14,034 |
| 14 | Insulation | | | | 4,230 | | 4,230 |
| 15 | Fireproofing | | | 530 | 1,110 | | 1,640 |
| 16 | Chemical cleaning | | | | | | |
| 17 | Testing | | | | | | |
| 18 | Const. equipment | | | | 35,666 | | 35,666 |
| 19 | Misc. directs | 1,008 | 10,608 | 2,050 | | 2,000 | 14,658 |
| 20 | Field extra work | | | | | | |
| Sub | Total Direct Cost | 7,862 | 89,953 | 76,924 | 180,416 | 72,207 | 419,500 |
| 21 | Con. tools/sup. | | | 7,361 | | | 7,361 |
| 22 | Field payroll/ burden | | | | | 16,580 | 16,580 |
| 23 | Start-up asst. | | | | | | |
| 24 | Ins. & taxes | | | | | 5,268 | 5,268 |
| 25 | Field sprvsn. | 480 | 7,200 | | | 2,038 | 9,238 |
| 26 | Home off. exp. | | | | | 2,454 | 2,454 |
| 27 | Field emp. ben. | | | | | 10,395 | 10,395 |
| Sub | Total Indirect Cost | 480 | 7,200 | 7,361 | | 36,735 | 51,296 |
| Adjustment Sheets | | | | | | | |
| Total Field Cost | | 8,342 | 97,153 | 84,285 | 180,416 | 108,942 | 470,796 |
| 28 | Escalation | | | | | | |
| 29 | Overhead & profit | | 8,342 | 5,057 | 9,021 | 10,190 | 32,610 |
| 30 | Contingency | | | | | | 18,076 |
| 31 | Total Project Cost | | | | | | 521,482 |

## 10.4   DEFINITION OF COST CENTERS

The subdivisions into which the project is divided for detailed cost estimation purposes are variously referred to as

1.   Estimating accounts
2.   Line items
3.   Cost accounts
4.   Work packages

As has been noted previously, the estimating account is typically defined so as to provide target values for the cost accounts that will be used to collect as-built costs while the job is in progress. Therefore, the end item that is the focus of cost development in the estimating account is linked to a parallel cost account for actual cost informatiron collection during construction. The cost account expenditures developed from field data are compared with the estimated cost as reflected by the estimating account to determine whether costs are exceeding, underrunning, or coming in on project values. Therefore, the use of the term "cost account" is not strictly correct during the preparation of bid, since this account is not active until the job is in progress and actual cost data are available.

The term *work package* has become current in the past few years and is commonly used to indicate a subdivision of the project that is used both for cost control and scheduling (i.e., time control). When both cost and time control systems are combined into an integrated project management system, work packages are controlled to determine cost versus estimate and time versus schedule.

The subdividing of the project into work packages results in the definition of a work breakdown structure (WBS).

A work package is a well-defined scope of work that usually terminates in a deliverable product. Each package may vary in size, but must be a measurable and controllable unit of work to be performed. It also must be identifiable in a numerical accounting system in order to permit capture of both budgeted and actual performance information. A work package is a cost center.*

The breakdown of a project into estimating accounts or work packages (depending upon the sophistication of the system) is aided by a comprehensive chart of cost accounts or listing of typical work packages which can be used as a checklist. This checklist or template can be matched to the project being estimated to determine what types of work are present. That is, accounts in the general chart are compared to the project being estimated to determine which ones apply.

---

* James N. Neil, *Construction Cost Estimating for Project Control* (Englewood Cliffs, N.J.: Prentice-Hall, 1982), p. 73.

## 10.5   QUANTITY TAKEOFF

The development of the quantities of work to be placed in appropriate units (e.g., square feet, cubic yards, etc.) is referred to as the *quantity takeoff (QTO)* or *quantity surveying*.† The procedures employed by the estimator to calculate these quantities should incorporate steps to minimize errors. Five of the most common errors experienced during quantity takeoff are

1. Arithmetic: Errors in addition, subtraction, and multiplication.
2. Transposition: Mistakes in copying or transferring figures, dimensions, or quantities.
3. Errors of Omission: Overlooking items called for or required to accomplish the work.
4. Poor reference: Scaling drawings rather than using the dimensions indicated.
5. Unrealistic waste or loss factors.

The first step in the quantity takeoff procedure is to identify the materials required by each estimating account or work package. Once the types of materials are identified, relevant dimensions are recorded on a spread sheet so that quantity calculations in the required unit of measure can be made. Calculation of quantities by estimating account or work package has several advantages, not the least of which is the fact that it allows the estimating process to be performed by several estimators, each with a well-defined area of responsibility. No matter how competent an estimator may be in his/her own field, it is not reasonable to expect him/her to have an intimate knowledge of all phases of construction. This method enables one estimator to check another estimator's work. It also facilitates computations required to develop the financial "picture" of the job and processing of progress payment requests. When changes occur, only those activities affected must be recalculated. Other procedures require a completely new takeoff.

Before the calculations for the quantity takeoff are performed, detailed working drawings are sometimes required to clarify the contract drawings and specifications or the chosen construction method (e.g., forming techniques). Such a drawing for a small wall is given in Figure 10.6. During construction these details are of tremendous value to the person in the field who is trying to perform the work within the cost guidelines provided. From these drawings and details a checklist should be developed to indicate all of the materials required for each work package. After this checklist has been made, it should be checked against a standard checklist to identify errors of omission.

The actual calculations should be performed on a standard worksheet in a clear and legible format so as to allow for independent checks and self-checks. As the calculations for those items shown on the plans progress, each item taken off should be highlighted by a color marker so that those items remaining to be considered are obvious at a glance. Arithmetic should be performed on a calculator that produces a tape. The tape

---

† This term is commonly used in the United Kingdom and the Commonwealth countries.

## FIGURE 10.6   Small Wall Construction

Prepare:   Activity list
           Activity material list (estimate)—include work sheets
           Material recap sheets

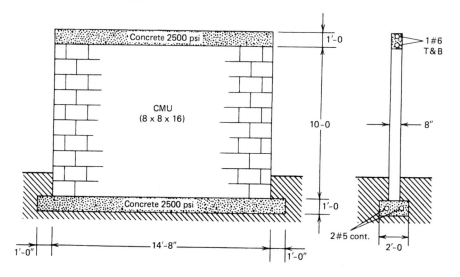

should be used by the estimator to identify errors. All tapes should be attached to the estimate to aid in checking by other sources, or at a later date. The quantities calculated should be exact. Waste and loss factors will be applied later. A materials takeoff sheet for the small wall (Figure 10.6) is given in Figure 10.7.

A summary or "recap sheet" should be made. This recap sheet should consist of a listing, by material type, of all the materials required for the entire work item or package. The listing should include total quantities as well as subquantities identified by activity code. The listing should also include appropriate waste and loss factor calculations. An example of a recap sheet is given in Figure 10.8. This example is simple, and is included only to demonstrate the nature of quantity development. In practice, such a detailed procedure may well be too cumbersome for a simple estimate item or work package. However, for more unique and special systems within the project, this approach may be the only means of achieving an accurate quantity estimate.

## 10.6   METHODS OF DETAILED COST DETERMINATION

After quantities have been determined for accounts that are relevant to the project at hand, the method by which costs will be assigned can be selected. The two methods

# FIGURE 10.7  Activity Material List

Activity material list

Project _____

| Activity Code | Activity Description | Material Description | Quantity | Unit | Cost Code |
|---|---|---|---|---|---|
| 1 | Layout | Stakes 2x4 x 24  8 ea. | 10.3 | BF | 0100 |
| | | | | | |
| 3 | Place rebar | #5 st.  2 PCS  16-2 | 32.3 | LF | 0320 |
| | | Tie wire | 1 | Roll | 0320 |
| | | | | | |
| 4 | Cast and Cure | footing | | | |
| | | Concrete | 1.23 | CY | 0330 |
| | | Curing compound | .25 | Gal | 0337 |
| | | | | | |
| 5 | Erect CMU wall | | | | |
| | | CMU 8x8 x16 strecher | 143 | Ea | 0412 |
| | | CMU 8x8 x16 corner | 14 | Ea | 0412 |
| | | CMU 8x8 x16 corner | 16 | Ea | 0412 |
| | | Scaffolding  4'x4'x6' | 2 | Sec. | 0100 |
| | | Mortar | .27 | CY | 0412 |
| 7 | Form bond beam | | | | |
| | | 2x4 (4-15'-0") | 43.5 | BF | 0310 |
| | | 2x2 | 12.7 | BF | 0310 |
| | | 1x2 | 2.0 | BF | 0310 |
| | | 3/4" ext ply | 60.3 | SF | 0310 |
| | | snapties 8" | 24 | Ea | 0310 |
| | | Nails 8d | 1.5 | Lb | 0310 |
| | | Nails 6d | .4 | Lb | 0310 |
| | | Form oil | .07 | Gal | 0310 |
| | | | | | |
| 8 | Place bond beam rebar | | | | |
| | | #6 rebar (str.) | 28.67 | LF | 0320 |
| | | | | | |
| 9 | Cast and cure | bond beam | | | |
| | | Concrete | .35 | CY | 0330 |
| | | Curing compound | .05 | Gal | 0337 |
| | | | | | |

(Continued)

# FIGURE 10.7   Activity Material List (Continued)

Activity material list

Project _____

| Activity Code | Activity Description | Material Description | Quantity | Unit | Cost Code |
|---|---|---|---|---|---|
| 10 | Strip forms and rub bond beam | | | | |
| | | Grout | 1 | CF | 0.339.2 |
| | | | | | |
| | | | | | |
| | | | | | |
| | | | | | |
| | | | | | |
| | | | | | |
| | | | | | |
| | | | | | |
| | | | | | |
| | | | | | |
| | | | | | |
| | | | | | |
| | | | | | |
| | | | | | |
| | | | | | |
| | | | | | |
| | | | | | |
| | | | | | |
| | | | | | |
| | | | | | |
| | | | | | |
| | | | | | |
| | | | | | |
| | | | | | |
| | | | | | |
| | | | | | |
| | | | | | |
| | | | | | |
| | | | | | |
| | | | | | |
| | | | | | |

# FIGURE 10.8 Construction Support Materials Recap Sheet

Construction support
materials recap sheet

Project _____ Wall _____

| Description | Activity code | Sub-quantity | Waste | Total quantity | Unit | | Cost code |
|---|---|---|---|---|---|---|---|
| | | | | | | | |
| | | | | | | | |
| 2X4 Lumber | Total | 53.8 | 10% | 60.0 | BF | | |
| | 1 | 10.3 | | | | | 0100 |
| | 7 | 43.5 | | | | | 0310 |
| | | | | | | | |
| 2X2 Lumber | 7 | 12.7 | 10% | 14.0 | BF | | 0310 |
| | | | | | | | |
| 1X2 Lumber | 7 | 2.0 | 10% | 2.25 | BF | | 0310 |
| | | | | | | | |
| 3/4" Exterior plywood | 7 | 60.3 | 10% | 66 | SF | | 0310 |
| | | | | | | | |
| Curing compound | Total | .30 | | 1 | Gal | | 0337 |
| | 4 | .25 | | | | | |
| | 9 | .05 | | | | | |
| | | | | | | | |
| Snap ties 8" | 7 | 24. | 5% | 25 | Ea | | 0310 |
| | | | | | | | |
| Nails 8d | 7 | 1.5 | | 3 | LB | | 0310 |
| | | | | | | | |
| Nails 6d | 7 | .4 | | 1 | LB | | 0310 |
| | | | | | | | |
| Form oil | 7 | .07 | | .25 | Gal | | 0310 |
| | | | | | | | |
| | | | | | | | |
| | | | | | | | |
| | | | | | | | |
| | | | | | | | |
| | | | | | | | |
| | | | | | | | |
| | | | | | | | |

of cost determination most frequently used are

1.  Unit pricing
2.  Resource enumeration

If the work as defined by a given estimating account is fairly standard, the cost can be calculated by simply taking *dollar per unit* cost from company records and applying this cost with a qualitative correction factor to the quantity of work to be performed. For instance, if the project calls for 100 lineal feet of pipe and historical data in the company indicate that the pipe can be placed for $65 a lineal foot to include labor and materials, the direct cost calculation for the work would yield a value of $6500. This value can then be adapted for special site conditions.

Unit pricing values are available in many standard estimating references.

The standard references normally give a nationally averaged price per unit. A multiplier is used to adjust the national price to a particular area. These references are updated on an annual basis to keep them current. Among the largest and best known of these services are

1.  *Dodge Construction Pricing and Scheduling Manual*
2.  R. S. Means, *Building Construction Cost Data*
3.  F. R. Walker's *The Building Estimator's Reference Book*
4.  The *Richardson General Construction Estimating Standards*

These references contain listings of cost *line items* similar to the cost account line items a contractor would maintain.*

A listing of the line items as given in the R. S. Means reference is shown in Figure 10.9. The development of direct costs to include overhead and profit for a particular line item using the R. S. Means system is shown in Figure 10.10.

The line items specified in the R. S. Means *Construction Cost Data* are defined by using the Uniform Construction Index numerical designators. The system assumes a given crew composition and production rate for each line item. In the case illustrated a standard crew designated C-1 can construct 190 SFCA (square foot contact area) of plywood column form per shift (daily output). This underlines the fact that unit pricing data must make some assumption regarding the resource group (i.e., crew, equipment fleet, etc.) and the production rate being used. That is, although unit pricing data are presented in dollars-per-unit format, the cost of the resource group and the rate of production achieved must be considered. The dollars-per-unit value is calculated as follows:

$$\frac{\text{Cost of resources per unit time}}{\text{Production rate of resources}} = \frac{\$/\text{hr}}{\text{unit/hr}} = \$/\text{unit}$$

---

* D. W. Halpin and R. W. Woodhead, *Construction Management* (New York: Wiley, 1980), pp. 271–272.

# FIGURE 10.9 Line Item Listing from R. S. Means

| 3.1 FORMWORK | | CREW | DAILY OUTPUT | UNIT | BARE COSTS MAT. | INST. | TOTAL | TOTAL INCL O&P |
|---|---|---|---|---|---|---|---|---|
| 320 | 30" diameter | C-1 | 70 | L.F. | 2 | 6.90 | 8.90 | 11.85 |
| 325 | 36" diameter | | 60 | | 2.25 | 8.05 | 10.30 | 13.75 |
| 330 | 48" diameter, heavy duty | | 50 | | 3.05 | 9.65 | 12.70 | 16.85 |
| 335 | 60" diameter, heavy duty | | 45 | | 4.40 | 10.70 | 15.10 | 19.85 |
| 400 | Column capitals, 4 use per mo., 24" column, 4' cap diameter | | 12 | Ea. | 10.75 | 40 | 50.75 | 68 |
| 405 | 5' cap diameter | | 11 | | 12.75 | 44 | 56.75 | 75 |
| 410 | 6' cap diameter | | 10 | | 17.75 | 48 | 65.75 | 87 |
| 415 | 7' cap diameter | | 9 | | 23 | 54 | 77 | 100 |
| 450 | For second and succeeding months, deduct | | | | 50% | | | |
| 500 (34) | Plywood, 8" x 8" columns, 1 use | C-1 | 165 | S.F.C.A. | 1.66 | 2.92 | 4.58 | 5.90 |
| 505 | 2 use | | 195 | | .89 | 2.47 | 3.36 | 4.45 |
| 510 (35) | 3 use | | 210 | | .64 | 2.30 | 2.94 | 3.92 |
| 515 | 4 use | | 215 | | .51 | 2.24 | 2.75 | 3.70 |
| 550 | 12" x 12" plywood columns, 1 use | | 180 | | 1.45 | 2.68 | 4.13 | 5.35 |
| 555 | 2 use | | 210 | | .80 | 2.30 | 3.10 | 4.10 |
| 560 | 3 use | | 220 | | .57 | 2.19 | 2.76 | 3.70 |
| 565 | 4 use | | 225 | | .46 | 2.14 | 2.60 | 3.51 |
| 600 | 16" x 16" plywood columns, 1 use | | 185 | | 1.36 | 2.61 | 3.97 | 5.15 |
| 605 | 2 use | | 215 | | .75 | 2.24 | 2.99 | 3.97 |
| 610 | 3 use | | 230 | | .54 | 2.10 | 2.64 | 3.53 |
| 615 | 4 use | | 235 | | .44 | 2.05 | 2.49 | 3.36 |
| 650 | 24" x 24" plywood columns, 1 use | | 190 | | 1.34 | 2.54 | 3.88 | 5.05 |
| 655 | 2 use | | 220 | | .74 | 2.19 | 2.93 | 3.89 |
| 660 | 3 use | | 235 | | .54 | 2.05 | 2.59 | 3.47 |
| 665 | 4 use | | 240 | | .44 | 2.01 | 2.45 | 3.30 |
| 700 | 36" x 36" plywood columns, 1 use | | 200 | | 1.58 | 2.41 | 3.99 | 5.10 |
| 705 | 2 use | | 230 | | .86 | 2.10 | 2.96 | 3.88 |
| 710 | 3 use | | 245 | | .62 | 1.97 | 2.59 | 3.44 |
| 715 | 4 use | | 250 | | .50 | 1.93 | 2.43 | 3.25 |
| 750 | Steel framed plywood, 4 use per mo., rent, 8" x 8" | | 290 | | .32 | 1.66 | 1.98 | 2.68 |
| 755 | 10" x 10" | | 300 | | .30 | 1.61 | 1.91 | 2.58 |
| 760 | 12" x 12" | | 310 | | .28 | 1.56 | 1.84 | 2.49 |
| 765 | 16" x 16" | | 335 | | .26 | 1.44 | 1.70 | 2.30 |
| 770 | 20" x 20" | | 350 | | .25 | 1.38 | 1.63 | 2.21 |
| 775 | 24" x 24" | | 365 | | .23 | 1.32 | 1.55 | 2.10 |
| 30 | FORMS IN PLACE, CULVERT 5' to 8' square or rectangular, 1 use | | 170 | | 1.27 | 2.84 | 4.11 | 5.35 |
| 005 | 2 use | | 170 | | .73 | 2.84 | 3.57 | 4.78 |
| 010 | 3 use | | 170 | | .55 | 2.84 | 3.39 | 4.58 |
| 015 (35) | 4 use | | 170 | | .46 | 2.84 | 3.30 | 4.48 |
| 35 | FORMS IN PLACE, ELEVATED SLABS | | | | | | | |
| 005 | See also corrugated form deck, division 5.2-30-610 | | | | | | | |
| 100 | Flat plate to 15' high, 1 use | C-2 | 470 | S.F. | 1.17 | 1.59 | 2.76 | 3.52 |
| 105 | 2 use | | 520 | | .66 | 1.44 | 2.10 | 2.75 |
| 110 | 3 use | | 545 | | .49 | 1.37 | 1.86 | 2.47 |
| 115 | 4 use | | 560 | | .40 | 1.34 | 1.74 | 2.31 |
| 150 | 15' to 20' high ceilings, 4 use | | 495 | | .50 | 1.51 | 2.01 | 2.67 |
| 160 | 21' to 35' high ceilings, 4 use | | 450 | | .60 | 1.66 | 2.26 | 2.99 |
| 200 | Flat slab with drop panels, to 15' high, 1 use | | 450 | | 1.50 | 1.66 | 3.16 | 3.98 |
| 205 | 2 use | | 505 | | .84 | 1.48 | 2.32 | 3 |
| 210 | 3 use | | 530 | | .62 | 1.41 | 2.03 | 2.66 |
| 215 | 4 use | | 545 | | .51 | 1.37 | 1.88 | 2.49 |
| 225 | 15' to 20' high ceilings, 4 use | | 480 | | .64 | 1.56 | 2.20 | 2.89 |
| 235 | 21' to 35' high ceilings, 4 use | | 435 | | .70 | 1.72 | 2.42 | 3.18 |
| 300 | Floor slab hung from steel beams, 1 use | | 485 | | 1.15 | 1.54 | 2.69 | 3.43 |
| 305 | 2 use | | 535 | | .65 | 1.40 | 2.05 | 2.68 |
| 310 | 3 use | | 550 | | .50 | 1.36 | 1.86 | 2.46 |
| 315 | 4 use | | 565 | | .41 | 1.33 | 1.74 | 2.31 |
| 350 (37) | Floor slab, with 20" metal pans, 1 use | | 350 | | 2.10 | 2.14 | 4.24 | 5.30 |
| 355 | 2 use | | 385 | | 1.25 | 1.94 | 3.19 | 4.10 |

*Source:* copyrighted by Robert Snow Means Co., Inc. It is reproduced from *Building Construction Cost Data 1981*, p. 53, with permission.

## FIGURE 10.10   Line Item Cost Development Using R. S. Means

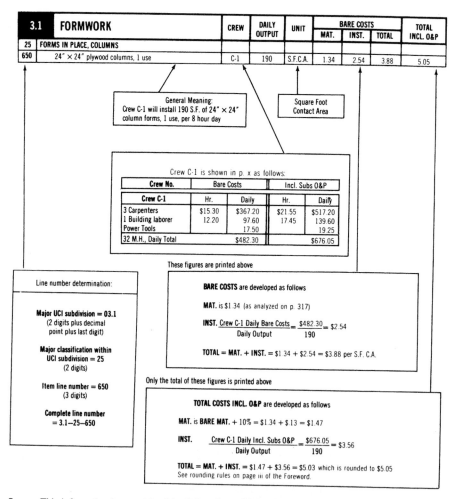

┌─────────────────────────────────────────────────────────────────┐
│                        **IMPORTANT**                              │
│                                                                   │
│       PRICES ARE LISTED TWO WAYS, ONE WITH BARE COSTS,            │
│      THE OTHER INCLUDES THE SUBCONTRACTOR'S OVERHEAD & PROFIT     │
│                                                                   │
│                     GENERAL RECOMMENDATION:                       │
│                                                                   │
│    When using total costs including Subcontractor's O & P, add 10%│
│               [see further discussion in Foreword]                │
└─────────────────────────────────────────────────────────────────┘

EXAMPLES OF HOW FIGURES ARE DEVELOPED
on p. 52 Under FORMWORK, column forms, 24″ × 24″ is shown as follows

| 3.1 | FORMWORK | CREW | DAILY OUTPUT | UNIT | BARE COSTS | | | TOTAL INCL. O&P |
|---|---|---|---|---|---|---|---|---|
| | | | | | MAT. | INST. | TOTAL | |
| 25 | FORMS IN PLACE, COLUMNS | | | | | | | |
| 650 | 24″ × 24″ plywood columns, 1 use | C-1 | 190 | S.F.C.A. | 1.34 | 2.54 | 3.88 | 5.05 |

General Meaning:
Crew C-1 will install 190 S.F. of 24″ × 24″ column forms, 1 use, per 8 hour day

Square Foot Contact Area

Crew C-1 is shown in p. x as follows:

| Crew No. | Bare Costs | | Incl. Subs O&P | |
|---|---|---|---|---|
| **Crew C-1** | Hr. | Daily | Hr. | Daily |
| 3 Carpenters | $15.30 | $367.20 | $21.55 | $517.20 |
| 1 Building laborer | 12.20 | 97.60 | 17.45 | 139.60 |
| Power Tools | | 17.50 | | 19.25 |
| 32 M.H., Daily Total | | $482.30 | | $676.05 |

These figures are printed above

Line number determination:

**Major UCI subdivision = 03.1**
(2 digits plus decimal point plus last digit)

**Major classification within UCI subdivision = 25**
(2 digits)

**Item line number = 650**
(3 digits)

**Complete line number = 3.1–25–650**

**BARE COSTS** are developed as follows

**MAT.** is $1.34 (as analyzed on p. 317)

**INST.** $\dfrac{\text{Crew C-1 Daily Bare Costs}}{\text{Daily Output}} = \dfrac{\$482.30}{190} = \$2.54$

**TOTAL = MAT. + INST.** = $1.34 + $2.54 = $3.88 per S.F. C.A.

Only the total of these figures is printed above

**TOTAL COSTS INCL. O&P** are developed as follows

**MAT.** is BARE MAT. + 10% = $1.34 + $.13 = $1.47

**INST.** $\dfrac{\text{Crew C-1 Daily Incl. Subs O&P}}{\text{Daily Output}} = \dfrac{\$676.05}{190} = \$3.56$

**TOTAL = MAT. + INST.** = $1.47 + $3.56 = $5.03 which is rounded to $5.05
See rounding rules on page iii of the Foreword.

*Source:* This information is copyrighted by Robert Snow Means Co., Inc. It is reproduced from *Building Construction Cost Data 1981*, p. 1, with permission.

The unit cost is the ratio of resource costs to production rate. The crew composition and assumed cost for the crew are shown in the middle of Figure 10.10.

In the R. S. Means system two costs are specified for each line item. The bare cost is the direct cost for labor and materials. The total cost includes the cost of burdens, taxes, and subcontractor overhead and profit (inclusive O&P). In Figure 10.9, the bare cost of the C-1 crew is calculated as $482.30 per shift. Therefore, the bare unit installation cost is

$$\frac{\$482.30/\text{shift}}{190 \text{ units/shift}} = \$2.54/\text{SFCA}$$

Combining this installation cost with the material cost per unit of $1.34 yields a bare unit cost for materials and installation of $3.88.

The overhead and profit charges associated with labor (as considered in the Means system) are

1. Fringe benefits (included in bare costs)
2. Workmen's compensation
3. Average fixed overhead
4. Subcontractor overhead
5. Subcontractor profit

These values, listed in Figure 10.11, vary from craft to craft. In order to adjust the bare costs of installation to include subcontractor's O&P (overhead and profit), the appropriate craft values for the members of the craft are located and applied. For the carpenters the total correction is 40.9 percent or $6.25 per hour. Therefore, the carpenter rate to include O&P is $21.55 per hour. Similarly, the laborer rate is adjusted to $17.45 per hour to include O&P. A markup of 10 percent is applied to the power tools, yielding a daily rate of $19.25. The new installation rate to include O&P is

$$\frac{\$676.05}{190} = \$3.56/\text{SFCA}$$

The 10 percent markup is applied to the material cost, resulting in a $1.47 charge per SFCA. The combined cost of materials and installation is $5.03 per square foot contact area (SFCA), which is rounded to $5.05.

## 10.7  PROBLEMS WITH UNIT-COST METHOD

The data that the contractor has available to him from company records are presented as dollars per unit, and in most cases no record of the crew composition, cost, and production rates is maintained. In fact, the dollars-per-unit value is typically an average of the values obtained on recent jobs. Since on each job the crew composition, costs, and production rates achieved are probably unique to the individual job, what the

# FIGURE 10.11   Subcontractor's Overhead and Profit

**⑥ SUBCONTRACTOR'S OVERHEAD & PROFIT** (Div. 1.1) p. 1 to 8

**1**

Listed below in the last two columns are average billing rates for Subcontractor's labor.

The Base Rate is for the building construction industry and includes the usual negotiated fringe benefits. These wage rates with trends are listed on the inside back cover. Workers' Compensation is a national average of states which have established rates in each trade. Average Fixed Overhead is a total of average rates for U.S. and State Unemployment, 4.95%; Social Security (FICA), 6.65%; Builders' Risk, .38%; and Public Liability, .80%. These are analyzed in ④ and ⑧. All the rates except Social Security

vary from state to state as well as from company to company.

The Subcontractor's Overhead presumes annual billing between 0.5 and 1.5 million dollars. A Subcontractor with lower annual billing usually has a higher overhead percentage.

Subcontractor's Overhead varies greatly within each trade. Some controlling factors are Annual Volume, Type Job, Size Job, Location, Local Economic Conditions, Engineering and Logistical Support Staff, and Equipment Requirements. These should be examined carefully for each job.

| Trade | Base Rate incl. fringes | | Workers' Comp. Ins. | Average Fixed Overhead | Subs Overhead | Subs Profit | Subs Total Overhead & Profit | | Rate with Subs O. & P. | |
|---|---|---|---|---|---|---|---|---|---|---|
| | Hourly | Daily | | | | | % | Amount | Hourly | Daily |
| Skilled Workers Average (35 trades) | $15.60 | $124.80 | 9.2% | 12.8% | 12.8% | 10% | 44.8% | $7.00 | $22.60 | $180.80 |
| Helpers Average (5 trades) | 12.10 | 96.80 | 8.8 | | 13 | | 44.6 | 5.40 | 17.50 | 140.00 |
| Foremen Average, Inside (50¢ over trade) | 16.10 | 128.80 | 9.2 | | 12.8 | | 44.8 | 7.20 | 23.30 | 186.40 |
| Foremen Average, Outside ($2.00 over trade) | 17.60 | 140.80 | 9.2 | | 12.8 | | 44.8 | 7.90 | 25.50 | 204.00 |
| Common Building Laborers | 12.20 | 97.60 | 9.4 | | 11 | | 43.2 | 5.25 | 17.45 | 139.60 |
| Asbestos Workers | 16.85 | 134.80 | 6.5 | | 16 | | 45.3 | 7.65 | 24.50 | 196.00 |
| Boilermakers | 17.20 | 137.60 | 7.8 | | 16 | | 46.6 | 8.05 | 25.25 | 202.00 |
| Bricklayers | 15.80 | 126.40 | 7.1 | | 11 | | 40.9 | 6.45 | 22.25 | 178.00 |
| Bricklayer Helpers | 12.50 | 100.00 | 7.1 | | 11 | | 40.9 | 5.10 | 17.60 | 140.80 |
| Carpenters | 15.30 | 122.40 | 7.1 | | 11 | | 40.9 | 6.25 | 21.55 | 172.40 |
| Cement Finishers | 15.05 | 120.40 | 5.3 | | 11 | | 39.1 | 5.90 | 20.95 | 167.60 |
| Electricians | 17.40 | 139.20 | 3.9 | | 16 | | 42.7 | 7.45 | 24.85 | 198.80 |
| Elevator Constructors | 16.95 | 135.60 | 7.0 | | 16 | | 45.8 | 7.75 | 24.70 | 197.60 |
| Equipment Operators, Crane or Shovel | 16.00 | 128.00 | 7.2 | | 14 | | 44.0 | 7.05 | 23.05 | 184.40 |
| Equipment Operators, Medium Equipment | 15.60 | 124.80 | 7.2 | | 14 | | 44.0 | 6.85 | 22.45 | 179.60 |
| Equipment Operators, Light Equipment | 14.75 | 118.00 | 7.2 | | 14 | | 44.0 | 6.50 | 21.25 | 170.00 |
| Equipment Operators, Oilers | 13.35 | 106.80 | 7.2 | | 14 | | 44.0 | 5.85 | 19.20 | 153.60 |
| Equipment Operators, Master Mechanics | 16.75 | 134.00 | 7.2 | | 14 | | 44.0 | 7.35 | 24.10 | 192.80 |
| Glaziers | 15.10 | 120.80 | 7.3 | | 11 | | 41.1 | 6.20 | 21.30 | 170.40 |
| Lathers | 14.85 | 118.80 | 4.5 | | 11 | | 38.3 | 5.70 | 20.55 | 164.40 |
| Marble Setters | 15.05 | 120.40 | 7.1 | | 11 | | 40.9 | 6.15 | 21.20 | 169.60 |
| Millwrights | 15.80 | 126.40 | 7.1 | | 11 | | 40.9 | 6.45 | 22.25 | 178.00 |
| Mosaic and Terrazzo Workers | 14.85 | 118.80 | 5.0 | | 11 | | 38.8 | 5.75 | 20.60 | 164.80 |
| Painters, Ordinary | 14.50 | 116.00 | 6.7 | | 11 | | 40.5 | 5.85 | 20.35 | 162.80 |
| Painters, Structural Steel | 15.10 | 120.80 | 28.6 | | 11 | | 62.4 | 9.40 | 24.50 | 196.00 |
| Paper Hangers | 14.65 | 117.20 | 6.7 | | 11 | | 40.5 | 5.95 | 20.60 | 164.80 |
| Pile Drivers | 15.40 | 123.20 | 17.9 | | 16 | | 56.7 | 8.75 | 24.15 | 193.20 |
| Plasterers | 14.85 | 118.80 | 6.5 | | 11 | | 40.3 | 6.00 | 20.85 | 166.80 |
| Plasterer Helpers | 12.70 | 101.60 | 6.5 | | 11 | | 40.3 | 5.10 | 17.80 | 142.40 |
| Plumbers | 17.45 | 139.60 | 4.6 | | 16 | | 43.4 | 7.55 | 25.00 | 200.00 |
| Rodmen (Reinforcing) | 16.60 | 132.80 | 16.9 | | 14 | | 53.7 | 8.90 | 25.50 | 204.00 |
| Roofers, Composition | 14.70 | 117.60 | 16.0 | | 11 | | 49.8 | 7.30 | 22.00 | 176.00 |
| Roofers, Tile & Slate | 14.80 | 118.40 | 16.0 | | 11 | | 49.8 | 7.35 | 22.15 | 177.20 |
| Roofer Helpers (Composition) | 11.00 | 88.00 | 16.0 | | 11 | | 49.8 | 5.50 | 16.50 | 132.00 |
| Sheet Metal Workers | 16.75 | 134.00 | 5.8 | | 16 | | 44.6 | 7.45 | 24.20 | 193.60 |
| Sprinkler Installers | 17.30 | 138.40 | 4.6 | | 16 | | 43.4 | 7.50 | 24.80 | 198.40 |
| Steamfitters or Pipefitters | 17.50 | 140.00 | 4.6 | | 16 | | 43.4 | 7.60 | 25.10 | 200.80 |
| Stone Masons | 15.75 | 126.00 | 7.2 | | 11 | | 41.0 | 6.45 | 22.20 | 177.60 |
| Structural Steel Workers | 16.70 | 133.60 | 22.2 | | 14 | | 59.0 | 9.85 | 26.55 | 212.40 |
| Tile Layers (Floor) | 14.85 | 118.80 | 5.0 | | 11 | | 38.8 | 5.75 | 20.60 | 164.80 |
| Tile Layer Helpers | 12.00 | 96.00 | 5.0 | | 11 | | 38.8 | 4.65 | 16.65 | 133.20 |
| Truck Drivers, Light | 12.35 | 98.80 | 5.9 | | 11 | | 39.7 | 4.90 | 17.25 | 138.00 |
| Truck Drivers, Heavy | 12.60 | 100.80 | 5.9 | | 11 | | 39.7 | 5.00 | 17.60 | 140.80 |
| Welders, Structural Steel | 16.70 | 133.60 | 22.2 | | 14 | | 59.0 | 9.85 | 26.55 | 212.40 |
| *Wrecking | 12.20 | 97.60 | 33.5 | | 11 | | 67.3 | 8.20 | 20.40 | 163.20 |

*Not included in Averages.

*Source:* This informaiton is copyrighted by Robert Snow Means Co., Inc. It is reproduced from *Building Construction Cost Data 1981* p. 301, with permission.

figure represents is an aggregate cost per unit. The actual number of man-hours used and the productivity achieved are masked by the dollar-per-unit figure. Unless the resource (i.e., man-hour, etc.) information is kept separately, it has been lost. Therefore, the unit price available from averaging values on previous jobs has to be treated with some caution. Since every job is unique, some of the estimator's intuition must be applied to insure that the value is adapted to the conditions of the job being estimated. If the conditions of jobs vary very little, however, the application of the unit pricing approach is both practical and efficient.

Clearly, the numerator (cost of resources per unit time) of the unit-cost ratio will vary significantly over time as the costs of labor and machines vary. The costs of all components of the construction process have risen sharply over the past 15 years. This is shown dramatically in the *Engineering News Record* construction and building cost indexes shown in Figure 10.12. In order to factor out the inflationary escalation inherent in resource costs, some contractors maintain the ratio of man-hours or resource-hours per hour to production. This establishes a company data base tied to *resource-hours required* rather than dollars per unit. Therefore, the contractor can retrieve a man-hour or resource-hour per-unit value for each line item. The value is calculated as

$$\frac{\text{Resource-hours per hour}}{\text{Units per hour}} = \text{RH/unit}$$

The cost per unit can then be calculated by multiplying the *resource-hours per unit* value by the average hourly cost per resource. If it takes 25 resource-hours per unit and the average cost of a resource-hour is $8.25, the unit cost will be $406.25 per unit. This method recognizes that the number of resource-hours required per unit is much more stable over the years than the cost per unit. Therefore, data on resource-hours per unit collected over several years will not be affected by inflationary trends and escalation in the cost of goods and services.

Use of the unit-pricing approach assumes that historical data have been maintained for commonly encountered cost accounts. Data are collected and linked to a reference unit such as a cubic yard or square foot. The costs of materials and installation are aggregated and then presented as a cost per unit. Companies typically accumulate such data either manually or on the computer as a by-product of the job cost system. On a typical job 80 to 90 percent of the work to be accomplished can be estimated by calculating the number of reference units and multiplying this number by the unit price. Typically the estimator will intuitively adjust this price to reflect special characteristics of the job, such as access restrictions, difficult management environment, and the like. One approach to the quantification of these site and job unique factors is proposed by Louis Dallavia. Although the Dallavia method is dated, it does reflect in an approximate way the factors that are considered by an estimator in adjusting general unit prices to a given project. The system defines a *percent efficiency factor* based on a production range index for each of eight job characteristics. The method of calculating the percent efficiency factor and the table production range indices are shown in Figure 10.13.

**FIGURE 10.12  Engineering News Record Construction Cost Indexes**

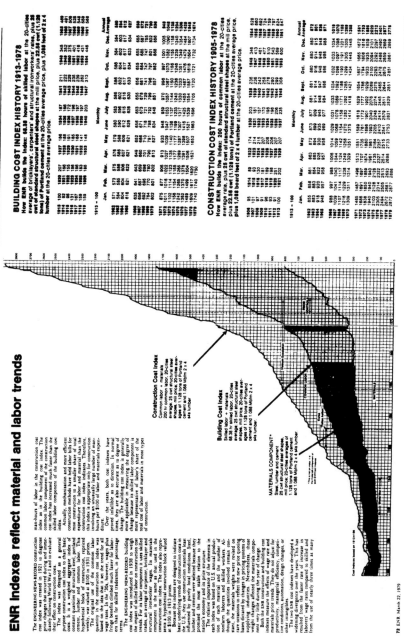

*Source:* Courtesy of *Engineering News Record* (March 22, 1979).

# FIGURE 10.13   Dallavia Method

## Production Range Index

### Production Effiency (%)

| Production Elements | 25   35 | 45   55   65   75   85   95 | 100 |
|---|---|---|---|
| | Low | Average | High |
| 1 *General Economy* | Prosperous | Normal | Hard times |
| Local business trend | Stimulated | Normal | Depressed |
| Construction volume | High | Normal | Low |
| Unemployment | Low | Normal | High |
| 2 *Amount of Work* | Limited | Average | Extensive |
| Design areas | Unfavorable | Average | Favorable |
| Manual operations | Limimted | Average | Extensive |
| Mechanized operations | Limited | Average | Extensive |
| 3 *Labor* | Poor | Average | Good |
| Training | Poor | Average | Good |
| Pay | Low | Average | Good |
| Supply | Scarce | Average | Surplus |
| 4 *Supervision* | Poor | Average | Good |
| Training | Poor | Average | Good |
| Pay | Low | Average | Good |
| Supply | Scarce | Average | Surplus |
| 5 *Job Conditions* | Poor | Average | Good |
| Management | Poor | Average | Good |
| Site and materials | Unfavorable | Average | Favorable |
| Workmanship required | First rate | Regular | Passable |
| Length of operations | Short | Average | Long |
| 6 *Weather* | Bad | Fair | Good |
| Precipitation | Much | Some | Occasional |
| Cold | Bitter | Moderate | Occasional |
| Heat | Oppressive | Moderate | Occasional |
| 7 *Equipment* | Poor | Normal | Good |
| Applicability | Poor | Normal | Good |
| Condition | Poor | Fair | Good |
| Maintenance, repairs | Slow | Average | Quick |
| 8 *Delay* | Numerous | Some | Minimum |
| Job flexibility | Poor | Average | Good |
| Delivery | Slow | Normal | Prompt |
| Expediting | Poor | Average | Good |

*Example:*

After studying a project on which he is bidding, a contractor makes the following evaluations of the production elements involved:

| Production Element | % Efficiency |
|---|---|
| 1. Present economy | 75 |
| 2. Amount of work | 90 |
| 3. Labor | 70 |
| 4. Supervision | 80 |
| 5. Job conditions | 95 |
| 6. Weather | 85 |
| 7. Methods and equipment | 55 |
| 8. Delays | 75 |
| Total | 625 |

As the total of the eight elements is 625, the average value will be 625/8, or 78 percent.

*(continued)*

## FIGURE 10.13   Dallavia Method (Continued)

Detailed estimates must be made for the individual operations required by that project. Each of these operations must be evaluated in a similar manner and adjusted against the overall rating before the production of the typical shift crew can be determined in actual units of scheduled work. In making such evaluations, those production elements which do not seem directly applicable to a particular operation may be ignored.

*Source:* Louis Dallavia, *Estimating General Construction Costs* (New York: F. W. Dodge, 1957). Used with permission of McGraw-Hill Book Company.

## 10.8   RESOURCE ENUMERATION

Although the unit-pricing approach is sufficiently accurate to estimate the common accounts encountered on a given project, almost every project has unique or special features for which unit-pricing data may not be available. Unusual architectural items that are unique to the structure and require special forming or erection procedures are typical of such work. In such cases, the price must be developed by breaking the special work item into its subfeatures and assigning a typical resource group to each subfeature. The productivity to be achieved by the resource group must be estimated by using either historical data or engineering intuition. The breakdown of the cost center into its subelements would occur much in the same fashion in which the wall of Section 10.5 was subdivided for quantity development purposes. The steps involved in applying the resource enumeration approach are shown in Figure 10.14.

An example of resource enumeration applied to a concrete-placing operation is shown in Figure 10.15.

In this example a concrete placement crew consisting of a carpenter foreman, two cement masons, a pumping engineer (for operation of the concrete pump), and seven laborers for placing, screeding, and vibrating the concrete has been selected. A concrete pump (i.e., an equipment resource) has also been included in the crew. Its hourly cost has been determined using methods described above. The total hourly rate for the crew is found to be $80.60. The average assumed rate of production for the crew is 12 cu yd/hr. This results in an average labor cost per cubic yard of concrete of $6.72. The line items requiring concrete are listed with the quantities developed from the plans and specifications. Consider the first item that pertains to foundation concrete. The basic quantity is adjusted for material waste. The cost per unit is adjusted to $7.47 based on an efficiency factor for placement of foundation concrete estimated as 90% (0.90).*

The resource enumeration approach has the advantage over unit pricing in that it allows the estimator to stylize the resource set or crew to be used to the work in question. The rates of pay applied to the resource group reflect the most recent pay and charge rates, and therefore incorporate inflationary or deflationary trends into the calculated price. The basic equation for unit pricing is

---

* D. W. Halpin, and R. W. Woodhead, *Construction Management* (New York: Wiley, 1980), p. 283.

**FIGURE 10.14   The Resource Enumeration Method of Estimating**

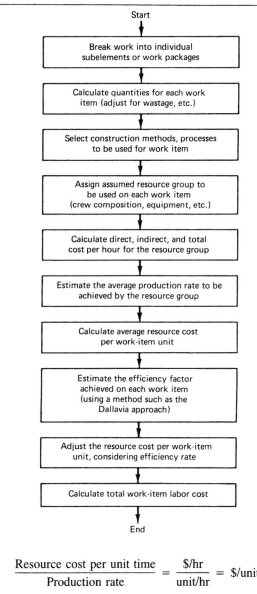

$$\frac{\text{Resource cost per unit time}}{\text{Production rate}} = \frac{\$/\text{hr}}{\text{unit/hr}} = \$/\text{unit}$$

In the unit-pricing approach the resource costs and the production rates are the aggregate values of resources and rates accumulated on a number of jobs over the period of historical data collection. With the resource enumeration approach, the estimator specifies a particular crew or resource group at a particular charge rate and a particular production level for the specific work element being estimated. This should yield a

## FIGURE 10.15   Labor Resource Enumeration

*Concrete Placing Crew*

| Quantity | Member | Rate | Total/ Hour |
|---|---|---|---|
| 1 | Carpenter foreman | $9.42 | $ 9.42 |
| 2 | Cement masons | 8.36 | 16.72 |
| 1 | Pumping engineer | 8.45 | 8.45 |
| 7 | Laborers | 5.43 | 38.01 |
| 1 | Concrete pump | 8.00 | 8.00 |
| | | Crew hourly rate | $80.60 |

Production rate of crew under normal circumstances (efficiency factor 1) = 12 cu yd/hr.
Average labor cost/cubic yard = $80.60/12 = $6.72.

| Area | Quantity | Percent Waste | Efficiency Factor | Labor Cost/ Cubic Yard | Activity Cost |
|---|---|---|---|---|---|
| 1. Foundation | 53.2 | 15 | 0.9 | $ 7.47 | $ 398 |
| 2. Wall to elevation 244.67 | 52.9 | 12 | 0.8 | 8.40 | 445 |
| 3. Slab 10 in. | 1.3 | 30 | 0.3 | 22.40 | 30 |
| 4. Beams elevated 244.67 | 10.5 | 15 | 0.7 | 9.60 | 101 |
| 5. Beams elevated 245.17 | 9.1 | 15 | 0.7 | 9.60 | 88 |
| 6. Slab elevation 244.67 | 8.7 | 10 | 0.7 | 9.60 | 84 |
| 7. Interior wall to 244.67 | 5.5 | 15 | 0.4 | 16.80 | 93 |
| 8. Slab elevation 254.17 | 6.3 | 10 | 0.75 | 8.96 | 57 |
| 9. Walls 244.67 − 254.17 | 57.2 | 10 | 0.8 | 8.40 | 481 |
| 10. Walls 254.17 − 267 | 42.0 | 10 | 0.8 | 8.40 | 353 |
| 11. Floors elevated 267 | 8.9 | 10 | 0.9 | 7.47 | 67 |
| 12. Manhole walls | 27.3 | 10 | 0.85 | 7.91 | 216 |
| 13. Roof | 14.0 | 15 | 0.7 | 9.60 | 135 |
| 14. Headwall | 8.5 | 10 | 0.8 | 8.40 | 72 |
| Total direct labor cost for concrete | | | | | $2620 |

*Source:* D. W. Halpin and R. W. Woodhead, *Construction Management,* copyright © 1980 by John Wiley and Sons, Inc. Reprinted by permission of John Wiley and Sons, Inc.

much more precise cost-per-unit definition. The disadvantage with such a detailed level of cost definition is the fact that it is time-consuming. Therefore, resource enumeration would be used only on (1) items for which no unit-cost data are available (2) "big-ticket" items, which constitute a large percentage of the overall cost of the job and for which such a precise cost analysis may lead to cost savings that may provide the winning margin at bid time, or (3) extremely complex work items on complicated and unique projects for which the use of the unit-pricing approach is deemed inadequate.

# FIGURE 10.16    Construction Systems Concept—Concrete Footing

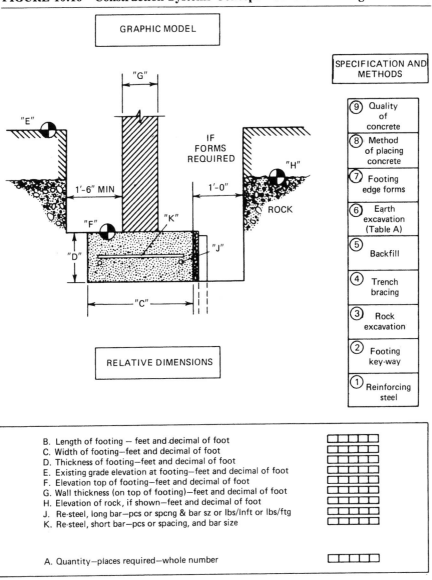

GRAPHIC MODEL

SPECIFICATION AND METHODS

9. Quality of concrete
8. Method of placing concrete
7. Footing edge forms
6. Earth excavation (Table A)
5. Backfill
4. Trench bracing
3. Rock excavation
2. Footing key-way
1. Reinforcing steel

RELATIVE DIMENSIONS

B. Length of footing — feet and decimal of foot
C. Width of footing—feet and decimal of foot
D. Thickness of footing—feet and decimal of foot
E. Existing grade elevation at footing—feet and decimal of foot
F. Elevation top of footing—feet and decimal of foot
G. Wall thickness (on top of footing)—feet and decimal of foot
H. Elevation of rock, if shown—feet and decimal of foot
J. Re-steel, long bar—pcs or spcng & bar sz or lbs/lnft or lbs/ftg
K. Re-steel, short bar—pcs or spacing, and bar size

A. Quantity—places required—whole number

## 10.9 WORK-PACKAGE OR ASSEMBLY-BASED ESTIMATING

In this approach to estimate development, a work package or assembly that is commonly encountered in construction is viewed as an estimating group, and appropriate dimensional and cost related parameters are defined for the package. The wall of Figure 10.6 could be considered an assembly. In this case, the height, width, and depth of the footer, block portion, and cap beam would be specified each time the assembly is encountered. Pricing information for the defined wall would be retrieved from a pricing catalogue. Since the reference subelement in this approach is the work package, an extensive listing of assemblies or packages into which the work can be subdivided is maintained.

A concrete footer assembly is shown in Figure 10.16. The relevant data required for takeoff are the dimensional values shown as items A through K. Data regarding

**FIGURE 10.17   Work Package Concept**

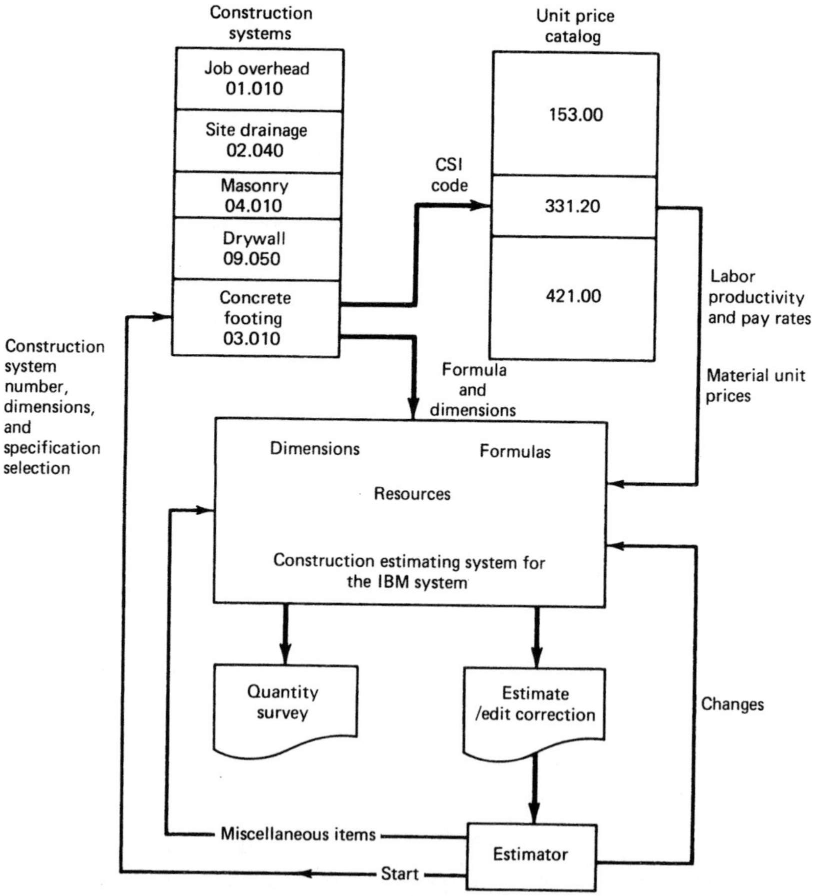

# FIGURE 10.18  Work Package Collection Sheet—Concrete Slab

**WORK PACKAGE COLLECTION SHEET**

System/Structure Identifier: 0 2 • 1 3 3
Crew Level Work Package Identifier: 0 3 • 1 3 1
Description: CONCRETE PLACEMENT, FLOAT FINISH — GROUND SLAB, BUILDING 2

**Productivity**

Base Unit for Productivity: CY CONCRETE

Total Quantity Base Unit: 128 CY

| Duration (Crew-Hours) | Low | Target | High |
|---|---|---|---|
| | | 8 | |

Escalation Rates (%)

Materials ☐
Labor ☐
Equipment ☐

## Permanent Materials (PM)

| Resource Code | Description | Unit | QTY | Unit Cost Low | Target | High | Extension |
|---|---|---|---|---|---|---|---|
| 1 3 2 5 | CONCRETE, 2500 PSI | CY | 135* | | 30 90 | | 4171 50 |
| | * 5% WASTE INCLUDED | | | | | | |
| | | | | | | | |
| | | | | | | | |

**Other Materials and Supplies (M&S)**

| | | | | | | | |
|---|---|---|---|---|---|---|---|
| | | | | | | | 4171 50 |

**Installed Equipment (IE)**

| | | | | | | | |
|---|---|---|---|---|---|---|---|

**Crew Labor (L)**

| Resource Code | Crew Labor (L) | NR. | Cost/Hour Low | Target | High | Extension |
|---|---|---|---|---|---|---|
| 0 2 9 1 | FORMAN | 1 | | 10 90 | | 87 20 |
| 0 2 9 3 | LABORERS | 4 | | 10 40 | | 332 80 |
| 0 2 9 2 | FINISHER | 1 | | 12 85 | | 411 20 |
| | | | | | | 831 20 |

**AVAILABILITY**

| Equipment Not Charged As Job Indirects | NR. | Periods | $/Period | Hours | USE | $/Hour | |
|---|---|---|---|---|---|---|---|
| 7 3 1 1  VIBRATOR, 6ED | 2 | ( × ) | + ( 8 × 3.25 ) = | | | | 52 00 |
| 7 3 1 9  HAND TOOLS | 1 | ( × ) | + ( 8 × 1.00 ) = | | | | 8 00 |
| 7 3 1 2  FINISHER, 6ED | 1 | ( × ) | + ( 8 × 4.50 ) = | | | | 36 00 |
| | | ( × ) | + ( × ) = | | | | 96 00 |

**Notes**

DURATION = $\frac{128 \text{ CY}}{22 \text{ CY/HR}}$ = 6 HRS

ALLOW 8 HRS WITH STARTUP AND CLEANUP

**Cost Summary:**
PM = 4171 50
M&S =
IE =
L = 831 20
CE = 96 00
TOTAL = 5098 70

*Source:* J. M. Neil, *Construction cost Estimating and Cost Control*, Englewood Cliffs, N.J.: Prentice-Hall, 1982), p. 231.

# FIGURE 10.19 Work Package Collection sheet—Excavation

## WORK PACKAGE COLLECTION SHEET

System/Structure Identifier: `01` • `300`  Crew Level Work Package `02` • `111` Identifier  Description: EXCAVATION, AREA 3, AND DISPOSAL W/O COMPACTION

### Permanent Materials (PM)

| Resource Code | Description | Unit | QTY | Unit Cost Low | Unit Cost Target | Unit Cost High | Extension |
|---|---|---|---|---|---|---|---|
| | | | | | | | |

### Other Materials and Supplies (M&S)

| | | | | | | | |
|---|---|---|---|---|---|---|---|

### Installed Equipment (IE)

| | | | | | | | |
|---|---|---|---|---|---|---|---|

### Crew Labor (L)

| | | Crew Labor (L) | NR. | Cost/Hour Low | Cost/Hour Target | Cost/Hour High | Extension |
|---|---|---|---|---|---|---|---|
| 0 8 | 1 1 | FOREMAN | 1 | | 13 90 | | 444 80 |
| 0 8 | 1 2 | EQUIP OPER, MEDIUM | 10 | | 13 40 | | 4288 00 |
| 0 8 | 1 3 | SPOTTER | 2 | | 10 40 | | 665 00 |
| | | | | | | | 5398 40 |

### Equipment Not Charged As Job Indirects

| | | | NR. | AVAILABILITY Periods | $/Period | Hours | USE $/Hour | |
|---|---|---|---|---|---|---|---|---|
| 7 3 | 2 2 | SCRAPER, ELEV 22 CY | 6 | ( 1 ) × ( 1850 — ) | + | ( 32 ) × ( 24.00 ) | = 15,708 00 |
| 7 2 | 0 7 | DOZER, D7 | 3 | ( 1 ) × ( 1600 — ) | + | ( 32 ) × ( 11.25 ) | = 5,808 00 |
| 7 4 | 1 2 | GRADER, 12' | 1 | ( 1 ) × ( 1000 — ) | + | ( 32 ) × ( 8.10 ) | = 1,259 20 |
| | | | | ( ) × ( ) | + | ( ) × ( ) | = |
| | | | | | | | 22,847 20 |

### Productivity

Base Unit for Productivity: BANK CY

Total Quantity Base Unit: 24,000 BCY

| Duration - (Crew-Hours) | Low | Target | High |
|---|---|---|---|
| | | 32 | |

### Escalation Rates (%)

| Materials | |
|---|---|
| Labor | |
| Equipment | |

### Notes

L.F. = 0.8

CYCLE = 6.3 MIN; ASSUME 45 MIN HOUR

$BCY/HR = \left(\frac{45 MIN}{6.3 MIN}\right)(22 CY)(0.8 LF)(6 SCRAPERS)$

= 754 BCY/HR

$DURATION = \frac{24,000 \; BCY}{754 \; BCY} = 32 \; HOURS$

ADD 1 DAY FOR BAD WEATHER

∴ PROJECT WILL LAST ONE WORK WEEK

### Cust Summary:

PM = 

M&S = 

IE = 

L = 5,398.40

CE = 22,847.20

TOTAL = 28,245.60

Source: J. M. Neil, *Construction Cost Estimating and Cost Control*, (Englewood Cliffs, N.J.: Prentice-Hall, 1982), p. 221.

the methodology of placement and the relevant specification are indicated by items (1) to (9) in the figure. Such work-package-based systems can be considered a structured extension of the resource enumeration approach and can be calculated manually. In general, most of these system-based (i.e., assembly-based) systems are computerized and are operated by presenting the estimator with individual assemblies. The estimator is interrogated by the computer and provides the dimensional and methodology information in a question-and-answer format. This procedure is shown schematically in Figure 10.17. The estimator goes through the construction systems sequentially, selecting those that are relevant and providing the required data. These data are integrated with information from a pricing catalogue. The pricing catalogue allows for price, resource, and productivity adjustment. The manual or computer program integration of these data produces the estimate reports required for bid preparation.

If a manual approach is used to estimate each work package, a work package takeoff sheet is helpful in organizing the collection of data. Such a work-package or assembly-collection sheet can be organized as shown in Figure 10.18. This form illustrates the development of an estimate for slab on grade in a building project. Material, labor, and equipment resources required for the package are shown on the left side of the sheet, together with target prices for each resource. It is interesting to note that the equipment resources are normally charged on a period basis, since partial-day allocation of equipment is not a common practice. The right side of the sheet considers the productivity rate to be used, special notes or characteristics relating to the package, and a total cost summary for the package. This sheet is quite versatile and can be used for earth work, masonry, and virtually any assembly encountered in the construction of a project. A similar sheet for earth work is shown in Figure 10.19.

## 10.10  SUMMARY

Chapters 9 and 10 have introduced concepts and topics related to the estimating and bidding processes. The estimate is the basis for the contractor's bid and, as such, has a significant effect upon whether or not a given project is profitable. In building construction the four levels of estimate preparation are

1. Conceptual
2. Preliminary
3. Engineer's
4. Bid

The first three of these estimates are typically prepared by the architect/engineer and reflect the increasing refinement of the design. Large and complex projects include a magnitude and definitive estimate in addition to those noted above. The bid estimate is a detailed estimate prepared by the contractor. The steps involved in preparing a detailed estimate are shown graphically in Figure 10.20. The project to be estimated

**FIGURE 10.20   Steps in the Estimating Process**

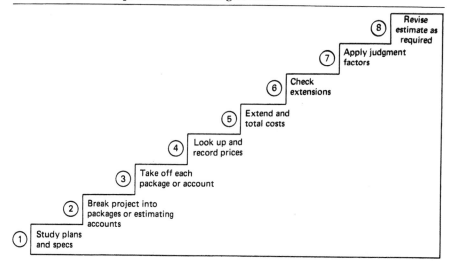

is subdivided for cost analysis purposes into estimating accounts or work packages. Quantities for each package or account are developed. These quantities are priced, and the extensions are calculated and checked for errors. At this stage professional judgment and engineering intuition are utilized to adjust the bid to reflect special or unique factors peculiar to the particular job. Profit margin is also applied at this point, and the bid is revised as required and finalized. Steps 3 through 6 are quantitative in nature and involve the application of formulas and arithmetical concepts. Steps 1 and 2 require professional expertise. Steps 7 and 8 require a good deal of experience and engineering judgment.

In the development of the estimate three methods are commonly used. These are:

1. Unit pricing or catalogue lookup method
2. Resource enumeration
3. Work-package/assembly method

The work-package method can be thought of as an extension of the resource-enumeration method. Different methods may be used on different parts of the job. On portions of the job that are very cost-sensitive and constitute a large portion of the overall project, methods 2 and 3 may be appropriate. On parts of the work that are standard and straightforward, the unit pricing approach may be acceptable. Selection of methods is a tradeoff between the need for accuracy and the cost of obtaining that accuracy. The keys to successful estimating are (1) the ability to assess the required level of accuracy and (2) the ability to achieve the required level of accuracy at minimal cost.

## REVIEW QUESTIONS AND EXERCISES

**10.1** Explain the difference between unit-cost estimating methods and resource-enumeration methods. When would you use unit cost? When would you use resource enumeration?

**10.2** What is meant by contractor O&P? Give three components that are considered in the O&P.

**10.3** What is the difference between labor cost and labor productivity? Use a sketch to illustrate.

**10.4** The partition wall shown is to be constructed of 8x16x6 block. Estimate the cost of the wall to include labor, materials, and contractor O&P using the R. S. Means building cost data or other appropriate estimating reference. The job is located in Cincinnati, Ohio.

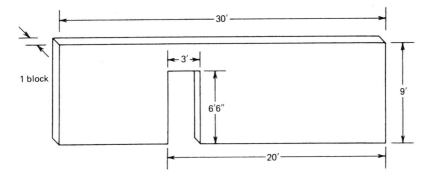

**10.5** You are excavating a location for the vault shown below. The top of the walls shown is one foot below grade. All slopes of the excavation are 3/4 to 1 to a toe one foot outside the base of the walls. The walls sit on a slab one foot in depth. Draw a sketch of the volume to be excavated, break it into components, and calculate the volume. Assume that a front loader (bucket) is supporting a fleet of trucks hauling material to an off-site fill location. The travel time to fill is six minutes and return travel takes five minutes. The dump time is one minute and the load time is 2.5 minutes. Two trucks (12 yard capacity each) are used. Assume that the material has a 30 percent swell (i.e., it bulks to a volume 30 percent larger than the in situ volume). What is the maximum hourly rate of production of this system (check whether loader or trucks limit the system)? How many hours are required to excavate the material for the vault? Disregard any backfill requirements.

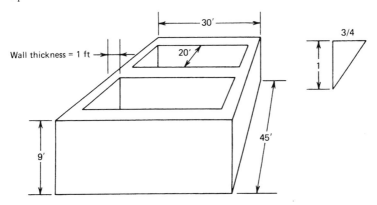

**10.6**  Given the bridge abutment shown below, determine the number of man-hours required to form the structure using data from R. S. Means or other appropriate reference.

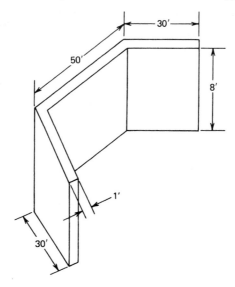

**10.7**  Develop a preliminary work-breakdown structure (WBS) for a small one-story commercial building to be constructed on the site of an existing small-frame structure. It is 30 by 60 feet in plan (see illustration below). The exterior and interior walls are of concrete block. The roof is constructed of bar joists on long-span bar joists covered with a steel roof deck, rigid insulation, and built-up roofing. The ceiling is suspended acoustical tile. The floor is a concrete slab on grade with an asphalt tile finish. Interior finish on all walls is paint.

(a)  Show the first level of this structure for the total project (WBS is developed from top to the bottom).

(b)  Select one work package (or building system) of the first level and develop the second level structure for this work package.

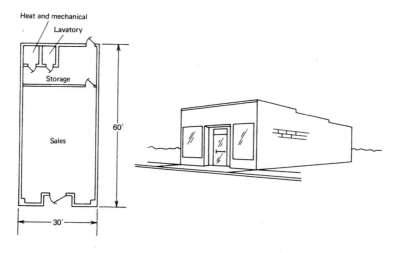

# Chapter 11

# TIME AND COST INTEGRATION

## 11.1 THE CONTROL BUDGET

The idea of cost planning implies the establishment of a framework within which costs can be controlled. This requires the restructuring of the cost data developed as part of the estimate. The framework for cost control that develops is commonly referred to as the *control budget* or the *project budget*. The budget contains the target values that will be used as guides to determine whether the project costs are developing as expected. In some cases, these target values are shown in the budget as values associated with each estimating account or work package without any relationship to the time at which the expenses will be incurred. In other cases the budget reflects not only the expected cost of project subelements but also the rate of expenditure over time. This approach implies the integration of cost and time. That is to say, it implies the integration of the estimate and the schedule.

Budgeting is the process of allocating the resources associated with the project (i.e., money, men, machines, and materials) to cost centers and comparing actual resource expenditure versus planned expenditure. The format in which the estimate has been developed may not be acceptable for this comparison. In such cases, the estimate is reworked to simplify the comparison of actual with projected cost values. The basis for control is usually the cost accounts that are relevant to the job. It will be recalled that the estimating accounts used for preparation of the estimate are essentially identical to the cost accounts. If the cost accounts are used as the basis for control, the reworking of the estimate developed in terms of estimating accounts is relatively straightforward. During recent years, the trend in estimating has been to reduce the amount of work required to restructure the estimate into an acceptable cost control vehicle so that the final estimate is, in fact, a control budget. An example of a project budget for a highway bridge is shown in Figure 11.1. In this case, cost targets associated with subelements of the project are organized in terms of the cost codes listed at the left side of the figure.

With increasing project complexity, scheduling has become more and more important. Increasing emphasis has been placed on the control of costs as a function of time. This has introduced a new element into the budgeting process—namely, that of the

**FIGURE 11.1   Highway Bridge Project**

| | PROJECT BUDGET | | | | | | |
|---|---|---|---|---|---|---|---|
| Job  Highway Bridge | | | | | | Estimator   G.A.S. | |
| Cost Code | Work Type | Quantity | Material Cost | Direct Labor Cost | Labor Unit Cost | Equipment Cost | Equipment Unit Cost |
| | Sitework | | | | | | |
| 02220.10 | Excavation, unclassified | 1,667 cy | $ — | $ 965 | $ 0.58 | $ 619 | $ 0.37 |
| 02222.10 | Excavation, structural | 120 cy | — | 1,112 | 9.27 | 390 | 3.25 |
| 02226.10 | Backfill, compacted | 340 cy | — | 1,190 | 3.50 | 323 | 0.95 |
| 02350.00 | Piledriving rig, mobilize and demobilize | job | 100 | 1,507 | 1,507.00 | 1,840 | 1,840.00 |
| 02361.10 | Piling, wood, driving | 2,240 lf | 6,944 | 2,122 | 0.95 | 1,912 | 0.85 |
| | Concrete | | | | | | |
| 03150.10 | Footing forms, fabricate | 360 sf | 137 | 468 | 1.30 | — | — |
| 03150.20 | Abutment forms, fabricate | 1,810 sf | 1,246 | 1,520 | 0.84 | — | — |
| 03157.10 | Footing forms, place and strip | 720 sf | 14 | 245 | 0.34 | — | — |
| 03157.20 | Abutment forms, place and strip | 3,620 sf | 290 | 3,819 | 1.05 | 1,320 | 0.36 |
| 03157.30 | Deck forms, place and strip | 1,800 sf | 1,044 | 2,124 | 1.18 | — | — |
| 03200.10 | Steel, reinforcing, place | 90,000 lb | 15,750 | 7,200 | 0.08 | 1,800 | 0.02 |
| 03251.10 | Concrete, deck, saw joints | 60 lf | — | 75 | 1.25 | 186 | 3.10 |
| 03311.10 | Concrete, footing, place | 120 cy | 4,473 | 401 | 3.34 | 295 | 2.46 |
| 03311.20 | Concrete, abutment, place | 280 cy | 10,437 | 3,107 | 11.10 | 1,834 | 6.55 |
| 03311.30 | Concrete, deck, place and screed | 200 sy | 2,095 | 362 | 1.81 | 420 | 2.10 |
| 03345.30 | Concrete, deck, finish | 1,800 sf | 342 | 1,314 | 0.73 | — | — |
| 03346.20 | Concrete, abutments, rub | 1,960 sf | 78 | 1,356 | 0.69 | — | — |
| 03370.20 | Concrete, abutments, curing | 3,820 sf | 48 | 88 | 0.023 | — | — |
| 03370.30 | Concrete, deck, curing | 1,800 sf | 23 | 47 | 0.026 | — | — |
| | Metals | • | | | | | |
| 05120.00 | Steel, structural, place | 65,500 lb | 20,305 | 1,310 | 0.02 | 880 | 0.013 |
| 05520.00 | Guardrail | 120 lf | 2,640 | 528 | 4.40 | 151 | 1.26 |
| 05812.00 | Bearing plates | 3,200 lb | 1,376 | 480 | 0.15 | 220 | 0.069 |

*Source:* G. A. Sears and R. Clough, *Construction Project Management,* copyright © 1979 by John Wiley and Sons, Inc. Reprinted by permission of John Wiley and Sons, Inc.

allocation of resources over time. As noted above, this implies the integration of schedule with estimate to establish the expected point in time at which expenditures occur. This time-based budgeting is central to the analysis of a project from a *cash flow* point of view.

A bar chart schedule for a small highway project is shown in Figure 11.2. The project consists of 19 activities, and the duration of the project is 55 working days. On the basis of an expectation of 20 working days per month, the duration of the project is just under three months. The project will begin on 1 June and end in August. The expenditure of resources to include dollars and "labor units" is given in Figure 11.3. In this figure the projected expenditure of cost by activity by month is shown. Two points are of interest. First, the cost breakdown in this example is based on schedule activities rather than cost accounts. Second, the projection of expenditures on a period-by-period basis over the life of the project is shown. No allocation of costs or resources over time is given in Figure 11.1. Whether or not a time distribution of expenses is required in the budget is a management decision. In most cases, the more complex the project and the longer its duration, the more desirable it is to integrate cost control with project time control. On a small project such as the highway bridge of Figure 11.1, a time-scaled budget may not be necessary.

On more complicated projects, where the allocation of cash expenditures over time is of interest, time-scaled budgets may be required. On complex projects, such as multiyear projects with values in excess of $100,000,000, more comprehensive bud-

**FIGURE 11.2   Bar Chart Schedule for Highway Project 101**

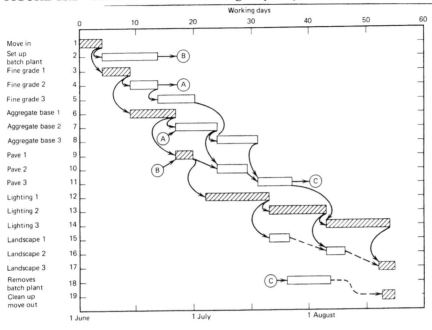

geting is utilized to include allocation of man-hours, equipment hours, and material procurement. Large and complicated industrial projects such as nuclear power plants and petrochemical plants tend to fall into this category. These large multiyear projects are typically constructed on a cost reimbursable basis, and design and construction are in progress at the same time. Budgeting focuses on "work packages," which are designed and then transferred to the construction force. In this sense, the overall project can be thought of as a large collection of miniprojects that rumble along the production line, first going through the design stages implied in Figure 10.4 and then being turned over to the construction force. The packages move through stages of evolving detail. As more detail becomes available, the estimating and hence the budgeting processes become more precise.

## 11.2   A SIMPLE TIME-SCALED BUDGET*

In order to illustrate the preparation of a cash flow budget, consider the highly simplified project shown in Figure 11.4. In this project, activities represented by bars are sched-

---

* This section is based on Section 8.1 of D. W. Halpin and R. W. Woodhead, *Construction Management* (New York: Wiley, 1980).

## FIGURE 11.3   Time-Scaled Budget for Highway Project of Figure 11.2

| Activity Number | Percent Complete | Labor Units to Date | Labor Cost to Date | Total Cost to Date |
|---|---|---|---|---|
| | | | | |
| **Cumulative Progress Report Through June** | | | | |
| * 1 | 100.00 | 159.70 | $ 1,142.40 | $ 1,142.40 |
| 2 | 100.00 | 375.67 | 2,592.00 | 2,592.00 |
| * 3 | 100.00 | 675.63 | 4,736.00 | 6,716.00 |
| 4 | 100.00 | 643.52 | 4,736.00 | 6,716.00 |
| 5 | 100.00 | 653.78 | 4,736.00 | 6,716.00 |
| * 6 | 100.00 | 842.03 | 6,386.80 | 8,786.80 |
| 7 | 76.91 | 665.99 | 4,562.00 | 6,407.91 |
| * 9 | 100.00 | 405.23 | 2,896.00 | 37,896.00 |
| | 50.45 | 4,425.54 | $33,587.20 | $78,773.11 |

| Activity Number | Percent Complete | Labor Units to Date | Labor Cost to Date | Total Cost to Date |
|---|---|---|---|---|
| | | | | |
| **Cumulative Progress Report Through July** | | | | |
| 7 | 100.00 | 865.90 | $ 6,386.80 | $ 8,786.80 |
| 8 | 100.00 | 868.55 | 6,386.80 | 8,786.80 |
| * 9 | 100.00 | 409.23 | 2,896.00 | 37,896.00 |
| 10 | 100.00 | 396.92 | 2,896.00 | 37,896.00 |
| 11 | 100.00 | 386.85 | 2,898.00 | 37,896.00 |
| * 12 | 100.00 | 78.52 | 548.00 | 948.00 |
| * 13 | 100.00 | 76.63 | 548.00 | 948.00 |
| 15 | 100.00 | 144.32 | 1,064.40 | 1,064.40 |
| 18 | 75.26 | 227.04 | 1,555.20 | 1,555.20 |
| | 92.94 | 6,804.32 | $53,106.40 | $172,046.40 |

| Activity Number | Percent Complete | Labor Units to Date | Labor Cost to Date | Total Cost to Date |
|---|---|---|---|---|
| | | | | |
| **Cumulative Progress Report Through August** | | | | |
| * 1 | 100.00 | 159.70 | $1,142.40 | $1,142.40 |
| 2 | 100.00 | 375.67 | 2,592.00 | 2,592.00 |
| * 3 | 100.00 | 675.63 | 4,736.00 | 6,716.00 |
| 4 | 100.00 | 643.52 | 4,736.00 | 6,716.00 |
| 5 | 100.00 | 653.78 | 4,736.00 | 6,716.00 |
| * 6 | 100.00 | 842.03 | 6,386.80 | 8,786.80 |
| 7 | 100.00 | 865.90 | 6,386.80 | 8,786.80 |
| 8 | 100.00 | 868.55 | 6,386.80 | 8,786.80 |
| * 9 | 100.00 | 409.23 | 2,896.00 | 3,7896.00 |
| 10 | 100.00 | 396.92 | 2,896.00 | 37,898.00 |
| 11 | 100.00 | 386.89 | 2,896.00 | 37,896.00 |
| * 12 | 100.00 | 78.52 | 548.00 | 948.00 |
| * 13 | 100.00 | 76.63 | 548.00 | 948.00 |
| * 14 | 100.00 | 72.95 | 548.00 | 948.00 |
| 15 | 100.00 | 144.32 | 1,064.40 | 1,064.40 |
| 16 | 100.00 | 149.40 | 1,064.40 | 1,064.40 |
| * 17 | 100.00 | 141.47 | 1,064.40 | 1,064.40 |
| 18 | 100.00 | 301.68 | 2,073.60 | 2,073.60 |
| * 19 | 100.00 | 78.83 | 542.40 | 542.40 |
| | 100.00 | 7,321.61 | $58,644.00 | $177,984.00 |

## FIGURE 11.4   Simple Time-Scaled Budget

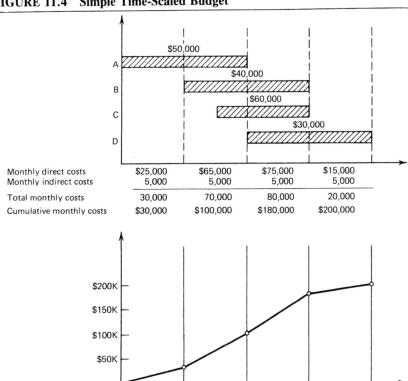

| | | | | |
|---|---|---|---|---|
| Monthly direct costs | $25,000 | $65,000 | $75,000 | $15,000 |
| Monthly indirect costs | 5,000 | 5,000 | 5,000 | 5,000 |
| Total monthly costs | 30,000 | 70,000 | 80,000 | 20,000 |
| Cumulative monthly costs | $30,000 | $100,000 | $180,000 | $200,000 |

The letter "K" is used to indicate thousands of dollars.

*Source:* D. W. Halpin and R. W. Woodhead, Construction Management, copyright © 1980 by John Wiley and Sons, Inc. Reprinted by permission of John Wiley and Sons, Inc.

uled across a four-month time frame. The direct costs associated with the individual activities are shown above the bars in the figure. It is assumed that project indirect costs will amount to $5000 per month. It has been assumed that the direct costs associated with each activity are even (linearly) distributed across the duration of the activity. The direct cost in the first month will be $25,000, since only activity A is in progress and half of it is scheduled for completion. In the second month expenses pertaining to activities A, B, and C will amount to $65,000 based on the following calculations:

$$\text{Activity A } 1/2 \times 50,000 = \$25,000$$

$$\text{Activity B } 1/2 \times 40,000 = \phantom{0}20,000$$

$$\text{Activity C } 1/3 \times 60,000 = \underline{\phantom{0}20,000}$$

$$\$65,000$$

**FIGURE 11.5   Project Total Cost S-Curve for Highway Paving Job 101 Based on Estimating Report**

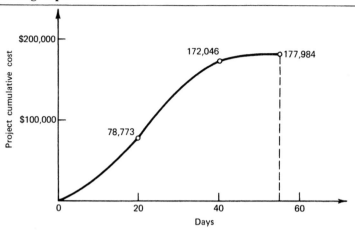

The total costs associated with each month are shown in Figure 11.4 below the bar chart. The monthly budgets for the four months of the project, as scheduled, are

Month 1   $30,000
Month 2   70,000
Month 3   80,000
Month 4   20,000

The cumulative expenditures on the project over time are shown in the form of an S-curve at the bottom of Figure 11.4.

> The name S-curve comes from the fact that the curve of cumulative expenditures has the appearance of a lazy S. The general shape characteristic results because early in the project activities are mobilizing and the expenditure curve is relatively flat. As many activities come on line, the level of expenditures increases and the curve has a steeper middle section. Toward the end of the project, activities are winding down and expenditures flatten again. . . ."*

The S-curve is a graphical representation of the cash budget for the project. The S-curve for the highway project of Figure 11.2 is shown in Figure 11.5.

## 11.3   A HIGHWAY PROJECT BUDGET

In order to consider the development of resource budgets in greater detail, consider a small highway project. A key plan, topographical plan, and typical section for this project are shown in Figure 11.6. The project entails the relocation of an 800-foot

* D. W. Halpin and R. W. Woodhead, *Construction Management* (New York: Wiley, 1980).

**FIGURE 11.6  Example Project Plan and Section of Relocated Road**

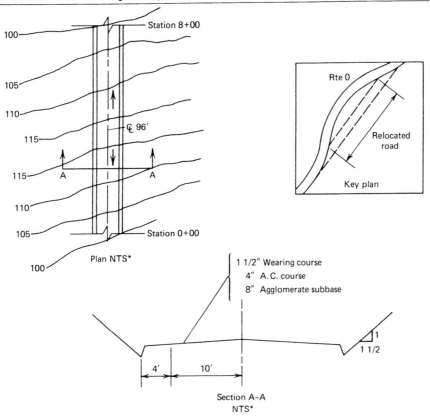

Plan NTS*

Section A–A
NTS*

**TABLE 11.1  Excavation Quantities**

| Station | End Area SF | Average End Area SF | Volume CY |
|---------|-------------|---------------------|-----------|
| 0 + 00  | 0           |                     |           |
|         |             | 170                 | 630       |
| 1 + 00  | 340         |                     |           |
|         |             | 411                 | 1522      |
| 2 + 00  | 482         |                     |           |
|         |             | 684                 | 2533      |
| 3 + 00  | 886         |                     |           |
|         |             | 919                 | 3404      |
| 4 + 00  | 952         |                     |           |
|         |             | 800                 | 2963      |
| 5 + 00  | 648         |                     |           |
|         |             | 425                 | 1574      |
| 6 + 00  | 202         | 160                 | 593       |
| 7 + 00  | 118         | 59                  | 219       |
| 8 + 00  | 0           |                     |           |

**TABLE 11.2   Production Rates Supplied by Estimating Department**

| | |
|---|---|
| Clearing | .6 acre/day |
| Excavation | 400 yd³/day |
| Fine grading | 1000 yd²/day |
| Place 8 in. base | 2200 yd²/day |
| Place 4 in. A.C. base course | 1200 yd²/day |
| Place 1½ in. A.C. wearing course | 1650 yd²/day |

**TABLE 11.3   Supplementary Conditions of the Contract**

| | |
|---|---|
| Project start | 1 June 19X2 |
| Traffic on new road | 1 October 19X2 |
| Sideslopes and ditches seeded by | 1 September 19X2 |
| Contractor off site | 15 October 19X2 |

**FIGURE 11.7   Arrow network Schedule for Example Highway Project**

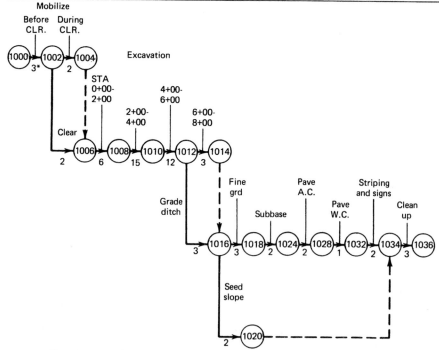

*Durations are given in working days.

## TABLE 11.4 Activity Listing with Durations and Cost Codes

| Activity Label | | Description | Duration | Cost Code | Nonlabor Expense |
|---|---|---|---|---|---|
| 1000 | 1002 | 'MOBILIZE BEFORE CLEARING' | 3 | 70200 | $1313. |
| 1002 | 1004 | 'COMPLETE MOBILIZATION' | 2 | 70200 | 875. |
| 1002 | 1006 | 'CLEAR AND GRUB' | 2 | 10000 | 3000. |
| 1006 | 1008 | 'EXCAVATE STA 0 + 00 TO 2 + 00' | 6 | 10300 | 5892. |
| 1008 | 1010 | 'EXCAVATE STA 2 + 00 TO 4 + 00' | 15 | 10300 | 14730. |
| 1010 | 1012 | 'EXCAVATE STA 4 + 00 TO 6 + 00' | 12 | 10300 | 11784. |
| 1012 | 1014 | 'EXCAVATE STA 6 + 00 TO 8 + 00' | 3 | 10300 | 2946. |
| 1013 | 1016 | 'GRADE DITCHES' | 3 | 10362 | 1962. |
| 1016 | 1020 | 'SEED DITCHES & SIDESLOPES' | 2 | 11000 | 1500. |
| 1016 | 1018 | 'FINE GRADE ROADWAY' | 3 | 10362 | 2946. |
| 1018 | 1024 | 'PLACE SUB-BASE' | 2 | 62016 | 6890. |
| 1024 | 1028 | 'PAVE 2 10FT. LANES' | 2 | 62039 | 6767. |
| 1028 | 1032 | 'PAVE WEARING COURSE' | 1 | 62044 | 3383. |
| 1032 | 1034 | 'PAVEMENT STRIPING & SIGNS' | 2 | 68400 | 1100. |
| 1034 | 1036 | 'CLEAN UP & DEMOBILIZE' | 3 | 70200 | 1312. |
| 1004 | 1006 | 'RESTRAINT' | | | |
| 1014 | 1016 | 'RESTRAINT' | | | |
| 1020 | 1034 | 'RESTRAINT' | | | |

$
$ EVENTS (MILESTONES)
$ ALL EVENTS IN THIS SCHEDULE WILL OCCUR AT ACTIVITY NODE NUMBERS.
$
1020 'DITCHES AND SLOPES SEEDED'
1024 'TRAFFIC ON RELOCATED ROAD'
1036 'PROJECT END—CONTRACTOR OFF SITE'

## TABLE 11.5 Activity Code Numbers for Relocated Road Project

*Standard Cost Code*

$$\underline{XX} \quad - \quad \underline{XX} \quad - \quad \underline{XXX} \quad - \quad \underline{XX}$$

| General ledger account number | Project number | Cost account number | Subaccount number |
|---|---|---|---|

*Cost Account Breakdown*
   000–099: subcontract item
   100–699: direct expenses
   700–999: project overhead (indirect) expenses
Activity code numbers will be the cost account number and subaccount number without the decimal. For example:
   Activity: fine grade
   Cost code: 80-76-103.62
   Activity code: 10362

*Note:* General activities not assigned a subaccount number will be given two zeros in the activity code. Dummy activities will be given activity code = 0.

## FIGURE 11.8    Event and Activity Schedule

### E V E N T    S C H E D U L E

| EVENT | DESCRIPTION | EARLY TIME | LATE TIME |
|---|---|---|---|
| 1020 | DITCHES AND SLOPES SEEDED | 3 AUG 1982<br>44 | 13 AUG 1982<br>52 |
| C 1024 | TRAFFIC ON RELOCATED ROAD | 6 AUG 1982<br>47 | 6 AUG 1982<br>47 |
| C 1036 | PROJECT END - CONTRACTOR OFF SITE | 17 AUG 1982<br>54 | 17 AUG 1982<br>54 |

### E N D    O F    E V E N T    S C H E D U L E

### A C T I V I T Y    S C H E D U L E

| ACTIVITY | | | DESCRIPTION | CODE |
|---|---|---|---|---|
| C | 1000 | 1002 | MOBILIZE BEFORE CLEARING | 70200 |
| C | 1002 | 1004 | COMPLETE MOBILIZATION | 70200 |
| C | 1002 | 1006 | CLEAR AND GRUB | 10000 |
| C | 1006 | 1008 | EXCAVATE STA 0 + 00 TO 2 + 00 | 10300 |
| C | 1008 | 1010 | EXCAVATE STA 2 + 00 TO 4 + 00 | 10300 |
| C | 1010 | 1012 | EXCAVATE STA 4 + 00 TO 6 + 00 | 10300 |
| C | 1012 | 1014 | EXCAVATE STA 6 + 00 TO 8 + 00 | 10300 |
| C | 1012 | 1016 | GRADE DITCHES | 10350 |
| C | 1016 | 1018 | FINE GRADE ROADWAY | 10350 |
| | 1016 | 1020 | SEED DITCHES AND SIDESLOPES | 11000 |
| C | 1018 | 1024 | PLACE SUB-BASE | 62016 |
| C | 1024 | 1028 | PAVE 2 10FT. LANES | 62039 |
| C | 1028 | 1032 | PAVE WEARING COURSE | 62044 |
| C | 1032 | 1034 | PAVEMENT STRIPPING & SIGNS | 68400 |
| C | 1034 | 1036 | CLEAN UP & DEMOBILIZE | 70200 |

### E N D    O F    S C H E D U L E

| DURA-TION | EARLY START | LATE START | EARLY FINISH | LATE FINISH | FREE FLOAT | TOTAL FLOAT |
|---|---|---|---|---|---|---|
| 3 | 1 JUN 82 | 1 JUN 82 | 3 JUN 82 | 3 JUN 82 | 0 | 0 |
|   | 1 | 1 | 3 | 3 |   |   |
| 2 | 4 JUN 82 | 4 JUN 82 | 7 JUN 82 | 7 JUN 82 | 0 | 0 |
|   | 4 | 4 | 5 | 5 |   |   |
| 2 | 4 JUN 82 | 4 JUN 82 | 7 JUN 82 | 7 JUN 82 | 0 | 0 |
|   | 4 | 4 | 5 | 5 |   |   |
| 6 | 8 JUN 82 | 8 JUN 82 | 15 JUN 82 | 15 JUN 82 | 0 | 0 |
|   | 6 | 6 | 11 | 11 |   |   |
| 15 | 16 JUN 82 | 16 JUN 82 | 7 JUL 82 | 7 JUL 82 | 0 | 0 |
|   | 12 | 12 | 26 | 26 |   |   |
| 12 | 8 JUL 82 | 8 JUL 82 | 26 JUL 82 | 26 JUL 82 | 0 | 0 |
|   | 27 | 27 | 38 | 38 |   |   |
| 3 | 27 JUL 82 | 27 JUL 82 | 29 JUL 82 | 29 JUL 82 | 0 | 0 |
|   | 39 | 39 | 41 | 41 |   |   |
| 2 | 27 JUL 82 | 28 JUL 82 | 28 JUL 82 | 29 JUL 82 | 1 | 1 |
|   | 39 | 40 | 40 | 41 |   |   |
| 3 | 30 JUL 82 | 30 JUL 82 | 3 AUG 82 | 3 AUG 82 | 0 | 0 |
|   | 42 | 42 | 44 | 44 |   |   |
| 2 | 30 JUL 82 | 11 AUG 82 | 2 AUG 82 | 12 AUG 82 | 0 | 8 |
|   | 42 | 50 | 43 | 51 |   |   |
| 2 | 4 AUG 82 | 4 AUG 82 | 5 AUG 82 | 5 AUG 82 | 0 | 0 |
|   | 45 | 45 | 46 | 46 |   |   |
| 2 | 6 AUG 82 | 6 AUG 82 | 9 AUG 82 | 9 AUG 82 | 0 | 0 |
|   | 47 | 47 | 48 | 48 |   |   |
| 1 | 10 AUG 82 | 10 AUG 82 | 10 AUG 82 | 10 AUG 82 | 0 | 0 |
|   | 49 | 49 | 49 | 49 |   |   |
| 2 | 11 AUG 82 | 11 AUG 82 | 12 AUG 82 | 12 AUG 82 | 0 | 0 |
|   | 50 | 50 | 51 | 51 |   |   |
| 3 | 13 AUG 82 | 13 AUG 82 | 17 AUG 82 | 17 AUG 82 | 0 | 0 |
|   | 52 | 52 | 54 | 54 |   |   |

**FIGURE 11.9   Highway Project Bar Chart**

```
                      P R O J E C T    R O A D A / A      B A R    C H A R T

ACTIVITY    NUMBER    DESCRIPTION
  1000       1002     MOBILIZE BEFORE CLEARING                      XXX
                                                          .     .        .        .
  1002       1004     COMPLETE MOBILIZATION                            XX
                                                          .     .        .        .
  1002       1006     CLEAR AND GRUB                                   XX
                                                          .     .        .        .
  1006       1008     EXCAVATE STA 0+00 TO 2+00                          XXXXXX
                                                          .     .        .        .
  1008       1010     EXCAVATE STA 2+000 TO 4+000                              XXXXXX
                                                          .     .        .        .
  1010       1012     EXCAVATE STA 4+000 TO 6+000          .     .        .        .

  1012       1014     EXCAVATE STA 6+00 TO 8+00            .     .        .        .

  1012       1016     GRADE DITCHES                        .     .        .        .

  1016       1018     FINE GRADE ROADWAY                   .     .        .        .

  1016       1020     SEED DITCHES & SIDESLOPES            .     .        .        .

  1020       1020     DITCHES AND SLOPES SEEDED            .     .        .        .

  1018       1024     PLACE SUB-BASE                       .     .        .        .

  1024       1024     TRAFFIC ON RELOCATED ROAD            .     .        .        .

  1024       1028     MOB EQUIP.,DRESS BASE & PAVE         .     .        .        .

  1028       1032     PAVE WEARING COURSE                  .     .        .        .

  1032       1034     PAVEMENT STRIPING & SIGNS            .     .        .        .

  1034       1036     CLEAN UP & DEMOBILIZE                .     .        .        .

  1036       1036     PROJECT END - CONTRACTOR OFF SITE    .     .        .        .

                                                          +.........+.......
                                             WORK DAYS    1          10
                                        CALENDAR DATES    1JUN82     14JUN82
```

section of two-lane road. The earthwork quantities required for the project are given in Table 11.1. Estimated production rates supplied by the estimating department are given in Table 11.2. The contractual required dates as given in the supplementary conditions of the construction contract are given in Table 11.3.

The first step in planning the budget for the project might be the preparation of a preliminary schedule for the project. A network schedule for the project consisting of 15 activities is shown in Figure 11.7. This schedule is shown in arrow notation, and the durations for each of the activities are shown below the activity arrows.†

A listing of the schedule activities and their durations is given in Table 11.4. The activity designation is defined by two numbers indicating the starting and ending events of the activity. This designation is followed by a description, the duration of the

---

† For information on network scheduling, consult D. W. Halpin and R. W. Woodhead, *Construction Management*, (New York: Wiley, 1980), Chapter 16. Many other excellent texts on the subject are available.

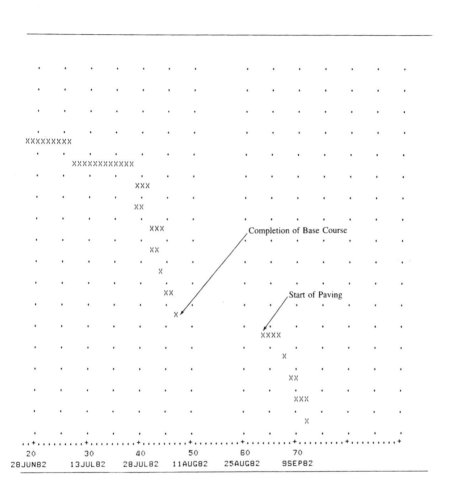

```
                                                                  .
         .     .    .    .    .    .    .        .    .    .    .    .    .    .    .
         .     .    .    .    .    .    .        .    .    .    .    .    .    .    .
         .     .    .    .    .    .    .        .    .    .    .    .    .    .    .
      XXXXXXXXX
         .     .    .    .    .    .    .        .    .    .    .    .    .    .    .
          XXXXXXXXXXXX
         .     .    .    .    .    .    .        .    .    .    .    .    .    .    .
                       XXX       .        .    .    .    .    .    .    .    .    .
         .     .    .    .    .    .    .        .    .    .    .    .    .    .    .
                      XX
         .     .    .    .    .    .    .        .    .    .    .    .    .    .    .
                          XXX        Completion of Base Course
         .     .    .    .    .    .    .        .    .    .    .    .    .    .    .
                      XX
         .     .    .    .    .    .    .        .    .    .    .    .    .    .    .
                        X
         .     .    .    .    .    .    .        .    .    .    .    .    .    .    .
                      XX           Start of Paving
         .     .    .    .    .    .    .        .    .    .    .    .    .    .    .
                    X
         .     .    .    .    .    .    .        .    .    .    .    .    .    .    .
                                    XXXX
         .     .    .    .    .    .    .        .    .    .    .    .    .    .    .
                                      X
         .     .    .    .    .    .    .        .    .    .    .    .    .    .    .
                                     XX
         .     .    .    .    .    .    .        .    .    .    .    .    .    .    .
                                     XXX
         .     .    .    .    .    .    .        .    .    .    .    .    .    .    .
                                      X
       ..+..........+..........+..........+..........+..........+..........+..........+
        20        30         40         50         60         70
      28JUN82   13JUL82    28JUL82    11AUG82    25AUG82    9SEP82
```

activity in days, and a five-digit group defining the cost account to which the activity is charged. The first excavation activity is 1006-1008. Its duration is six days. This is based on the required quantity of 2152 cubic yards and the excavation rate of 400 yd$^3$/day. The associated cost account is 10300. The structure of the cost codes used is given in Table 11.5. The mobilization activities are coded with cost account label 70200. The excavation activities are linked to the cost code 10300.

The contract constraints establish that the project may begin on 1 June and must end not later than 15 October. Landscaping must be completed by 1 September. Traffic must be routed onto the relocated road by 10 October. The schedule must comply with these constraints. The activity and event schedules based on a 1 June start are shown in Figure 11.8. This schedule can be considered as a planning document. After examining this planning schedule, it is clear that the project will be completed considerably in advance of the required 1 October date. Assume that in order to accommodate other projects, the paving equipment will not be available for the project until Monday 3 August. The excavation work is to be completed as soon as possible to release that equipment to other projects.

It is decided that the work should proceed on an early start basis up to and including placement of the base course. Once the paving machines become available, it will take two days to mobilize them on the job and to dress the base course for paving. An adjustment of the schedule to reflect these changes is shown in the bar chart schedule of Figure 11.9.

Resource planning for this project entails determining expenditure levels for the project by:

1. Cost account
2. Schedule activity
3. Month and day

The purpose of the time-based budgeting is to achieve an improved pattern of cash flow and resource consumption. The purpose of the cost account and schedule activity budgets is to establish target levels that can be used to detect overrun and underrun of projected values. This allows the manager to manage "by exception." That is, the manager can locate areas that require attention by noting significant variation from planned expenditure levels.

The budgeted amounts for non-labor-related expense are shown by schedule activity in Table 11.4. The non-labor budgeted amounts by cost account are as follows:

*Relevant Cost Codes*

| Activity | Cost Code | Non-Labor Budgeted Amount |
| --- | --- | --- |
| Mobilization | 702.01-99 | $ 3,500 |
| Clearing and grubbing | 100.01-99 | 3,000 |
| Excavation | 103.01-49 | 35,352 |
| Grading | 103.50-99 | 4,908 |
| Landscaping | 110.01-99 | 1,500 |
| Base course and paving | 620.01-99 | 17,040 |
| Pavement striping and signs | 684.00 | 1,100 |

Two labor resources are budgeted in connection with this schedule. These are the equipment operators, coded 6.79, and the laborers, coded 6.74. The labor resource requirements by activity are as shown in Table 11.6.

The plot of cost versus time for the operators' wages is shown in Figure 11.10. The cumulative (S-curve) plot of nonlabor expenses as a function of time is shown in Figure 11.11.

Clearly, the types of resource plans that the manager can establish are quite varied. The manager must decide what parameters are to be monitored and in what fashion this monitoring is to be conducted. The variety of budgets and plans to be established

**TABLE 11.6    Activity Labor Requirements**

| Activity | 6.74 Laborers (per day) | 6.79 Operators (per day) |
| --- | --- | --- |
| Mobilization | 7 | 1 |
| Clearing | 10 | 2 |
| Excavation | 1 | 2 |
| Grading | 1 | 1 |
| Landscaping | 5 | 0 |
| Sub-base | 3 | 1 |
| Pave | 8 | 2 |
| Paint and signs | 3 | 0 |
| Demobilization | 7 | 1 |

**FIGURE 11.10    Plot of Operators' Wages Versus Time**

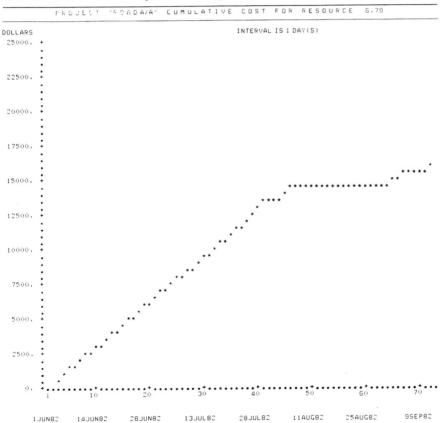

**FIGURE 11.11   S-Curve Plot of Nonlabor Expenses**

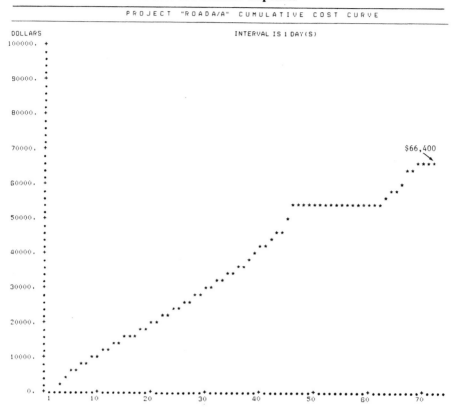

and maintained must be balanced with their informational value and the cost of establishing and maintaining them.

## 11.4   CASH FLOW TO THE CONTRACTOR*

The flow of money from the owner to the contractor is in the form of progress payments. Estimates of work completed are made by the contractor periodically (usually monthly), and are verified by the owner's representative. Depending on the type of contract (e.g., lump-sum, unit-price, and the like), the estimates are based on evaluations of the percentage of total contract completion or actual field measurements of quantities placed. This process is best demonstrated by further consideration of the four-activity

---

* Material in this and the following sections is taken from D. W. Halpin and R. W. Woodhead, *Construction Management* (New York: Wiley, 1980), Chapter 8.

example described in Figure 11.4. Assume that the contractor originally included a profit or markup in his bid of $10,000 (i.e., 5 percent) so that the total bid price was $210,000. The owner retains 10 percent of all validated progress payment claims until one-half of the contract value (i.e., $105,000 × 0.10) has been built and approved, as incentive for the contractor to complete the contract. The progress payments will be billed at the end of the month, and the owner will transfer the billed amount minus any retainage to the contractor's account 30 days later. The amount of each progress payment can be calculated as

$$\text{Pay} = 1.05 \times (\text{indirect expense} + \text{direct expense})$$

$$- 0.10 \times 1.05 \, (\text{indirect expense} + \text{direct expense})$$

The minus term for retainage drops out of the equation when 50 percent of the contract has been completed. Because of the delay in payment of billings by the owner and the retainage withheld, the income profile lags behind the expense S-curve, as shown in Figure 11.12.

The income profile has a stair-step appearance, since the progress payments are transferred in discrete amounts based on the above equation. The cross-hatched area in the figure between the income and expense profiles indicates the need on the part of the contractor to finance part of the construction until such time as he is reimbursed by the owner. The difference between income and expense makes it necessary for the contractor to obtain temporary financing. Usually, a bank extends a line of credit against which the contractor can draw to buy materials, make payments, and pay other expenses while waiting for reimbursement. This is similar to the procedure used by major credit card companies, in which they allow credit card holders to charge expenses and carry an outstanding balance for payment. Interest is charged by the bank (or

**FIGURE 11.12  Expenses and Income Profiles**

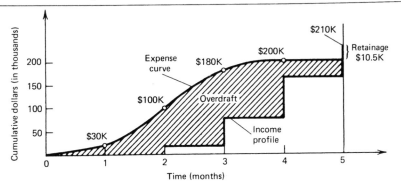

*Source:* D. W. Halpin and R. W. Woodhead, Construction Management, copyright © 1980 by John Wiley and Sons, Inc. Reprinted by permission of John Wiley and Sons, Inc.

**FIGURE 11.13   Influence of Front or Mobilizaiton Payment or Expense and Income Profiles**

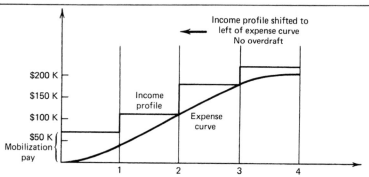

*Source:* D. W. Halpin and R. W. Woodhead, Construction Management, copyright © 1980 by John Wiley and Sons, Inc. Reprinted by permission of John Wiley and Sons, Inc.

credit card company) on the amount of the outstanding balance or overdraft.* It is, of course, good policy to try to minimize the amount of the overdraft and, therefore, the interest payments. The amount of the overdraft is influenced by a number of factors, including the amount of markup or profit the contractor has in his bid, the amount of retainage withheld by the owner, and the delay between billing and payment by the owner.

Interest on this type of financing is usually quoted in relationship to the *prime rate*. The prime rate is the interest rate charged preferred customers who are rated as very reliable and who represent an extremely small risk of default (e.g., General Motors, Exxon, and the like). The amount of interest is quoted in the number of points (i.e., the number of percentage points) above the prime rate. The higher-risk customers must pay more points than more reliable borrowers. Construction contractors are normally considered high-risk borrowers; if they default, the loan is secured only by some materials inventories and partially completed construction. In the event that a manufacturer of household appliances defaults, the inventory of appliances is available to cover part of the loss to the lender. Additionally, since construction contractors have an historically high rate of bankruptcy, they are more liable to be charged additional interest in most of their financial borrowings.

Some contractors offset the overdraft borrowing requirement by requesting front or mobilization money from the owner, thus shifting the position of the income profile so that no overdraft occurs (Figure 11.13). Since the owner is normally considered a

---

* Similar examples of this type of inventory financing can be found in many cyclic commercial undertakings. Automobile dealers, for instance, typically borrow money to finance the purchase of inventories of the new car models and then repay the lender as cars are sold. This kind of financing is often achieved by a plan whereby a major distributor guarantees a specific-purpose loan or overdraft with the dealer's bankers. Further, clothing stores buy large inventories of spring or fall fashions with borrowed money and then repay the lender as sales are made.

better risk than the contractor, he can borrow short-term money at a lower interest rate. If the owner agrees to this approach, he essentially takes on the interim financing requirement normally carried by the contractor. Front money may be made available on cost-reimbursable contracts when the owner has great confidence in the contractor's ability to complete the project. In such cases it represents an overall cost savings to the owner, since otherwise he will ultimately be back-billed for the contractor's higher financing rate if the contractor must carry the overdraft.

## 11.5 OVERDRAFT REQUIREMENTS

In order to know how much credit must be made available at the bank, the contractor needs to know what the maximum overdraft will be during the life of the project. With the information given regarding the four-activity project, the overdraft profile can be calculated and plotted. For purposes of illustration, the interest rate applied to the overdraft will be assumed to be one percent per month. That is, the contractor may pay the bank 1 percent per month for the amount of the overdraft at the end of the month. More commonly, daily interest factors may be employed for the purpose of calculating this interest service charge. Month-end balances might otherwise be manipulated by profitable short-term borrowings at the end of the month. The calculations required to define the overdraft profile are summarized in Table 11.7. The table indicates that the payment by the owner occurs at the end of a month, based on the billing at the end of the previous month. It is assumed that the interest is calculated on the overdraft and added to obtain the amount financed. This amount is then reduced by the amount received from the owner for previous billings. To illustrate: The overdraft at the bank at the end of the second month is $100,300. The interest on this amount is $1003 and is added to the overdraft to obtain the total amount financed ($101,303). To obtain the overdraft at the end of the third month, the progress payment of $28,350 is applied to reduce the overdraft at the beginning of the third month to $72,953. The overdraft at the end of the period is, then, $72,953 plus the costs for the period. Therefore, the overdraft is $72,953 plus $80,000 or $152,953. The information in the table is plotted in Figure 11.14. The overdraft profile appears as a sawtooth curve plotted below the base line. This profile shows that the maximum requirement is $154,483. Therefore, for this project the contractor must have a line of credit that will provide at least $155,000 at the bank plus a margin for safety, say $175,000 overall to cover expenses.

Requirements for other projects are added to the overdraft for this project to get a total overdraft or cash commitment profile. The timing of all projects presently under construction by the contractor leads to overlapping overdraft profiles that must be considered to find the maximum overdraft envelope for a given period of time.

Bids submitted that may be accepted must also be considered in the projection of total overdraft requirement. The plot of total overdraft requirements for a set of projects is shown in Figure 11.15.

**TABLE 11.7 Overdraft Calculations**

| | Month | | | | | |
|---|---|---|---|---|---|---|
| | 1 | 2 | 3 | 4 | 5 | |
| Direct cost | $25,000 | $65,000 | $75,000 | $15,000 | | |
| Indirect cost | 5,000 | 5,000 | 5,000 | 5,000 | | |
| *Subtotal* | 30,000 | 70,000 | 80,000 | 20,000 | | |
| Markup | 1,500 | 3,500 | 4,000 | 1,000 | | |
| *Total Billed* | 31,500 | 73,500 | 84,000 | 21,000 | | |
| Retainage withheld | 3,150 | 7,350 | 0 | 0 | | |
| *Payment Received* | | $28,350 → | $66,150 → | $84,000 → | $31,500 → | |
| Total cost to date | 30,000 | 100,000 | 180,000 | 200,000 | 200,000 | |
| Total amount billed to date | 31,500 | 105,000 | 189,000 | 210,000 | 210,000 | |
| Total paid to date | | 28,350 | 94,500 | 178,500 | 210,000 | |
| Overdraft end of month | 30,000 | 100,300 | 152,953 | 108,333 | 25,416 | (−)5830 |
| Interest on overdraft balance[a] | 300 | 1,003 | 1,530 | 1,083 | 254 | |
| Total amount financed | $30,300 | $101,303 | $154,483 | $109,416 | $25,670 | 0 |

[a] A simple illustration only. Most lenders would calculate interest charges more precisely on the amount/time involved, employing daily interest factors.

232

**FIGURE 11.14    Plot of Maximum Overdraft**

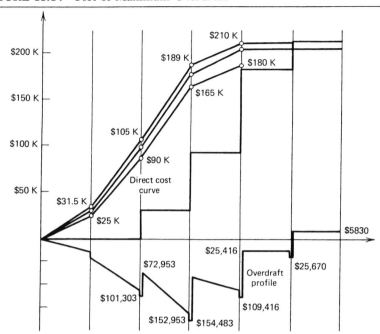

*Source:* D. W. Halpin and R. W. Woodhead, Construction Management, copyright © 1980 by John Wiley and Sons, Inc. Reprinted by permission of John Wiley and Sons, Inc.

## 11.6    CASH-FLOW MANAGEMENT CONSIDERATIONS

Cash-flow management involves all of the techniques described in this chapter and much more. It is fairly true to say, for example, that you cannot budget the other fellow's payments. That is, cash inflows are affected by a significant degree of uncertainty. A cash-flow management model of a relatively simple kind involves making provision for a set of at least 50 variables and requires a computer program to secure sufficiently timely and usable decision-making information.

A simplified illustration of an estimate of bank overdraft requirements is given in Table 11.7. It is generally necessary in practice to prepare a much more detailed analysis that takes account of the expected levels of working capital and the extent of the support likely to be received by way of credit from suppliers and subcontractors. In addition, it is necessary to estimate and allow for the variations that are likely to occur, especially lags and delays in the flow of moneys from progress-payment billings.

A more detailed system for gathering relevant data for meaningful cash flow forecasts will begin at the individual project level; a summary and a time schedule for the various elements and items making up total construction costs moving from time of incurrence to expected time of payment will be required. Similarly, a matching schedule for

**FIGURE 11.15   Composite Overdraft Profiles**

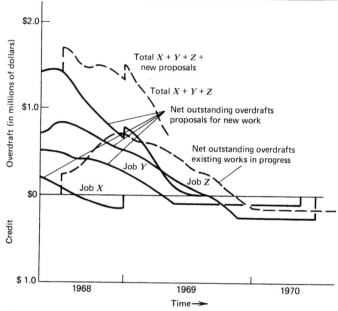

*Source:* D. W. Halpin and R. W. Woodhead, Construction Management, copyright © 1980 by John Wiley and Sons, Inc. Reprinted by permission of John Wiley and Sons, Inc.

anticipated progress payments moving from the time of billing to the expected time of receipt must be prepared.

These data can be combined for all current projects and can incorporate capital outlays and general administrative cost schedules to yield a total organization-wide forecast of cash flow. It must be remembered that at any time it may become necessary to adjust this forecast in the light of actual events as they unfold, and to call up contingency plans to deal with unexpected shortfalls or an unexpected funds surplus. One of the primary purposes of the cash-flow forecast is, of course, to make it possible to plan ahead, either to ensure that any additional funds requirements may be met by advance provisions or to ensure that a cash surplus does not remain idle when it should be earning some reasonable return. The importance of organization-wide cash-flow forecasting lies in the lead time provided by competent forecasting and the improved efficiency of cash management.

## 11.7   SUMMARY

Most of the discussion and the illustrations in this chapter reflect fairly simplistic manual system-based summaries, but increasing use is being made of computers and various modes of input related to programs of varying degrees of sophistication. The cash-flow-based computer model for determining the level of project markup described in Appendix C is indicative of the sophistication in using cash-flow methods that can be achieved. Before these computer financing models can be activated, a great deal of detailed information is required. Information needed ranges from field details on individual projects to top management-level inputs relating to planned capital outlays and to a variety of financing packages. At various times, bond and debenture issues, share issues, mortgage loans, and bankers' loans including, of course, standby arrangements and contingency plans are also incorporated into such models.

## REVIEW QUESTIONS AND EXERCISES

**11.1**  Given the following cost expenditures for a small warehouse project (to include direct and indirect changes), calculate the peak financial requirement.

(a) Assume 12 percent markup
Retainage is 10 percent throughout project
Finance charge is 1.5 percent per month

Payments are billed at the end of the month, and are received one month later. Sketch a diagram of the overdraft profile.

| Month | 1 | 2 | 3 | 4 |
|---|---|---|---|---|
| Monthly Indirect plus direct cost | 69,000 | 21,800 | 17,800 | 40,900 |

What would be the impact on the maximum overdraft of giving the contractor a $25,000 mobilization payment? (Show calculation.)

**11.2**  The contract between Ajax Construction Co. and Mr. Jones specifies the contractor will bill Mr. Jones at the end of each month for the amount of work finished that month. Mr. Jones will then pay Ajax a specified percentage of the bill the same day. The accumulated retainage is to be paid one month after project completion. The latest cumulative billing was $5 million, of which Ajax has actually received $4.5 million. The project is to be finished two months from now. Ajax estimates the bill for the remaining two months will be $100,000 and $50,000. Mr. Jones, being short of cash at present, proposed the following alternative: Rather than follow the contract and make the three payments required, he will make one final payment (for the two months' work plus the retainage) five months from now. Mr. Jones will also pay a 4-percent monthly interest rate because of the delay in payment.

(a) Find what would be the total final payment according to the actual contract and the new final payment according to the new proposal.
(b) Should the contractor accept the new proposal? Why?

**11.3**  Direct costs of a project are shown on the following bar chart. Assume $500 per month indirect cost and 10% retainage with payment time lag of one month.

(a) Calculate the project's net cash flow for 10-percent markup.
(b) Estimate the proper markup for 1 percent, 1.5 percent, and 2 percent rate of return per month, using methods described in Appendix C

| Month / Activity | 1 | 2 | 3 | 4 | 5 |
|---|---|---|---|---|---|
| 1. Survey | $2,000 | | | | |
| 2. Mobilization | | 1200 | | | |
| 3. Trench | | | 3000 | 6000 | |
| 4. Lay pipe | | | | 3000 | |
| 5. Concrete | | | | 500 | 500 |
| 6. Backfill | | | | 500 | |
| 7. Move out | | | | 500 | |
| 8. Prepare valves | | | | 1000 | |
| 9. Install valves | | | | 1000 | |
| 10. Test | | | | | 1500 |

**11.4**  Given the following cost expenditures for a small bridge job to include direct and indirect charges (but not bank interest), calculate the peak financial requirement for the job.

Assume 8-percent markup
Retainage is 6 percent throughout the project
Finance charge is 1 percent per month
Payments are billed at the end of the month, and are received one month later
Sketch a diagram of the overdraft profile.

| Month | 1 | 2 | 3 |
|---|---|---|---|
| Direct cost plus indirect cost | $29,000 | $48,900 | $16,400 |

**11.5**  The following table and graph represent a contractor's overdraft requirements for a project. Complete the table for costs, markup, total worth, retainage, and pay received. Retainage is 10 percent, markup is 10 percent, and interest is 1 percent per month. The client is billed at the end of the month. Payment is received the end of the next month, to be deposited in the bank the first of the following month.

| Overdraft | −50,000 | −120,500 | −82,205 | −13,727 |
|---|---|---|---|---|
| Interest | 500 | 1,205 | 822 | 137 |
| Cumulative overdraft | −50,500 | −121,705 | −83,027 | −13,864 |

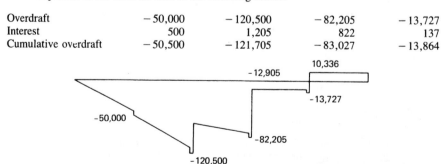

| | 1 | 2 | 3 | 4 | 5 |
|---|---|---|---|---|---|
| Direct cost | | | | | |
| Indirect cost | 10,000 | 10,000 | 5,000 | | |
| Total cost | | | | | |
| Markup @10% | | | | | |
| Total worth | | | | | |
| Retainage @10% | | | | | |
| Pay received | | | | | |

**11.6** A contractor is preparing to bid for a project. He has his cost estimate and the work schedule. The following table gives his expected expenses and their time of occurrence. Other expenses such as insurance, bonds, and payroll taxes are included. For simplicity of analysis he assumed that all expenses are recognized at the end of the month in which they occur.

## Table of Expenses

| Month | Mobiliza-tion Demobiliza-tion | Subcontractors | Materials | Payroll | Equipment | Field Overhead |
|---|---|---|---|---|---|---|
| 0 | $ 40,000 | $ 0 | $ 0 | $ 0 | $ 0 | $ 0 |
| 1 | 0 | 10,000 | 10,000 | 10,000 | 20,000 | 1,000 |
| 2 | 0 | 30,000 | 20,000 | 15,000 | 10,000 | 5,000 |
| 3 | 0 | 30,000 | 30,000 | 20,000 | 20,000 | 6,000 |
| 4 | 0 | 40,000 | 30,000 | 20,000 | 30,000 | 6,000 |
| 5 | 0 | 50,000 | 40,000 | 40,000 | 20,000 | 6,000 |
| 6 | 0 | 50,000 | 40,000 | 40,000 | 15,000 | 6,000 |
| 7 | 0 | 40,000 | 30,000 | 40,000 | 10,000 | 6,000 |
| 8 | 0 | 40,000 | 10,000 | 20,000 | 10,000 | 6,000 |
| 9 | 0 | 70,000 | 10,000 | 10,000 | 10,000 | 6,000 |
| 10 | 0 | 30,000 | 5,000 | 5,000 | 10,000 | 6,000 |
| 11 | 0 | 30,000 | 5,000 | 5,000 | 5,000 | 6,000 |
| 12 | 20,000 | 50,000 | 0 | 5,000 | 5,000 | 5,000 |
| Total | $ 60,000 | $ 470,000 | $ 230,000 | $ 230,000 | $ 165,000 | $ 65,000 |

Total cost = 60,000 + 470,000 + 230,000 + 230,000 + 165,000 + 65,000 = $1,220,000
Profit plus overhead = (10%) (total cost) = $122,000
Bid price = total cost + profit + overhead = $1,342,000

(a) The contractor is planning to add 10 percent to his estimated expenses to cover profits and office expenses. The total will be his bid price. He is also planning to submit for his progress payment at the end of each month. Upon approval the owner will subtract 5 percent for retainage and pay the contractor one month later. The accumulated retainage will be paid to the contractor with the last payment (i.e., end of month 13).
What is the monthly rate of return?
What is the annual rate of return?
What is the peak financial requirements and when does it occur?
(b) Assume the same as in part (a) except that the owner will retain 10 percent instead of 5 percent.
What is the monthly rate of return?
What is the annual rate of return?
What is the peak financial requirement?

(c) With a 10-percent retainage, assume that the owner pays the contractor only twice: the first time at the end of the sixth month for the work done for first six months; the last with the retainage being at the end of month 13.

What is the monthly rate of return?

What is the annual rate of return?

What is the peak financial requirement?

(d) Assume the owner will delay his payments by an extra month. Repeat part (b) (i.e., contractor will receive his progress payment, two months after he submits them).

If the contractor delays the payments to his subcontractors so that they are paid when he gets paid, would the monthly rate of return remain the same as in part (b), or would it be higher or lower? Explain.

(e) Plot the S-curve for part (a) against time.

# Chapter 12

# DATA COLLECTION AND REPORTING

## 12.1    FINANCIAL VERSUS MANAGEMENT DATA

The beginning point in collecting data for management purposes is the financial accounting system. The financial accounting system performs mandatory functions such as the payment of workers, the payment of vendors and subcontractors, and the generation of reports to creditors and stockholders. Collection of the data required by these functions is not discretionary. Therefore, these data are available to be used to detect the status of the job. In the raw data form in which they are used within the financial accounting system, worker pay amounts and vendor payments may have only limited value from the management point of view. To illustrate, if the amount of the payroll for the past two months on a particular job is available, this tells us what the labor cost is on the job as a whole. It does not, however, indicate how much labor was used for placement of concrete, excavation, and other work elements within the job. In this sense, the payroll amounts do not directly tell us the labor costs associated with work items within the job. Payroll amounts do not reflect whether concrete, for instance, is costing more or less than was originally budgeted.

By judicious restructuring of the data available from parts of the financial accounting system, information that is significant from the standpoint of management can be developed. For instance, if the number of hours worked by the individual workers is not only reported for pay purposes but is also distributed against the actual work items (e.g., cost accounts) on which they worked, information as to the labor cost status of each cost subaccount can be maintained. This kind of management report is a simple and straightforward by-product of the payroll subsystem within financial accounting.

The distribution of the hours to the appropriate accounts requires effort on the part of the foreman, the superintendent, the timekeeper, or the cost engineer. This additional effort yields a potential benefit, but management must accept the cost involved in this restructuring of the mandatory data collected by the payroll system. The effort to distribute labor costs is minimal and, therefore, the management data generated from financial data come rather cheaply. As management's need for more sophisticated data becomes greater, the costs to acquire the needed information and restructure it will become greater and greater.

In fact, the operation of the job cost system is based on the restructuring of data from the financial accounting system (payroll and accounts payable) and the collection of additional data not required by this financial accounting system. This is shown schematically in Figure 12.1. A company may decide that the cost and effort involved in restructuring data from the financial area and the acquisition of additional data (e.g., quantities of production) for discretionary management reports is too great. These companies satisfy themselves that the data available in the financial accounting system are sufficient for management purposes. In small and uncomplicated firms, this may be true. The amount of data required by most individuals to organize their personal affairs is normally limited to the maintenance of a checkbook balance.

**FIGURE 12.1  Relationship of Required Financial Systems to Management Systems**

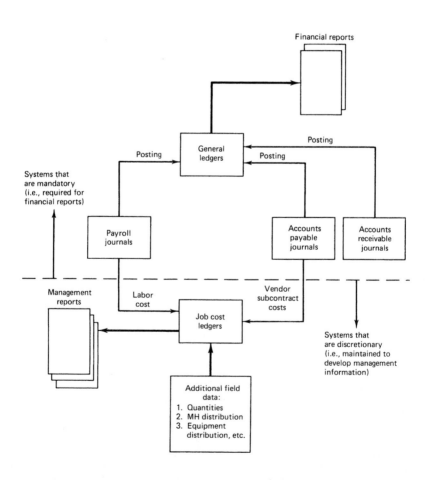

Construction companies of any size, however, require more management data and monitoring capability than that provided by the financial accounting system. Therefore, a job cost system becomes the data acquisition and monitoring system needed to cost-manage a job and compare actual costs with budgeted costs. The design of the job cost system starts with data available from payroll and accounts payable and adds additional data to aid in the interpretation of these financial data from a management point of view. Since payroll and accounts payable provide "free" data to the job costing function, they will be examined more closely in the next several sections.

## 12.2 DATA COLLECTION FROM PAYROLL

The purpose of the payroll system is to (1) determine the amount of and disburse wages to the labor force, (2) provide for payroll deductions, (3) maintain records for tax and other purposes, and (4) provide information regarding labor expenses. The source document used to collect data for payroll is a daily or weekly time card for each hourly employee similar to those shown in Figures 12.2 and 12.3. These cards are usually prepared by foremen, checked by the superintendent or field office engineer, and transmitted via the project manager to the head office payroll section for processing. The makeup of the cards is such that the foreman or timekeeper has positions next to the name of each employee for the allocation of the time worked on appropriate cost subaccounts. The foreman in the distribution made in Figure 12.2 has charged four hours of A. Apple's time to an earth excavation account and four hours to rock excavation. Apple is a code 15 craft, indicating that he is an operating engineer (equipment operator). As noted, this distribution of time allows the generation of management information aligning work effort with cost center. If no allocation is made, these management data are lost.

The flow of data from the field through preparation and generation of checks to cost accounts and earnings accumulation records is shown in Figure 12.4.

This data structure establishes the flow of raw data or information from the field to management. Raw data enter the system as *field entries* and are processed to service both payroll and cost accounting functions. Temporary files are generated to calculate and produce checks and check register information. Simultaneously, information is derived from the field entries to update project cost accounts. These quantity data are not required by the financial accounting system and can be thought of as management data only.

From the time card, the worker's ID (badge number), pay rate, and hours in each cost account are fed to processing routines that cross-check them against the worker data (permanent) file and use them to calculate gross earnings, deductions, and net earnings. Summations of gross earnings, deductions, and net earnings are carried to service the legal reporting requirements placed on the contractor by insurance carriers (PL and PD, workmen's compensation), the unions, and government agencies (e.g., Social Security and Unemployment). Figure 12.5 shows a typical certified payroll sent to such agencies for verification purposes. It will be noted that the certified payroll is a financial report in the sense that cost accounts and the effort associated with them

## FIGURE 12.2 Foreman's Daily Labor Distribution Report

Dewey, Cheatum, and Howe
Company

Report No. 16

**DAILY LABOR DISTRIBUTION REPORT**

Date 12 September 83

Job No. 101

Don Duck
Foreman's Signature

Location Peachtree Corners Shopping Mall

Description of Work Performed and Cost Account Number

| Employee or Badge Number | Name | Code | Craft or Union | Rate | Hours (80.103) | Hours (80.104) | Hours (80.260.01) | Hours (80.260.07) | Hours | Hours | Total Hours ST | Total Hours PT |
|---|---|---|---|---|---|---|---|---|---|---|---|---|
| 65 | Adam Apple | ST | 15 | 16.50 | 4 | 4 | | | | | 8 | 0 |
| | | PT | | | | | | | | | | |
| 14 | Ella Del Fabbro | ST | 10 | 12.50 | | | 8 | | | | 8 | 0 |
| | | PT | | | | | | | | | | |
| 22 | Charles Hoarse | ST | 10 | 12.50 | | | 6 | 2 | | | 8 | 0 |
| | | PT | | | | | | | | | | |
| | | ST | | | | | | | | | | |
| | | PT | | | | | | | | | | |
| | | ST | | | | | | | | | | |
| | | PT | | | | | | | | | | |
| | | ST | | | | | | | | | | |
| | | PT | | | | | | | | | | |
| | | ST | | | | | | | | | | |
| | | PT | | | | | | | | | | |
| | | ST | | | | | | | | | | |
| | | PT | | | | | | | | | | |
| | | ST | | | | | | | | | | |
| | | PT | | | | | | | | | | |
| | | | | Totals | 4 | 4 | 14 | 2 | | | 24 | 0 |

Approved by Dwf

**FIGURE 12.3   Time Card**

JOB #

WEEK ENDING                              CODE          HOURS

| EMP # | Name | CLASS CODE | CLASSIFICATION | RATE CODE | RATE | S | M | T | W | T | F | S | S | M | T | W | T | F | S | Total Hrs. |
|---|---|---|---|---|---|---|---|---|---|---|---|---|---|---|---|---|---|---|---|---|
| | | | | | | | | | | | | | | | | | | | | |
| | | | | | | | | | | | | | | | | | | | | |

| EMP # | Name | CLASS CODE | CLASSIFICATION | RATE CODE | RATE | S | M | T | W | T | F | S | | | | | | | | Total Hrs. |
|---|---|---|---|---|---|---|---|---|---|---|---|---|---|---|---|---|---|---|---|---|
| | | | | | | | | | | | | | | | | | | | | |
| | | | | | | | | | | | | | | | | | | | | |

| EMP # | Name | CLASS CODE | CLASSIFICATION | RATE CODE | RATE | S | M | T | W | T | F | S | | | | | | | | Total Hrs. |
|---|---|---|---|---|---|---|---|---|---|---|---|---|---|---|---|---|---|---|---|---|
| | | | | | | | | | | | | | | | | | | | | |
| | | | | | | | | | | | | | | | | | | | | |

| EMP # | Name | CLASS CODE | CLASSIFICATION | RATE CODE | RATE | S | M | T | W | T | F | S | | | | | | | | Total Hrs. |
|---|---|---|---|---|---|---|---|---|---|---|---|---|---|---|---|---|---|---|---|---|
| | | | | | | | | | | | | | | | | | | | | |
| | | | | | | | | | | | | | | | | | | | | |

*Source:* Courtesy of Bellamy Brothers, Inc., Ellenwood, Ga.

are not noted. The reporting requirements to agencies outside of the contractor's organization generally require only financial data.* Management reports are generated primarily for internal use.

Hourly data reported on the daily or weekly time cards are combined with *quantity installed* entries and are routed to the cost account files for the project. The cost account productivity can be developed by dividing quantity of material installed by the number of worker hours. Further, unit costs can be developed by using the relation

$$\$/unit = \frac{\sum_{i=1}^{m} (\text{craft rate}) \times (\text{total craft hours})}{\text{total quantity installed}}$$

where $m$ is the number of crafts working the cost account. Project unit prices are valuable in estimating and budgeting future projects, as described in Section 10.6. They are, of course, critical in determining whether cost account estimates (target values) from budget are being achieved on the present project. A typical labor cost report generated by the payroll system is shown in Figure 12.6. This report, organized by cost account, is a management report designed to compare actual with projected

---

* On cost reimbursable jobs the client may require copies of management (job cost) reports. Federal agencies monitoring certain types of construction (e.g., Federal Energy Administration) sometimes also require cost reports and go as far as stipulating the cost codes to be used in the report.

**FIGURE 12.4   Payroll Data Structure**

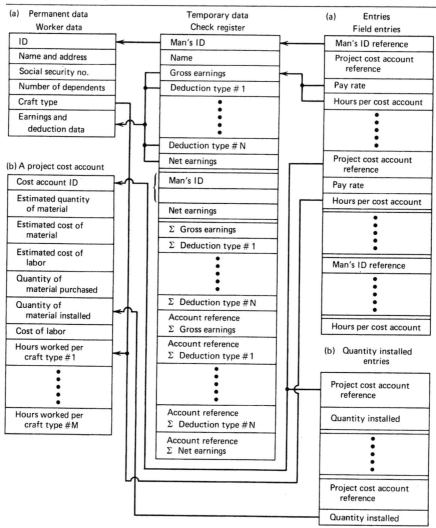

progress. Since it derives from the distribution of worker hours and the estimated quantities placed for each cost account, it is normally generated by computerized payroll programs.

The labor cost report highlights work units performed during the week of the report as well as cumulative units. The actual and estimated (target) units costs as well as actual versus estimated total labor costs are reported. The report also reflects the under/over value to date and makes a projection of the under/over value for the account as of the end of the project. This projection allows the manager to identify accounts

# FIGURE 12.5   Certified Payroll

```
        DEWEY, CHEATUM, AND HUME, INCORPORATED
        1111 WEST COLUMBIA AVE
        OAK PARK, ILLINOIS
           61600

XXXXXXXXXXXXXXXXXXXXX......................................XXXXXXXXXXXXXXXXXXXX

                        CERTIFIED PAYROLL

            JOB 2000    MANVILLE Y.W.C.A.           ILL
            WEEK ENDING  2/16/69   CURRENT DATE   3/ 2/69
---------------------------------------------------------------------------
102        HOARSE             CHARLES      A            MAR / 4      223975177
1141 N CENTRAL AVE          CHICAGO ILL          60637               wIS
    CRAFT    1011   LABOR        FOREMAN
       2/10    2/11    2/12    2/13    2/14    2/15    2/16   TOT HRS   RATE    PAY
 TOT  8.000   8.000   8.000   0.000   0.000   0.000   0.000   24.000   5.600  134.40
 BON  0.000   0.000   0.000   0.000   0.000   0.000   0.000            5.600    0.00
       TOT PAY    FIT      FICA    UNION    OTHER   TOT DED   PAID    JB GRS
        242.00   33.18    11.62    0.00     7.75    52.55   189.45   134.40
---------------------------------------------------------------------------
103        SHUE              JAMES       T            SINGLE / 1     350722195
96 S AUSTIN AVE             CHICAGO ILL          60640               ILL
    CRAFT    1021   LABOR      JOURNEYMAN
       2/10    2/11    2/12    2/13    2/14    2/15    2/16   TOT HRS   RATE    PAY
 TOT  8.000   9.000   8.000   0.000   0.000   0.000   0.000   25.000   4.225  105.62
 BON  0.000   0.500   0.000   0.000   0.000   0.000   0.000            0.500    2.11
       TOT PAY    FIT      FICA    UNION    OTHER   TOT DED   PAID    JB GRS
        107.73    0.00     0.00    0.00     0.00     0.00   107.73   107.73
---------------------------------------------------------------------------
101        APPLE              ADAM                    MAR / 3      257257257
401 MAPLE ST               OAK PARK ILLINOIS      61601               ILL
    CRAFT    2011   CARPENTER    FOREMAN
       2/10    2/11    2/12    2/13    2/14    2/15    2/16   TOT HRS   RATE    PAY
 TOT 11.000   8.000   8.000   8.000   8.000   8.000   0.000   51.000   6.250  318.75
 BON  0.000   0.000   0.000   0.000   0.000   0.000   0.000            6.250    0.00
       TOT PAY    FIT      FICA    UNION    OTHER   TOT DED   PAID    JB GRS
        318.75   53.03    15.30    0.00    57.33   125.66   193.09   318.75
---------------------------------------------------------------------------

XXXXXXXXXXXXXXXXXXXXXX......................................XXXXXXXXXXXXXXXXXXXXX
XXXXXXXXXXXXXXXXXXXXX.......................................XXXXXXXXXXXXXXXXXXXXX
XXXXXXXXXXXXXXXXXXXXX.......................................XXXXXXXXXXXXXXXXXXXXX



            TOTALS FOR      3  MEN ON THIS JOB THIS WEEK
TOTAL HRS WORKED ON THIS JOB THIS WEEK =    100.000

   TOT PAY      FIT       FICA      UNION     OTHER    TOT DED     PAID     JOB GRS

   668.48     86.21      26.92      0.00     65.08    178.21     490.27    569.88
----------------------------------------------------------------------------
                 END OF CERTIFIED PAYROLL JOB  2000


              MEMO FOR CERTIFIED PAYROLL

LAYOFF OF MAN 103 FOR PERSONAL REASONS
```

# FIGURE 12.6　Labor Cost Report

| CENTURY CENTER BLDG #5 | | | | | LABOR COST REPORT | |
| ATLANTA, GA | | | | | MCDEVITT & STREET COMPANY ATLANTA DIVISION | |

| COST CODE INFORMATION | | | | QUANTITY | | UNIT |
|---|---|---|---|---|---|---|
| COST CODE | DESCRIPTION | UNITS | % COMP | ESTIMATED | ACTUAL | ESTIMATED |
| 10 | **THIS WEEK** | | 2 | | 1 | |
| | SUPV | WKS | 86 | 65 | 56 | 461.538 |
| 11 | **THIS WEEK** | | 2 | | 1 | |
| | CLERK | WKS | 82 | 65 | 53 | 153.846 |
| 21 | SIGNS | LS | | | | |
| 22 | **THIS WEEK** | | | | | |
| | TEMP BLDG | LS | | | | |
| 34 | PARTS & REPAIR | LS | 50 | | | |
| 60 | **THIS WEEK** | | 2 | | 1 | |
| | FIELD ENG | WKS | 100 | 43 | 43 | 383.721 |
| 61 | OFFICE ENG | WKS | 68 | 60 | 41 | 233.333 |
| 70 | HOIST FDN | EA | 100 | 2 | 2 | 200.000 |
| 71 | ER & DISMT HOIST | EA | 90 | | | |
| 72 | LANDINGS | EA | 100 | 40 | 40 | 111.875 |
| 73 | MATL HOIST OPER | WKS | 89 | 35 | 31 | 257.143 |
| 74 | **THIS WEEK** | | 3 | | 1 | |
| | PERS HOIST OPER | WKS | 77 | 35 | 27 | 285.714 |
| 75 | EQUIP OPERATOR | LS | 85 | | | |
| 93 | TEMP WINTER | LS | | | | |
| 100 | **THIS WEEK** | | 1 | | 1 | |
| | ICE WATER | WKS | 64 | 69 | 44 | 25.362 |
| 110 | FINAL CLEAN | SF | | 300,000 | | .050 |
| 111 | **THIS WEEK** | | 2 | | 1 | |
| | HAUL TRASH | WK | 68 | 50 | 34 | 200.000 |
| 112 | **THIS WEEK** | | 1 | | 1 | |
| | DAILY CLEAN | WK | 68 | 69 | 47 | 217.391 |
| 115 | | | | | | |
| 130 | **THIS WEEK** | | 1 | | 1 | |
| | SAFETY | WK | 81 | 69 | 56 | 217.391 |
| 131 | PROTECT TREES | LS | | | | |
| 132 | SHORING | LS | 100 | 18 | 18 | 1,388.889 |
| 200 | **THIS WEEK** | | | | | |
| | BKCHRG | | | | | |
| 304 | PROT BNKS | LS | 1 | | | |
| 307 | **THIS WEEK** | | 1 | | 5 | |
| | HAND EXC | CY | 97 | 725 | 705 | 18.793 |
| 310 | DE WATER | LS | 100 | | | |
| 312 | BKFL HAND | CY | 83 | 6,000 | 5,000 | 1.500 |
| 316 | FINE GR | SF | 99 | 15,000 | 14,830 | .167 |

*Source:* Courtesy of McDevitt and Street Company, Atlanta Division.

| PRICE | COST | | PROJECTED COST | |
|---|---|---|---|---|
| ACTUAL | ESTIMATED | ACTUAL | TO DATE OVER/UNDER | TO COMPLETE OVER/UNDER |
| 435.000 | | 435 | 26- | |
| 440.375 | 30,000 | 24,661 | 1,185 | 191 |
| 150.000 | 10,000 | 150 | 4- | |
| 142.585 | | 7,557 | 597 | 135- |
| | 300 | | | |
| | 1,500 | 21 | 21 | |
| | | 787 | 113- | 75- |
| | 1,500 | 242 | 508- | 508- |
| 219.000 | | 219 | 165- | |
| 386.930 | 16,500 | 16,638 | 138 | |
| 246.878 | 14,000 | 10,122 | 3878- | COMPLETE |
| 210.500 | 400 | 421 | 21 | COMPLETE |
| | 6,000 | 8,350 | 2,350 | COMPLETE |
| 51.975 | 4,475 | 2,079 | 2,396- | COMPLETE |
| 293.129 | 9,000 | 9,087 | 87 | COMPLETE |
| 446.000 | | 446 | 160 | |
| 333.185 | 10,000 | 8,996 | 1,282 | 239 |
| | 4,000 | 4,179 | 179 | COMPLETE |
| | 2,000 | 73 | 1,927- | COMPLETE |
| 26.000 | | 26 | 1 | |
| 21.136 | 1,750 | 930 | 186- | 106- |
| | 15,000 | | | |
| 248.000 | | 248 | 48 | |
| 92.765 | 10,000 | 3,154 | 3,646- | 1,716- |
| 543.000 | | 543 | 326 | |
| 311.596 | 15,000 | 14,645 | 4,428 | 2,072 |
| | | 1,747 | 1,747 | NO BUDGET |
| 13.000 | | 13 | 204- | |
| 206.304 | 15,000 | 11,553 | 621- | 144- |
| | 500 | 31 | | |
| 1,345.389 | 25,000 | 24,217 | 783- | COMPLETE |
| | | 257 | | NO BUDGET |
| | | 2,650 | | |
| | 700 | 138 | 562- | COMPLETE |
| 19.600 | | 98 | 4 | |
| 19569 | 13,625 | 13,796 | 547 | 15 |
| | 2,000 | 2,060 | 60 | COMPLETE |
| 1.291 | 9,000 | 6,453 | 1,047- | 209- |
| .135 | 2,500 | 1,999 | 501- | COMPLETE |

## Fig. 12.6   (cont.)

CENTURY CENTER BLDG #5

ATLANTA, GA

**LABOR COST REPORT**
MCDEVITT & STREET COMPANY
ATLANTA DIVISION

| COST CODE INFORMATION | | | | QUANTITY | | UNIT |
|---|---|---|---|---|---|---|
| COST CODE | DESCRIPTION | UNITS | % COMP | ESTIMATED | ACTUAL | ESTIMATED |
| 320 | FINE GR SITE | SF | 23 | 20,000 | 4,650 | .100 |
| 324 | 4 DR TILE | LF | 81 | 400 | 324 | 3.250 |
| 324 | 2 DRAIN TILE | LF | 83 | 1,200 | 1,000 | .500 |
| 325 | PEA GR @ DRAIN | CY | 80 | 70 | 56 | 6.000 |
| 326 | SPRD TOP SOIL | CY | | 1,000 | | 1.000 |
| 330 | FINE GR SW | SF | 47 | 10,000 | 4,700 | .100 |
| 401 | FOOTING | CY | 138 | 135 | 186 | 3.000 |
| 402 | PILE CAPS | CY | 83 | 150 | 124 | 4.000 |
| 403 | GR BMS | CY | 64 | 170 | 108 | 4.000 |
| 404 | DRILL DOWELS | EA | 48 | 80 | 38 | 4.500 |
| 500 | STAIRS & FL 2-6 | LS | 98 | | | |
| 501 | COOL TOW SLAB | CY | 80 | 5 | 4 | 25.000 |
| 502 | **THIS WEEK** | | | | 3 | |
| | WALLS & BMS | CY | 92 | 835 | 772 | 4.000 |
| 505 | PLACE SLABS | CY | 93 | 3,000 | 2,788 | 3.500 |
| 510 | CANOPIES | CY | 43 | 40 | 17 | 15.000 |
| 603 | GRAVEL UNDER SOG | CY | 112 | 225 | 253 | 2.956 |
| 605 | EXP JT | LF | 72 | 1,250 | 900 | .500 |
| 606 | WATERSTOP | lf | 74 | 1,650 | 1,217 | 1.000 |
| 607 | VAPOR BARRIER | SF | 26 | 30,000 | 7,650 | .010 |
| 611 | SIDEWALKS | CY | 58 | 165 | 95 | 6.970 |
| 612 | **THIS WEEK** | | 97 | | 32 | |
| | CONC DRIVE | CY | 109 | 33 | 36 | 4.545 |
| 616 | | | | | | |
| 618 | CONC SEAL SLAB | CY | 80 | 50 | 40 | 8.000 |
| 619 | CONC TOPPING | SF | 100 | 700 | 700 | .357 |
| 620 | PLACE CONC SPLASH | CY | 40 | 10 | 4 | 60.000 |
| 704 | **THIS WEEK** | | | | 50 | |
| | FORM WALLS | SF | 99 | 33,250 | 32,844 | 1.244 |
| 706 | FORM BEAMS | SF | 100 | 5,000 | 4,999 | 2.140 |
| 709 | **THIS WEEK** | | | | | |
| | F SLABS & STEPS | SF | 78 | 3,350 | 2,602 | 1.642 |

| PRICE | COST | | PROJECTED COST | |
| --- | --- | --- | --- | --- |
| ACTUAL | ESTIMATED | ACTUAL | TO DATE OVER/UNDER | TO COMPLETE OVER/UNDER |
| .095 | 2,000 | 442 | 23- | 76- |
| 4.210 | 1,300 | 1,364 | 64 | COMPLETE |
| .211 | 600 | 211 | 389- | COMPLETE |
| 6.964 | 420 | 390 | 30- | COMPLETE |
|  | 1,000 |  |  |  |
| .165 | 1,000 | 776 | 306 | 345 |
| 4.817 | 405 | 896 | 491- | COMPLETE |
| 3.798 | 600 | 471 | 129- | COMPLETE |
| 5.907 | 680 | 638 | 42- | COMPLETE |
| 3.921 | 360 | 149 | 22- | 24- |
|  | 7,550 | 6,739 | 660- | 13- |
| 10.750 | 125 | 43 | 82- |  |
| 8.000 |  | 24 | 12 |  |
| 3.645 | 3,340 | 2,814 | 274- | 22- |
| 3.788 | 10,500 | 10,533 | 33 | COMPLETE |
| 30.235 | 600 | 514 | 259 | 350 |
| 2.040 | 665 | 516 | 149- | COMPLETE |
| .170 | 625 | 153 | 297- | 116- |
| .261 | 1,650 | 318 | 1,332- | COMPLETE |
| .017 | 300 | 128 | 172- | COMPLETE |
| 2.926 | 1,150 | 278 | 384- | 283- |
| 4.750 |  | 152 | 20 |  |
| 7.222 | 150 | 260 | 110 |  |
|  |  | 54 | 54 | NO BUDGET |
| 5.100 | 400 | 204 | 196- | COMPLETE |
| .153 | 250 | 107 | 143- | COMPLETE |
| 14.000 | 600 | 56 | 184- | 276- |
| 1.760 |  | 88 | 26 |  |
| 1.285 | 41,350 | 42,190 | 1,345 | 17 |
| 2.123 | 10,700 | 10,613 | 87- | COMPLETE |
|  |  | 502 | 502 |  |
| 1.836 | 5,500 | 4,778 | 506 | 146 |

## FIGURE 12.6 (cont.)

| CENTURY CENTER BLDG #5 | LABOR COST REPORT |
|---|---|
| | MCDEVITT & STREET COMPANY |
| ATLANTA, GA | ATLANTA DIVISION |

| | COST CODE INFORMATION | | | QUANTITY | | UNIT |
|---|---|---|---|---|---|---|
| COST CODE | DESCRIPTION | UNITS | % COMP | ESTIMATED | ACTUAL | ESTIMATED |
| 710 | **THIS WEEK** | | 4 | | 60 | |
| | F 4 SIDEWALK | LF | 44 | 1,500 | 660 | .667 |
| 711 | F KEY KOLD | LF | | 1,000 | | .400 |
| 712 | KEYWAY | LF | 61 | 1,650 | 1,013 | .197 |
| 716 | INSERT & BLKOUTS | LS | 96 | | | |
| 719 | F PADS | EA | 100 | 8 | 8 | 62.500 |
| 720 | F ELEV THRES | EA | 100 | 75 | 75 | 8.000 |
| 725 | CONST JTS | LS | 100 | | | |
| 727 | F DRAINS | SF | 75 | 200 | 150 | 2.000 |
| 728 | F SCISSOR LIFT | EA | 100 | 3 | 3 | 33.333 |
| 729 | F TOPPING EDGE | LF | 160 | 100 | 160 | 1.000 |
| 730 | F ROOF CURBS | LF | 117 | 175 | 204 | .571 |
| 731 | F SPLASH | LF | 55 | 600 | 330 | 1.000 |
| 805 | **THIS WEEK** | | 2 | | 300 | |
| | RUB FINISH | SF | 82 | 15,000 | 12,300 | .242 |
| 806 | GROUT ELEV THRES | EA | 100 | 75 | 75 | 10.000 |
| 807 | FIN WALKS & SPLASH | SF | | 700 | | .071 |
| 811 | CURE WALLS | SF | 13 | 33,000 | 4,328 | .010 |
| 820 | REMOVE DWL & PATCH | LS | | | | |
| 1201 | **THIS WEEK** | | | | | |
| | MISC BLOCKING | LS | | | | |
| 1215 | TREADED BLK | FBM | 43 | 3,000 | 1,280 | .750 |
| 1302 | **THIS WEEK** | | | | | |
| | WD DOORS | EA | | 925 | | 21.757 |
| 1307 | WD CAB & CTR | EA | | 34 | | 25.000 |
| 1402 | HM DOORS | EA | | | | |
| 1403 | HM FRAMES | EA | 100 | 7 | 7 | 28.571 |
| 1500 | **THIS WEEK** | | | | | |
| | | LS | 50 | | | |
| 1601 | SET ABS | EA | 62 | 400 | 248 | 3.000 |
| 1602 | | EA | | 50 | | 15.000 |
| 1608 | MISC IRON | LS | 45 | | | |
| 1615 | **THIS WEEK** | | | | | |
| | TOIL PART SUPPORT | EA | 100 | 34 | 34 | 100.000 |

WEEK ENDING 10/11/

PROJECT NUMBER 13-5265

| PRICE | COST | | PROJECTED COST | |
| --- | --- | --- | --- | --- |
| ACTUAL | ESTIMATED | ACTUAL | TO DATE OVER/UNDER | TO COMPLETE OVER/UNDER |
| .483 | | 29 | 11- | |
| .723 | 1,000 | 477 | 37 | 47 |
| | 400 | | 400- | COMPLETE |
| .305 | 325 | 309 | 16- | COMPLETE |
| | 1000 | 1,198 | 238 | 10 |
| 57.625 | 500 | 461 | 39- | COMPLETE |
| 13.827 | 600 | 1,037 | 437 | COMPLETE |
| | 1,500 | 438 | 1,062- | COMPLETE |
| 1.273 | 400 | 191 | 109- | 36- |
| 23.333 | 100 | 70 | 30- | COMPLETE |
| .875 | 100 | 140 | 40 | COMPLETE |
| 1.931 | 100 | 394 | 294 | COMPLETE |
| .418 | 600 | 138 | 192- | 157- |
| .847 | | 254 | 182 | |
| .439 | 3,625 | 5,403 | 2,431 | 533 |
| 4.520 | 750 | 339 | 411- | COMPLETE |
| | 50 | | | |
| .032 | 330 | 139 | 191- | COMPLETE |
| | 300 | | | |
| | | 167 | | |
| | 2,000 | 167 | | |
| .901 | 2,250 | 1,153 | 1,097- | COMPLETE |
| | | 10 | | |
| | 20,125 | 3,065 | | |
| | 850 | | | |
| | 270 | 510 | | |
| 61.143 | 200 | 428 | 228 | COMPLETE |
| | | 216 | 216 | |
| | 4,200 | 1,385 | 715- | 715- |
| 5.790 | 1,200 | 1,436 | 236 | COMPLETE |
| | 750 | | 750- | COMPLETE |
| | 1,400 | 1,596 | 966 | 1,181 |
| | | 170 | 170 | |
| 68.882 | 3,400 | 2,342 | 1,058- | |

# Fig. 12.6   (cont.)

| CENTURY CENTER BLDG #5 | **LABOR COST REPORT** |
| ATLANTA, GA | MCDEVITT & STREET COMPANY |
| | ATLANTA DIVISION |

| | COST CODE INFORMATION | | | | QUANTITY | |
|---|---|---|---|---|---|---|
| COST CODE | DESCRIPTION | UNITS | % COMP | ESTIMATED | ACTUAL | |
| | **THIS WEEK** | | | | | |
| 1616 | | LS | 90 | | | |
| 1901 | GROUT | LS | 100 | | | |
| 1902 | | | | | | |
| 1907 | SIGNS | LS | | | | |
| 1908 | WIRE MESH FENCE | | | | | |
| | **THIS WEEK** | | | | | |
| 3000 | | | | | | 1 |
| 3001 | | | | | | |

| CENTURY CENTER BLDG #5 | **LABOR COST REPORT** |
| ATLANTA, GA | MCDEVITT & STREET COMPANY |
| | ATLANTA DIVISION |

| | COST CODE INFORMATION | | | QUANTITY | | UNIT PRICE | |
|---|---|---|---|---|---|---|---|
| COST CODE | DESCRIPTION | UNITS | % COMP | ESTIMATED | ACTUAL | ESTIMATED | ACTUAL |

| ***PROJECT TOTALS*** | REGULAR | PREMIUM | TOTAL |
|---|---|---|---|
| CURRENT WEEK | 4,673.00 | 155.00 | 4,828.00 |
| JOB TO DATE | 288,182.00 | 7611.00 | 295,793.00 |

| | | |
|---|---|---|
| EFFECTIVE LABOR ACCOMPLISHED TO-DATE | 301,717 | TOTAL LABOR EXCLUDING 200-299 CODES |
| EFFECTIVE LABOR ACCOMPLISHED THIS WEEK | 2,533 | |
| PCT LABOR OVERRUN TO-DATE | 4.098DS- | |
| PROJECTED TOTAL LABOR OVERRUN | 15,298- | |
| PERCENT COMPLETE | 80.825 | |
| | | EMPLOYEES THIS WEEK 21 |
| PCT LABOR OVERRUN THIS WEEK | 70.154 | |

| | | WEEK 57 | WEEK ENDING 10/11/ | | |
| | | PAGE 4 | PROJECT NUMBER 13-5265 | | |

| UNIT PRICE | | COST | | PROJECTED COST | |
| ESTIMATED | ACTUAL | ESTIMATED | ACTUAL | TO DATE OVER/UNDER | TO COMPLETE OVER/UNDER |
|---|---|---|---|---|---|
| | | | 499 | 499 | 499 |
| | | 10,000 | 6,307 | 2693- | 2,693- |
| | | 650 | 630 | 20- | COMPLETE |
| | | | 51 | 51 | NO BUDGET |
| | | 250 | | | |
| | | 500 | | | |
| | | | 261 | | NO BUDGET |
| | | | 3,670 | | |
| | | | 119 | | NO BUDGET |

WEEK 57    WEEK ENDING 10/11/

PAGE 5    PROJECT NUMBER 13-5265

| COST | | PROJECTED COST | |
| ESTIMATED | ACTUAL | TO DATE OVER/UNDER | TO COMPLETE OVER/UNDER |
|---|---|---|---|
| 373,295 | 295,793 | 12,363- | 6- |

293,142.97

**FIGURE 12.7  Payroll Information Flow**

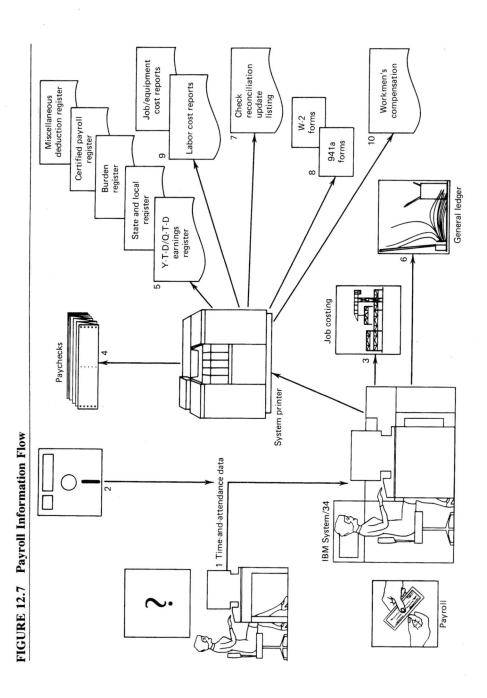

that vary considerably from the budgeted values and to utilize the variation information to take corrective actions as required. This approach of identifying accounts that vary greatly from budget values is called *management by exception*. Methods of analyzing these report data are discussed in Chapter 16.

An overview of the flow of information in the payroll module of the IBM Construction Management and Accounting System (CMAS) is shown in Figure 12.7. This diagram highlights the information flow into the system from field data and the various reports that are generated by the system. The figure also indicates the flow of information from payroll to the general ledger and job costing modules. The CMAS system generates a wide variety of output documents including

1. Paychecks (labeled 4)
2. Various earnings and deduction registers (labeled 5)
3. Labor cost and equipment cost reports (9)
4. Check reconciliation listing (7)
5. Workmen's compensation and worker tax forms (8 and 10)

**FIGURE 12.8   Labor Cost Report**

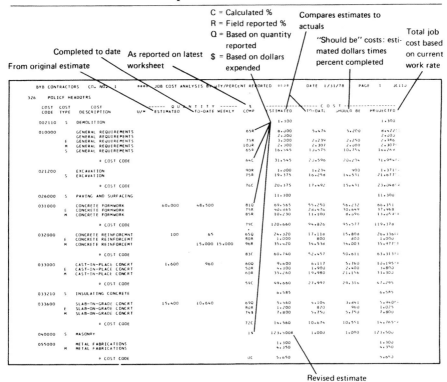

*Source:* Reprinted by permission from *Introducing the Construction Management and Accounting System for the IBM System/34.* © 1978 by International Business Machines Corporation.

## 12.3 QUANTITY REPORTING

Payroll data alone without information concerning the quantity of work produced by a given number of man-hours is of no assistance in determining the unit cost for purposes of comparison with the control budget. The acquisition of valid and correct quantities to match with resource (man-hour, equipment-hour, etc.) expenditures is key to the operation of an effective cost control system.

Reporting of quantities from the field is usually handled in two ways. Certain items of work (cubic yards of concrete poured, for example) lend themselves to daily reporting by the supervisor in the daily time report. Other work types, such as square feet of form-work placed, are best suited to weekly reporting by the job engineer or superintendent. The same cutoff date should be used for both reported quantities and costs, if these figures are to reflect actual unit costs.

In the CMAS system, the quantity estimates are submitted as input to the job cost module (see Figure 12.1). The labor cost by quantity report similar to Figure 12.8 is generated by the job cost module. Required quantity information is collected by using a computer-generated form that is filled out by field personnel and submitted on a weekly basis. This *job field report worksheet* is shown in Figure 12.9. A similar form for quantity reporting in a simple manual system is shown in Figure 12.10. This *job*

**FIGURE 12.9 Job Field Report Worksheet**

Prepared by job

As reported on preceding worksheet

Includes last week's

To be completed by field personnel and returned

XYB CONTRACTORS CO. NO. 1 **** JOB FIELD REPORT WORKSHEET **** DATE 1/31/78 PAGE 1 JC090

| JOB NUMBER | JOB NAME | COST CODE | COST TYPE | COST DESCRIPTION | PERCENT COMP TO-DATE | QUANTITY TO DATE | QUANTITY LAST WEEK | TOTAL PERCENTAGE | THIS WEEK QUANTITY | U/M |
|---|---|---|---|---|---|---|---|---|---|---|
| 326 | POLICE HEADQTRS | | | TOTAL JOB PERCENTAGE | 33 | | | .......... | TOTAL JOB | |
| | | 002110 | S | DEMOLITION | | | | .......... | .......... | |
| | | 010000 | | GENERAL REQUIREMENTS | | | | .......... | .......... | |
| | | | F | GENERAL REQUIREMENTS | 75 | | | .......... | .......... | |
| | | | M | GENERAL REQUIREMENTS | 100 | | | .......... | .......... | |
| | | | S | GENERAL REQUIREMENTS | 65 | | | .......... | .......... | |
| | | 021200 | S | EXCAVATION | 75 | | | .......... | .......... | |
| | | 026000 | S | PAVING AND SURFACING | | | | .......... | .......... | |
| | | 031000 | E | CONCRETE FORMWORK | 75 | | | .......... | .......... | |
| | | | M | CONCRETE FORMWORK | 85 | | | .......... | .......... | |
| | | 032000 | E | CONCRETE REINFORCEMT | 80 | | | .......... | .......... | |
| | | | M | CONCRETE REINFORCEMT | 96 | 15,000 | 15,000 | .......... | .......... | |
| | | 033000 | E | CAST-IN-PLACE CONCRT | 50 | | | .......... | .......... | |
| | | | M | CAST-IN-PLACE CONCRT | 60 | | | .......... | .......... | |
| | | 033210 | S | INSULATING CONCRETE | | | | .......... | .......... | |
| | | 033600 | F | SLAB-ON-GRADE CONCRT | 80 | | | .......... | .......... | |
| | | | M | SLAB-ON-GRADE CONCRT | | | | .......... | .......... | |
| | | 040000 | S | MASONRY | | | | .......... | .......... | |
| | | 055000 | M | METAL FABRICATIONS | | | | .......... | .......... | |
| | | 057000 | M | ORNAMENTAL METAL | 100 | 800 | 800 | .......... | .......... | |
| | | 062200 | M | MILLWORK | | | | .......... | .......... | |
| | | 071000 | S | WATERPROOFING | 45 | | | .......... | .......... | |
| | | 075000 | S | MEMBRANE ROOFING | | | | .......... | .......... | |
| | | 076000 | S | FLASHING & SHEET MTL | | | | .......... | .......... | |

*Source:* Reprinted by permission from *Introducing the Construction Management and Accounting System for the IBM System/34*. © 1978 by International Business Machines Corporation.

*progress report* requires the foreman or superintendent filling it out to write in the cost code and description for each work item. This form has the disadvantage that errors in cost coding can be made, thus causing a mistake in the number of hours recorded for a given cost account.

The source documents should be designed so that they are as simple as possible and require a minimum of time to prepare. Foremen normally have more than a full-time job in directing the work for which they are responsible, and by tradition they dislike paper work with a passion. If accurate reporting is made as easy as possible, the foreman can have more time available for directing the work at hand.

If the field reporting system is to operate properly, it must be enforced. Supervisors and others who are responsible for input and preparation of cost data must be made aware of management's commitment to the system. If late, careless, or incomplete reporting is either tolerated or halfheartedly criticized by management, the system will stand little chance of success. In this particular situation, top management's vocal support for the cost system is of great importance, since anything less could, and most likely will, be taken as support for the opposite position.

Finally, management should use discretion in confronting a foreman whose job has high cost reports. The purpose of cost reporting is to provide feedback from projects rather than a "club" to use on foremen. If caution is not exercised in such a situation, the foreman could decide to falsify the report as much as is necessary to escape management's wrath. A good solution to such a problem rarely exists. However, some type of action must be taken to prompt the foreman to improve field efficiency. If the foreman is on an incentive or profit-sharing plan, he has some motivation to take an interest in the cost reports and to try to correct cost overruns. Second, if a foreman's good performance is stressed along with the bad, the overall effect may be to prompt the necessary corrective action without producing motivation to falsify field data.

Should these approaches to the problem be unsuccessful, it may be necessary to

**FIGURE 12.10   Job Progress Report—Quantities of Work Performed**

| Cost Code | Contract Item No. | Description | Unit | Quantity | | | |
|---|---|---|---|---|---|---|---|
| | | | | Previous Report | This Period | Adjustment | To Date |
| | | | | | | | |
| | | | | | | | |
| | | | | | | | |
| | | | | | | | |
| | | | | | | | |

place a timekeeper, who is responsible to the home office, in charge of the cost reports. While this procedure will most likely eliminate bias in field reporting, it will also increase project overhead.

"Garbage in, garbage out" is a phrase commonly used in the computer business to indicate that if erroneous input is submitted to the computer, erroneous information will be generated. This is particularly true in a cost control system. A mistake by the supervisor in allocation of man-hours to a particular account can cause strong fluctuation in the cost per unit value. Similarly, an incorrect estimate of the number of units placed on a particular account during a particular period will cause sharp variation of cost per unit figures.

Consider the following simple illustration in the context of the placement of concrete for a flat slab in a high-rise building. Assume that the crew size for direct labor is six workers and the foreman reports 48 man-hours, or eight crew-hours, against cost account 80.260.10. Assume that the hourly rate for the crew is $120, yielding a charge of $960 for the shift. If the quantity estimate for the amount of concrete placed during the shift on the slab is 20 cubic yards (based on the dimensions of the slab) the cost per yard is $48. If the foreman is careless and enters the delivered quantity of 32 cubic yards as the amount placed, the labor cost is incorrectly reduced to $30 per yard. This is a substantial difference (60 percent) in cost and emphasizes how radically unit costs can be influenced by inaccurate or faulty field reports.

If through a mistake the foreman had allocated half of the crew's time to beam rather than slab concrete but used the 20-cubic-yard slab quantity, the cost per unit of slab concrete would become $24 per cubic yard. This simple illustration indicates that mistakes in field reporting generate a range of prices from $24 to $48 per cubic yard. Radical shifts of this type render the data generated unreliable at best. To improve reliability, a well-established reporting procedure defining the basis upon which field reports are to be made is needed. The first requirement in field reporting is to require that those individuals who are responsible for filling out the source documents understand the correct method to prepare them. This problem can largely be solved through proper training.

## 12.4   PURCHASING MATERIALS AND SERVICES

Transactions that involve the purchase of materials and services (other than internal payroll) flow through the accounts payable system and generate data that provide a source of additional management information. Contractors typically purchase materials and equipment with a purchase order. Services are procured by means of a subcontract.

The purchase order is a purchase contract between the contractor and the supplier. This document depicts the material to be supplied, their quantities, and the dollar amount of the purchase order.

Purchase orders vary in complexity; they can be as simple as a mail-order house (e.g., Sears) order form or almost as complex as the construction contract itself. When complex and specially fabricated items are to be included in the construction, very detailed specifications and drawings become part of the purchase order. Some typical purchase order forms are shown in Figures 12.11 and 12.12. Figure 12.11 shows a

**FIGURE 12.11    Field Purchase Order**

## Special Purchase Order

**HCB**    HENRY C BECK COMPANY

VENDOR:

MAIL INVOICE TO:
HENRY C. BECK COMPANY
1210 S. Old Dixie Highway
Jupiter, Florida 33468

DATE:

CHG. TO JOB    # 21330 ____

SHIP TO: 1210 S. Old Dixie Highway / Jupiter, Florida 33468

| QUANTITY | ARTICLE | U.P. | AMOUNT | COST CODE |
|----------|---------|------|--------|-----------|
|          |         |      |        |           |
|          |         |      |        |           |
|          |         |      |        |           |
|          |         |      |        |           |
|          |         |      |        |           |
|          |         |      |        |           |
|          |         |      |        |           |

**STATE AND LOCAL SALES TAXES MUST BE SET OUT SEPARATELY ON INVOICE**

Invoice in Triplicate
To Above Address
No Later Than 25th of Month —— Vendor's Acceptance (when required)
Show S.P.O. Number On Invoice    WHITE (ORIGINAL) - VENDOR'S COPY
CANARY    —JOB OFFICE COPY
(MAIL TO DALLAS WITH INVOICE)

**SUPT. OR PROJECT MGR.**
PINK    - SUPERINTENDENT'S COPY
GOLDENROD - PROJECT MANAGER'S COPY

*Source:* Courtesy of Henry C. Beck Company.

form for field-purchased items procured from locally available sources. Such items are usually purchased on a cash-and-carry basis. The purchase order in this case is used primarily to document the purchase for record-keeping and cost-accounting purposes (rather than as a contractual document). A more formal purchase order used in a contractual sense is shown in Figure 12.12. It is used in the purchase of more complex items from sources that are remote to the site.

When the contract for construction is awarded, the contractor immediately begins awarding subcontracts and purchase orders for the various parts of the work. How much of the work is contracted depends on the individual contractor. Some contractors subcontract virtually all of the work in an effort to reduce the risk of cost overruns and to have every cost item assured through stipulated-sum subcontract quotations. Others perform almost all the work with their own field forces.

The subcontract agreement defines the specialized portion of the work to be performed and binds the contractor and subcontractor to certain obligations. The subcontractor, through the agreement, must provide all materials and perform all work described in the agreement. The Associated General Contractors (AGC) of America publish a Standard Subcontract Agreement for use by their members.

A sample of this agreement can be found in Appendix D. Most contractors either

## FIGURE 12.12 Formal Purchase Order

Letter or transmittal form accompanying this order when mailed to Vendor should show the number of shop drawings and/or samples to be furnished and the address to which they must be sent; also the address to which Vendor is to mail correspondence relating to this order.

PURCHASE ORDER

**HCB** HENRY C BECK COMPANY          **No.**

VENDOR

ADDRESS

_____**19**_____

JOB:

Job Mailing Address:

Please ship the following to HENRY C. BECK COMPANY, at

SHIP VIA:

It is agreed that shipment will be made on or before                    or right is reserved to cancel order.

IMPORTANT NOTE: It is IMPERATIVE in the interest of prompt payment that all invoices be rendered in the original with two (2) copies. Mail together with two (2) copies of bills of lading and/or other papers to JOB at address above.

| ITEM NO. | QUANTITY | DESCRIPTION | UNIT | AMOUNT |
|---|---|---|---|---|
| | | | | |
| | | | | |
| | | | | |
| | | | | |
| | | | | |
| | | | | |
| | | | | |
| | | | | |

SALES or USE TAX (is) (is not) included in amounts shown above.          HENRY C. BECK COMPANY

F.O.B.

TERMS:                                                        By_____

See above IMPORTANT NOTE for invoicing instructions. They MUST be complied with.

Accepted:_____

Show above order number on invoices, and on the outside of each package containing Shipment.

By_____

*Soure:* Courtesy of Henry C. Beck Company.

adopt a standard agreement, such as that provided by the AGC, or implement their own agreement. In most cases, a well-defined and well-prepared subcontract is used for subcontracting work.

The sequence of actions involved in procuring materials and services is shown in Figure 12.13. The sequence begins during the estimate preparation, at which time the need for materials and services from specialty subcontractors is identified. For work to be undertaken by the prime or general contractor, a formal *bill of materials* (similar to the bill in Figure 10.8) may be prepared for each work package, delineating the types and amounts of materials to be procured. If the firm has a purchasing group or department, all bills of materials are referred to the purchasing group by a *requisition*. The requisition is a request to purchase and instructs the purchasing group as to what is to be purchased and the specifications and characteristics of the supplied item.

**FIGURE 12.13    Sequence of Actions Relating to material and Subcontract Procurement**

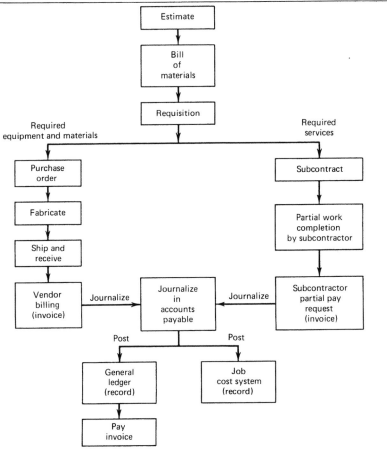

Procedural aspects of the procurement, such as the need to submit drawings for the engineer's approval, must also be reflected in the requisition. On the basis of the requisition, the purchasing department studies manufacturer's catalogues, solicits quotations, selects the lowest or most responsive quotation, prepares the purchase order, and enters into a contract. If a subcontract is involved, the negotiation of the subcontract is coordinated by the purchasing group in cooperation with the project manager and his staff.

Some companies prefer to assign the purchasing function for all but the most major equipment items to the field staff. In such cases, the procurement activity is divided between project staff and purchasing specialists answering to a senior executive for procurement in the home office.

Actions taken by field staff in arriving at the point of issuing a purchase order or subcontract may be less formal than those just described. Nevertheless, the actions described to arrive at issuance of a purchase order are accomplished either directly or implicitly. Once a purchase order or subcontract is signed, the basis for cost is fixed. Such costs associated with the purchase orders and subcontracts on a particular job are called *committed costs*. That is, under the contractual provisions of the purchase order or subcontract, the contractor is committed to pay a given price for a defined item or service.

Receipt of items under a purchase order triggers the incurring of an expense. To be more precise, receipt of an invoice (i.e., billing statement) for an item received results in the accrual of an expense. Some purchase orders provide for the partial payment of the vendor for work accomplished toward completion of an end item. In such cases, invoices for partial payment result in incurring expense. Under subcontracts, it is typical for the subcontractor to submit partial payment requests, which result in the accrual of expense. All such invoiced expenses are processed through accounts payable and provide raw data for use by management.

When journalized, these costs are captured both for the general ledger as well as the job cost systems. The transactions are ultimately posted to both systems. The G/L system picks up the transaction as a posting against the appropriate vendor or subcontractor account (e.g., Jones Glass Co., Smith Plumbing, Inc.). A parallel posting to the appropriate job cost ledger (or ledgers) results in the generation of management data. These accounts payable data, combined with labor cost data from payroll, provide the basis for developing unit cost information to be compared with that used in the original estimate. The assembly of these data to generate a project cost status report is shown schematically in Figure 12.14.

## 12.5   DISTRIBUTION OF COST DATA TO JOB COST SYSTEM (FROM ACCOUNTS PAYABLE)

In much the same fashion that labor costs are distributed to appropriate accounts in the job cost system, materials and subcontract costs can be distributed to the job cost system. This distribution is normally carried out by the job cost or site office engineer.

**FIGURE 12.14   Flows to Project Cost Status Report**

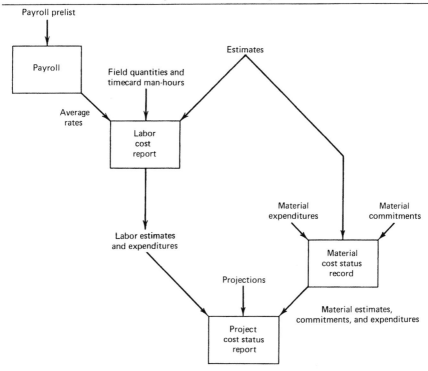

In the case of a subcontract, the job cost accounts are typically set up so that one account covers a particular subcontract, and vice versa. That is, the subcontract covers one cost account, and the decision to subcontract this cost account is made during the preparation of bid. To illustrate, if it has been decided to subcontract glass installation on the job, a single cost account for glazing accumulates the charges from the glazing subcontractor in the job cost ledger. In effect, all requests from the subcontractor for payment are recorded against the glass installation subaccount in the job cost ledgers. The target value is the amount of the subcontract. The committed cost associated with this account will, at any time, be the amount of the original subcontract amount still to be paid out.

In some cases, the general contractor may desire to distribute work under a single subcontract to several job cost accounts. This may occur if a single subcontractor is handling several major subelements of the project (e.g., heating, ventilating, and air conditioning) and it is desired to maintain cost records on each subelement. It may also occur when the general contractor's subaccount structure is so detailed as to require distribution of effort from a single subcontractor. Such as procedure may be required of the general contractor under his prime contract. As noted previously, certain work, such as power plant construction, requires that the contractor submit job effort and cost reports in a predefined cost code format. When such codes are detailed, the work of a single subcontractor may have to be distributed to several cost codes.

Fabricated equipment items procured by using a purchase order are normally charged to a single job cost account. However, bulk and semi-finished items often result in charges to several cost accounts. The obvious example is the procurement of transit mix concrete from a single concrete supplier. The concrete purchased may be used for footers, slabs both on grade and elevated, columns, beams, and a multitude of other cost centers. In such cases, the quantity and cost of the concrete must be distributed to the appropriate job cost account. Items such as electrical cable, cable trays, pipe, hangers, and the like are similar, in that they must be distributed to a number of differing cost centers.

The basis and method for distributing material and service (i.e., subcontract) costs to job cost accounts is extremely important, since, as was illustrated in the discussion of labor costs, unit costs are very sensitive to the quantity and cost values distributed. Small variations in quantities can lead to large variations in unit cost.

## 12.6   SUMMARY

This chapter has addressed the topic of data collection and reporting for the purpose of managing construction progress. The primary source of data on costs is the financial accounting system. These data, combined with field reports allocating man-hours to appropriate cost accounts and indicating quantities placed, allow the development of unit costs. These unit costs can be compared with estimated costs to detect overruns and underruns. The reported costs are very sensitive to the allocation of effort and the accuracy of reported quantities. In many systems, allocation of effort and quantity reporting is handled by field supervision. Therefore, failure to impress upon field personnel the importance of consistency and accuracy in reporting can lead to unreliable status reports and cost information. Top management must insure that these field data are as accurate as possible.

## REVIEW QUESTIONS AND EXERCISES

**12.1**   Why does a contractor have to "certify" his payroll, and who requires this? Name at least three interested agencies.

**12.2**   Name and discuss the types of reports produced by a typical payroll system. Give at least
(a) Two developed on a weekly basis
(b) One developed on a quarterly basis
(c) Two developed on an as-required basis

**12.3**   Name and discuss two types of reports produced by a typical computerized cost-control system.

**12.4**   Is it more important to be on schedule or to be within the budget? Elaborate on your ideas, using different examples.

**12.5** What meaningful data (besides cost data) should be collected on a major concrete-paving job or a new airfield? Of what value are they to the manager?

**12.6** The following planned figures for a trenching job are available:

| Excavation—second hauling | Quantity yd³ 100,000 | Resources | | Cost |
|---|---|---|---|---|
| | | Machines | 1000 hours | $100,000 |
| | | Labor | 5000 hours | $100,000 |
| | | Trucks | 2000 hours | $ 62,500 |

At a particular time during the construction, the site manager realizes that the actual excavation will be in the range of 110,000 yd³. Based on the new quantity he figures that he will have 30,000 yd³ left.

From the main office, the following job information is available.

| Resources | | Cost |
|---|---|---|
| Machines | 895 hours | 85,000 |
| Labor | 6011 man-hours | 79,000 |
| Trucks | 1684 hours | 50,140 |

What would concern you as manager of this job?

# Chapter 13

# ACCOUNTING FOR EQUIPMENT COSTS

## 13.1 THE NATURE OF EQUIPMENT AND JOB OVERHEAD COSTS

Cost accounting as it relates to equipment and overhead expenses differs somewhat from its application in accounting for labor and materials. Labor and material charges are linked directly to portions of the work and, therefore, can be allocated or distributed directly to a physical cost center within the project. Such items are directly linked to production units or items that, taken as a whole, represent the total project. Equipment and overhead charges *support* the production process. In this support function, they may be more difficult to distribute to individual cost centers representing concrete beams and columns, or installed air conditioning ducts (i.e., physical end items). To illustrate, the generator powering a lighting set used for night operations may support a wide variety of operations. To which accounts should its costs be distributed? The project manager and job superintendent supervise the project as a whole. To which physical cost centers should their time and salaries be charged? Distributing these support and management resources to physical subelements of the project in a manner that throws light on the cost of a particular portion of the project is virtually impossible.

One straightforward way of addressing this problem is to set up individual cost accounts for each of the support resources. By so doing, expenses associated with the resource are charged to the individual account established for that resource. This is the approach utilized in the labor cost report shown in Figure 12.6. It will be noted that separate items have been set up for the field engineer (item 60), the office engineer (item 61), and the job site trailer (item 22). Other job site overhead items are represented by control account line items in this report (e.g., ice water and hauling trash). Instead of distributing the charges associated with these support items to production accounts, separate line items are established.

In accounting for major equipment items, the establishment of individual line items for each machine is recommended. This means that a separate account for the generator and lighting set mentioned above would be established. Equipment items are looked upon as "profit centers" in support of the overall construction cost accounts established for each work category or item. One account number is assigned

. . . to each type of equipment used, with a subaccount number for each equipment unit. For an unusually large equipment fleet . . . sub-subaccount numbers may also be used.

Recommended groupings are as follows:

1. Automotive Equipment
   Personnel Vehicles
   Trucks
   Trailers
2. Construction Equipment
3. Fixed Plant
4. Marine Equipment
5. Aviation Equipment

Each equipment unit should be identified by a number that identifies the type as well as unit. For example, Bulldozers may be assigned type number 12, with each unit numbered from one chronologically as purchased, allowing sufficient numbers for each type for a ten-year period before reusing numbers previously assigned. A Supplementary designation should be used for rented/leased equipment which should be included for costing.*

This procedure amounts to setting up a separate subsidiary ledger for each major equipment and overhead item in the job cost system. As expenses are incurred, the appropriate job cost account is debited to reflect the actual charge. For instance, if a new carburetor is installed on a scraper tractor 601, the entry would indicate a debit to the equipment account and a credit to the vendor account:

|                                        | Dr     | Cr     |
|----------------------------------------|--------|--------|
| Scraper tractor 13.601                 | XXXXX  |        |
| Yancy Brothers, Inc. (Acct. payable)   |        | XXXXX  |

In addition to direct costs such as the repair noted above, equipment items lose value over time due to wear and tear as well as obsolescence. Industry practice provides for recovery of this value through a charge to the client. The basis for recovering this ownership cost as well as other categories of equipment-associated costs are discussed in the following sections.

## 13.2  EQUIPMENT COSTS

Costs associated with equipment are of two types. One cost is the loss of value of a piece of equipment due to obsolescence or usage. This is accounted for in terms of depreciation. Amounts to amortize the equipment may be charged to the client. That is, it is typical to charge the client an amount that will be escrowed to provide funds to replace the equipment at the end of its service life.

---

* Dan S. Brock, "Cost Accounting for Highway Contractors—A System for Cost Control," Washington, D.C.: American Road Builders' Association, 1971).

The second cost category is related to other expenses incurred in owning and operating the equipment.* The other (nondepreciation) ownership costs typically relate to payment of

1. Insurance
2. Interest on the note (if the equipment is purchased by borrowing)
3. Taxes
4. Storage

Operational costs relate to:

1. Replacement of tires (if the vehicle has rubber tires)
2. Fuel, oil, and grease
3. Repair and maintenance (to include replacement of expendable items—e.g., batteries, rubber covers, and the like)

All of these costs are discussed in detail in Chapter 9 of *Construction Management* by Halpin and Woodhead.

Accounting for equipment costs is complicated because part of the charge is for annual costs (equipment replacement, insurance, taxes, and so on) that are projected and must be prorated to individual equipment pieces. The other part is for actual expenditures for fuel, repair parts, and other operational support items that are associated with the equipment's operation. Actual billings for fueling several pieces of equipment must be distributed to each machine. They may be further distributed to the cost account on which the machine is working. Many contractors choose to avoid the second distribution, and, as described in Section 13.1, establish individual equipment accounts that collect all charges on an equipment piece-by-piece basis. On equipment-intensive work such as excavation, however, the only way of obtaining a good unit price is to distribute operational cost (e.g., fuel, oil, lubrication) to the machine and then to the cost account. Distribution to the machine allows management to determine the relevant cost effectiveness of various machines. Distribution to the cost account allows for comparison with budgeted costs and generation of pricing data for future estimates.

The accounting difficulties are simplified for rented pieces of equipment. The rental amount is simply invoiced by the vendor providing the equipment. This charge is then distributed as required to the appropriate cost accounts. Both rental costs as well as operational costs are charged against accounts payable and the appropriate cost account within the job cost system in the same manner as a material or service purchase. Since the tax and financial accounting for ownership cost is the responsibility of the equipment rental company, the contractor can handle the problem in a neat and straightforward manner. Some contractors establish separate companies to hold their equipment and,

---

* Depreciation is normally considered to be part of the ownership cost.

in effect, "rent" their own equipment. That is, the contractor rents equipment from a separate company owned by himself. This method simplifies equipment costing and has other tax-related advantages.

For contractors who own their equipment and charge it to the client within the framework of the job cost system, the industry has adopted the practice of using precalculated *charge rates* to allocate cost to the individual jobs and cost centers. These charge rates typically cover ownership costs and may cover both ownership and operating costs as described above. The rates are updated on a job-by-job or periodic basis to reflect recent cost experience. A reconciliation is made as required between the actual billed costs for operational expenses and the portion of the charge rate covering operational expenses. If the charge rate is too low, the contractor is incurring costs on the job that are not being expensed to the job. If the charge rate is too high, the contractor is posting costs to the job that are in excess of actual incurred cost. From a tax standpoint, it is advantageous to overcharge, since doing so adds apparent expense to the job and reduces taxable income. Charge rates used by contractors can be referenced to an evaluation of ownership and operational costs for each individual equipment item. Some contractors use a percentage of the rental rates published annually by the Associated Equipment Distributors (AED) as the basis for their charge.

## 13.3  REPORTING EQUIPMENT COSTS

Reporting of equipment costs is similar in many respects to labor cost reporting. Some record of hours worked must be maintained as a basis for expensing the equipment to a job. It is possible to record equipment time either with labor on the labor time sheet or separately on another source document. If separate equipment time cards are used, it becomes possible to distribute some of the responsibility for field reporting to the equipment operator. The actual data that are recorded depend largely on the method used by the contractor to account for equipment costs.

Regardless of how these costs are accounted for, an equipment time distribution report generally includes the following information:

1. Job name and number
2. Cost code and time worked
3. Idle time
4. Downtime for repairs
5. Repairs, maintenance supplies, and the like needed

A typical equipment time sheet is shown in Figure 13.1.

The treatment and recovery of equipment costs varies in accordance with the difference in operational conditions of various companies and differences in opinion among contractors themselves. In an effort toward reaching a common ground, some authorities and contractor organizations have prepared guidelines to be used as a basis for recovering expenses associated with equipment.

**FIGURE 13.1  Foreman's Daily Equipment Distribution Report**

DAILY TIME DISTRIBUTION
EQUIPMENT

Report No. _____

Foreman's Daily Report

Station

Cost Code

Company

Job Name _____ No. _____

Foreman _____

Date _____ 19 ___

Day*  M  T  W  T  F  S  S
       1  2  3  4  5  6  7

Describe Work Done

Report work done, repairs needed, repairs made, equipment transfers and receipts, diary etc. on reverse side

| Equipment No. | Description | Use Rate | Productive | | | | | Hours | | | | | Total | | Idle Hours | | | Total Hours Available | Down for Repairs | Total Shift Hrs. |
|---|---|---|---|---|---|---|---|---|---|---|---|---|---|---|---|---|---|---|---|---|
| | | | Hrs. | Amt. | Hrs. | Amt. | Hrs. | Amt. | Hrs. | Amt. | Hrs. | Amt. | Hrs. | Amt. | Weather | No Work | Other | | | |
| | | | | | | | | | | | | | | | | | | | | |
| | | | | | | | | | | | | | | | | | | | | |
| | | | | | | | | | | | | | | | | | | | | |
| | | | | | | | | | | | | | | | | | | | | |

*Circle appropriate letter

Totals

Approved by _____

Foreman's Daily Equipment Distribution Report

*Source:* Dan S. Brock, *Cost Accounting Manual for Highway Contractors* (Washington, D.C.: American Road Builder's Association, 1971).

Annual equipment costs that are commonly considered include the following items:

1. Depreciation \
2. Interest on investment /     Ownership costs (fixed)
3. Repairs
4. Maintenance     Operating costs
5. Lubrication     (variable)
6. Fuel

The Associated General Contractors has developed a method for obtaining annual equipment ownership expense that is widely used as a framework for charging equipment to a job. The AGC method is described in the publication *Contractors' Equipment Ownership Expense Manual*. The annual equipment ownership expense, as identified in this manual, includes depreciation, replacement cost escalation, interest on investment, taxes, insurance, and storage. In addition, the total annual equipment expense includes ownership expense plus four other factors:

1. Repair and maintenance expense
2. Operating expenses
3. Operating labor expenses
4. Mobilization and demobilization expenses

AGC gives recommended values for average hourly ownership expense and average hourly repair and maintenance expense expressed as a percentage of the new acquisition cost of each unit of equipment. The remaining elements of the total equipment expense rate can generally be determined from the contractor's cost records.

This method makes provision for replacement cost escalation. The replacement cost factor is designed to generate sufficient capital during the economic life of a piece of equipment to insure its replacement at the end of its depreciation period. The replacement unit is considered to be equal to the original unit in all respects. Users of this manual are cautioned that funds generated by this factor will not be sufficient to produce a replacement unit of equipment that has been "updated" through innovations and changes. An explanation of the concept used in the AGC method is given in Appendix E.

To illustrate the AGC method of calculating the hourly charge rate, consider the following situation. A highway contractor has purchased a wheeled tractor. The first cost for the tractor is $54,800. The "new value" under the AGC method is calculated by subtracting the value of the tires, which are expensed on a different basis than the machine is. The cost of the machine includes the ad valorem tax paid as well as the costs to ship the machine to the contractor's yard. These costs are summarized as follows:

| | |
|---|---|
| First cost | $54,800 |
| Less tires | 9,400 |
| | $45,400 |
| Tax @ 5% | 2,270 |
| Freight | 1,330 |
| "New Value" | $49,000 |

**FIGURE 13.2  AGC Table for Cost Recovery of Ownership Costs**

| Equipment | Average Annual Expense Percent of Capital Investment Without Field Repairs | | | | Average Use Months Per Year | Expense Per Working Month Percent | Application of A.G.C. Schedule to Owner's Values | |
|---|---|---|---|---|---|---|---|---|
| | Depreciation | Overhauling, Major Repairs Painting | Interest Taxes Storage Insurance | Total Ownership Expense | | | Value Dollars | Expense Per Working Month Dollars |
| | | | | | | | (Fill in your own values) | |
| TOWERS (Cont'd.) | | | | | | | | |
| Steel, tubular (cont'd.) | | | | | | | | |
| Heavy: (Max. live load 5,000 lbs; or 35 cu. ft. concrete bucket; 2 or 3 wheelbarrows) | | | | | | | | |
| Single tower, 39–300 ft. | 20 | 8 | 11 | 39 | 8 | 4.9 | —— | —— |
| Double tower, 39–300 ft. | 20 | 8 | 11 | 39 | 8 | 4.9 | —— | —— |
| Concrete apparatus for any of above towers | 25 | 15 | 11 | 51 | 7 | 7.3 | —— | —— |
| Steel, portable | | | | | | | | |
| Including hoist and elevator, complete | | | | | | | | |
| 20–40 ft. height | 20 | 10 | 11 | 41 | 8 | 5.1 | —— | —— |
| 10 ft. section | 20 | 10 | 11 | 41 | 8 | 5.1 | —— | —— |
| TOWER EQUIPMENT | | | | | | | | |
| Buckets, 7–36 cu. ft. | 20 | 16 | 11 | 47 | 8 | 5.9 | —— | —— |
| Hopper, vertical back 10–60 cu. ft. | 20 | 20 | 11 | 51 | 8 | 6.4 | —— | —— |
| Sheaves, top tower, 12–18 in. | 50 | 15 | 11 | 76 | 8 | 9.5 | —— | —— |
| Swivel, bottom, 12–18 in. | 50 | 15 | 11 | 76 | 8 | 9.5 | —— | —— |

# TRACTORS

**Crawler, diesel engine, gear drive**

Drawbar h.p.

| From | To | | | | | | |
|------|------|----|----|----|----|----|-----|
| 20 | 52 ...... | 25 | 15 | 11 | 51 | 9 | 5.7 |
| 53 | 265 ..... | 20 | 15 | 11 | 46 | 8 | 5.8 |

torque converter drive

| *30 | 331 ..... | 20 | 15 | 11 | 46 | 8 | 5.8 |
|-----|-----------|----|----|----|----|---|-----|

**Crawler, gasoline engine**

Drawbar h.p.

| From | To | | | | | | |
|------|------|----|----|----|----|----|-----|
| 20 | 32 ...... | 33 | 15 | 11 | 59 | 9 | 6.6 |
| 33 | 41 ...... | 30 | 15 | 11 | 56 | 9 | 6.2 |
| 42 | 52 ...... | 28 | 15 | 11 | 54 | 9 | 6.0 |
| 53 | 66 ...... | 25 | 15 | 11 | 51 | 9 | 5.7 |
| 67 | 105 ..... | 20 | 15 | 11 | 46 | 8 | 5.8 |

**Four-Wheeled, rubber tired**

Diesel engine

Belt h.p.

| From | To | | | | | | |
|------|------|----|----|----|----|----|-----|
| 38 | 47 ...... | 30 | 15 | 11 | 56 | 9 | 6.2 |
| 48 | 60 ...... | 25 | 15 | 11 | 51 | 9 | 5.7 |
| 93 | 300 ..... | 20 | 15 | 11 | 46 | 8 | 5.8 |
| *301 | 600 ..... | 20 | 15 | 11 | 46 | 8 | 5.8 |

*Source:* Reprinted by permission from *Contractors Equipment Ownership Expense Manual.* © 1966 by The Associated General Contractors of America.

By referring to a table similar to that shown in Figure 13.2, we find the annual ownership recovery percentage to be 46 percent. The average use in months per year is 7. The recovery percentage per month is 6.6 percent. The number of straight hours available for the equipment to work per month is taken as 176 (22 days/month × 8 hours/day) when the AGC method is used. Therefore the straight time rate to recover ownership costs is calculated as

$$\text{S.T. rate} = \$49,000 \times \left(\frac{0.066}{176}\right) = \$18.40$$

For hours worked in excess of 8 per day, an adjusted overtime rate is calculated by using the AGC method. The assumption is that the rate for hours in excess of 8 are charged at half (50 percent) of the straight-time rate. Therefore, the charge rate for overtime hours is $9.20 per hour.

Assume that the hourly operator's rate and the operating expenses are as follows:

|  | Straight Time | Overtime |
|---|---|---|
| Operator | $ 20.00 | $ 30.00 |
| Operating expense* | 25.00 | 25.00 |

The hourly charge rate would then be as follows:

|  | Straight Time | Overtime |
|---|---|---|
| Ownership | $ 18.40 | $ 9.20 |
| Operator and Operating Expense | 45.00 | 55.00 |
| Total | $ 63.40 | $ 64.20 |

The "new value" used in the AGC method includes (1) the cost of the equipment plus taxes, (2) freight to the point of use, (3) sales and use taxes, and (4) license. The cost of tires, if any, is deducted. The *convention* of charging 50 percent of the straight recovery rate for hours worked over 8 per day allows for the increased wear and tear on the equipment due to extra use.

## 13.4   DEPRECIATION OF EQUIPMENT

The method by which depreciation is calculated for tax purposes must conform to standards established by the Internal Revenue Service. Under the Tax Reform Act of 1981, all equipment placed in service after 1980 must be depreciated by either the Accelerated Cost Recovery System (ACRS) method or the alternate ACRS.† Table 13.1 gives ACRS depreciation amounts. Figure 13.3 shows graphically the depreciation

---

* This includes expenses incurred while machine is in operation (e.g., fuel, oil, etc.).

† The alternate ACRS is essentially a linear or straight-line method. See U.S. Congress Joint Committee on Taxation, *General Explanation of the Economic Recovery Tax of 1981*, H.R. 4242, 97th Congress, Public Law 97-34, pp. 7582, December 1981.

**TABLE 13.1   For Property Placed in Service 1981–84 (Recovery Percentage)**

|  | Class of Property | | | |
|---|---|---|---|---|
|  | *3-Year* | *5-Year* | *10-Year* | *15-Year Public Utility Property* |
| Recovery year | | | | |
| 1 | 25 | 15 | 8 | 5 |
| 2 | 38 | 22 | 14 | 10 |
| 3 | 37 | 21 | 12 | 9 |
| 4 |  | 21 | 10 | 8 |
| 5 |  | 21 | 10 | 7 |
| 6 |  |  | 10 | 7 |
| 7 |  |  | 9 | 6 |
| 8 |  |  | 9 | 6 |

for a $100,000 asset with five-year service life. The revised ACRS amounts effective in 1985 are shown in Table 13.2, and those effective after 1985 are shown in Table 13.3. Equipment placed in service before 1981 can be depreciated by using one of the following methods:

1. Straight-line
2. Declining-balance
3. Sum-of-years-digits
4. Production

**FIGURE 13.3   Depreciation Using ACRS method**

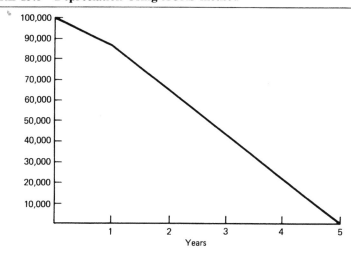

**TABLE 13.2   For Property Placed in Service in 1985 (Recovery Percentage)**

|  | Class of Property | | | |
|---|---|---|---|---|
|  | 3-Year | 5-Year | 10-Year | 15-Year Public Utility Property |
| Recovery year |  |  |  |  |
| 1 | 29 | 18 | 9 | 6 |
| 2 | 47 | 33 | 19 | 12 |
| 3 | 24 | 25 | 16 | 12 |
| 4 |  | 16 | 14 | 11 |
| 5 |  | 8 | 12 | 10 |
| 6 |  |  | 10 | 9 |
| 7 |  |  | 8 | 8 |
| 8 |  |  | 6 | 7 |
| 9 |  |  | 4 | 6 |
| 10 |  |  | 2 | 5 |
| 11 |  |  |  | 4 |
| 12 |  |  |  | 4 |
| 13 |  |  |  | 3 |
| 14 |  |  |  | 2 |
| 15 |  |  |  | 1 |
| Total | 100 | 100 | 100 | 100 |

**TABLE 13.3   For Property Placed in Service After 31 December 1985 (Recovery Percentage)**

|  | Class of Property | | | |
|---|---|---|---|---|
|  | 3-Year | 5-Year | 10-Year | 15-Year Public Utility Property |
| Recovery year |  |  |  |  |
| 1 | 33 | 20 | 10 | 7 |
| 2 | 45 | 32 | 18 | 12 |
| 3 | 22 | 24 | 16 | 12 |
| 4 |  | 16 | 14 | 11 |
| 5 |  | 8 | 12 | 10 |
| 6 |  |  | 10 | 9 |
| 7 |  |  | 8 | 8 |
| 8 |  |  | 6 | 7 |
| 9 |  |  | 4 | 6 |
| 10 |  |  | 2 | 5 |
| 11 |  |  |  | 4 |
| 12 |  |  |  | 3 |
| 13 |  |  |  | 3 |
| 14 |  |  |  | 2 |
| 15 |  |  |  | 1 |
| Total | 100 | 100 | 100 | 100 |

These methods are described in detail in Chapter 10 *Construction Management* by Halpin and Woodhead. Declining-balance and sum-of-years-digits are referred to as *accelerated methods,* since they allow larger amounts of depreciation to be taken in the early years of the life of the asset. The contractor usually selects a method that offsets or reduces the reported profit for tax purposes as much as possible. If taxable profits are anticipated to be large in a given year, then the company will try to take as much depreciation as possible and will use an accelerated method on as many asset items as possible. Most heavy construction contractors attempt to utilize the depreciation available on each piece of equipment to reduce or alter the incidence of the taxable profit or income it will generate. Each individual asset may be depreciated separately under IRS rules. The contractor may establish as many accounts as desired. He can use a separate account for each item, or two or more items may be included in one account. Asset items may also be grouped by common useful lives or common uses. In a similar approach, a composite account may be used without special regard for the character of the depreciable assets involved.

The major factors to be considered in calculating the depreciation of an asset are shown in Figure 13.4. Three major factors from the three sides of the depreciation "box" are linked by the method of depreciation selected. They are

1. Initial cost or basis in dollars
2. Service life in years or hours
3. Salvage value in dollars

The amount that can be depreciated or claimed by way of a tax deduction is the difference between the initial net value of the asset and its residual or salvage value. This *depreciable amount* establishes the maximum number of depreciation dollars available in the asset during its service life.

**FIGURE 13.4  Factors in Depreciation**

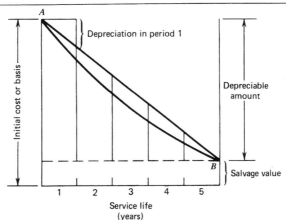

The declared initial cost of the asset must be acceptable in terms of the IRS definition of depreciable cost. For instance, suppose a $75,000 scraper is purchased. The tires on the scraper cost $15,000. They are considered a current period expense and therefore are not depreciable. That is, they are not part of the capital asset for purposes of depreciation, just as a typewriter ribbon is not a depreciable portion of a typewriter. Such items are considered current period expenses. The initial cost of the scraper for depreciation purposes is $60,000. The initial depreciable cost or basis is often referred to as the net first cost or "new value," as described in connection with the AGC method. In addition to the purchase price minus major expense items such as tires, items such as freight costs and taxes are included in the net first cost and are part of the amount depreciable.

Each asset has a residual or scrap value that is referred to as the *salvage value*. The concept is that, toward the end of its useful life, the value of the asset will not decline below this residual value. In the case of machines, the residual value is normally assumed to be the sale or trade-in value at the end of its normal service life or a scrap value, if the machine is sold as scrap for recycling. Equipment vendors can provide information concerning a reasonable and acceptable level of salvage value for a given piece of equipment. There is a tendency to use a zero value for salvage because of the provision in the Internal Revenue Code that salvage value up to a 10 percent level may be ignored for tax depreciation purpose. A reasonable salvage value, if it exists, must be recognized.

## 13.5   TAX-RELATED EQUIPMENT COST FACTORS

The tax implications of equipment ownership relate to the following areas:

1.   The amount of income that a company can shelter from taxation through judicious use of depreciation.
2.   The long-term capital gain that a company can incur through sale of equipment.
3.   The investment tax credit a company is entitled to claim by means of equipment ownership.

Depreciation makes it possible for a company to reduce its income. Less income results in less money being taxed. The amount of taxable income that is offset by depreciation constitutes a savings to the company.

Deviation of the market value of a piece of equipment from its book value creates a capital gain (loss) for the company at the time of sale. The amount of this capital gain, once taxes are paid, represents a benefit to the company.

A company that invests in new equipment is allowed to deduct some of the money invested in it in the form of a tax credit. Thus, the company saves an amount equal to the taxes on the amount deducted.

Tax benefits for a particular piece of equipment can be calculated in dollars for the year under consideration. As the value of money changes from year to year, it is necessary to refer to benefits obtained for a base year so that various strategies of ownership can be compared. In determining the time at which an equipment should be sold, each alternative (in this case the year of disposal) involves different periods of time.

In order to determine the equivalent annual benefits to be derived from the sale of a piece of equipment in a given year of its service life, the following steps are necessary:

1. Determination of the net benefit due to depreciation for year $j$, where $j$ varies from 1 to $n$ and $n$ is the year of sale.

$$A(j) = B(j) \times C$$

$$B(j) = \text{depreciation of asset in year } j$$

$$C = \text{actual tax rate}$$

2. Determination of net capital gains at time of sale, year $n$.

$$D(n) = (E(n) - F(n)) \times (1 - G)$$

where

$$E(n) = \text{market value of asset at year } n$$

$$F(n) = \text{book value of asset at year } n$$

$$D(n) = \text{net capital gain at year } n \text{ if asset disposed of}$$

$$G = \text{alternative (capital gains) tax rate}$$

3. Determination of investment tax credit based on year of sale $H(n,j)$.
4. Determination of total benefits per year

$$T(n, j) = A(j) + D(n, j) + H(n, j)$$

where

$$j = 1, n \qquad D(n, j) = 0 \text{ for } j \neq n$$

and

$$T(n, j) = \text{total benefits at year } j \text{ when asset disposed of in year } n$$

5. Determination of total benefits in terms of present worth.

$$\text{PW}(n) = \sum_{j=1}^{n} T(n, j) \times \frac{(1)}{(1 + i)^j}$$

where $i$ = minimum attractive rate of return

$n$ = year of disposal of piece of equipment

## FIGURE 13.5   Program Flow Chart

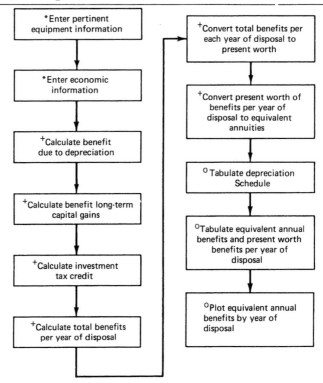

## TABLE 13.4   Equivalent Annual Benefits According to Year of Disposal

| Disposal Year | Depreciation Method | | | | |
|---|---|---|---|---|---|
| | SL | SOYD | DDB | SDB | ACRS |
| 1 | −5199 | 7230 | 11373 | 3087 | −2370 |
| 2 | 2712 | 12576 | 14070 | 8122 | 9186 |
| 3 | 6036 | 13911 | 13852 | 9520 | 12451 |
| 4 | 7341 | 12971 | 12105 | 8891 | 13894 |
| 5 | 8627 | 12540 | 11366 | 8924 | 14648 |
| 6 | 8258 | 10691 | 9532 | 7626 | |
| 7 | 9254 | 10420 | 9476 | 7941 | |

SL  = straight line
SOYD  = sum of years digits
DDB  = double declining balance
SDB  = 150 declining balance
ACRS  = accelerated cost recovery system

6. Conversion of present worth benefits to equivalent annual benefits.

$$EAB(n) = PW(n) \times \frac{(1 + i)^n}{(1 + i)^n - 1}$$

$EAB(n)$ = equivalent annual benefit if machine disposed of at year $n$

A flow diagram showing the different steps required to evaluate the three tax related factors (depreciation, capital gain or loss, and investment tax credit) is shown in Figure 13.5. A simple analysis of the various strategies of depreciation and replacement can be developed within the context of this diagram.

Table 13.4 shows the results of comparison of various strategies for a $100,000 asset.

## 13.6   SUMMARY

Accounting for the cost of equipment is complicated by the fact that fixed costs associated with equipment are commonly period costs that are incurred in a distributed fashion throughout the year. Variable costs such as fuel, oil, and operator's salary are incurred at random times and in direct conjunction with the production process and a particular project. To gain a total cost value for the equipment, these two dissimilar cost structures must be integrated. Industry practice is to use a charge rate, which is multiplied times the hours of utilization of a particular piece of equipment to determine how much cost to bill to the project. Periodically, the charge rates for all equipment are verified by comparing actual costs with the charge rate amount. On the basis of this comparison, the charge rates can be adjusted. A framework for this comparison is presented in Chapter 14.

Fixed costs that are commonly recovered through the application of the charge rate are

1. Depreciation
2. Interest on investment
3. Insurance
4. Taxes
5. Storage

Variable or operational costs pertain to recovery of expenses related to operator's salary, repairs, and maintenance, as well as items that are consumed during operation (e.g., tires, fuel, oil).

Replacement of equipment is strongly influenced by tax considerations. Principal tax items influencing the time at which a piece of equipment should be replaced are (1) method of depreciation utilized, (2) capital gains considerations (tax amount paid upon resale), and (3) the impact of investment tax credit availability. Tax rates under the Accelerated Cost Recovery System (ACRS) are to be applied to equipment placed in service after 1980.

## REVIEW QUESTIONS AND EXERCISES

**13.1** You have just bought a used track-type tractor to add to your production fleet. The estimated service life of this piece of equipment is 7000 hours and it is anticipated it will operate 2000 hours per year. The initial capitalized value of the tractor is $90,000. The salvage value is less than 10 percent of the purchase value.

(a) Using the straight-line method, what amount of depreciation would you claim for the first two years of the equipment's life?

(b) If you resell the equipment at the end of the second year for $50,000, will you pay a capital gains tax? What is the approximate amount of tax (if any) you will pay?

(c) Assume you do not sell, but rebuild the engine and transmission at a cost of $20,000. If this establishes a new service life of three years from the time of rebuild, how much depreciation will you claim for each of these three years, using the straight-line method?

(d) Using the ACRS method of depreciation, what amount of depreciation would you claim for the first two years of the equipment's life?

**13.2** You have just purchased a portable steel tower including hoist and elevator for $42,000. Using the AGC method, calculate the hourly rate for straight time and overtime you would charge for use of this tower and hoist. Assume the operator's rate is $18 per hour and that he receives time and a half for overtime hours. Sales tax is 3 percent and the fuel, oil, and grease costs are estimated at $30 per hour. Freight cost to your location is $860.

**13.3** Compute the average hourly cost to a contractor of a 110 horsepower four-wheeled tractor with cable scraper based on the AGC method. The first cost is $50,000. The sales tax is 5 percent and the tire cost is $12,000. The shipping weight of the combination is 62,500 pounds. The shipping cost is $3/cwt.

(a) Calculate the straight-time and overtime ownership costs.

(b) Calculate the total cost per hour (with operation) based on the following information:

| | |
|---|---|
| Life of tires | 3000 hr |
| Fuel cost | 0.06 gal/hr/bhp @ $0.80/gal |
| Oil and grease | 10% of fuel cost |
| Operator's salary | $16/hr (straight time); $24 (overtime) |

# Chapter 14

# A CONCEPTUAL EQUIPMENT COST-CONTROL SYSTEM

## 14.1  INTRODUCTION

The fundamental purpose of an equipment control system is to provide management with a formalized framework for the development of cost control targets and the recording of actual performance to be compared against original projections.

From an organizational point of view the equipment control system can be viewed as a separate entity within the structure of the company. This approach simplifies the recordkeeping within the organization and facilitates the tracking of expenses related to equipment operations. In order to have an idea of how this separate entity interacts with the recordkeeping operations of the company, a schematic diagram of this interaction is given in Figure 14.1. The equipment cost-control activity can be viewed as an autonomous equipment division within the company and is best thought of as a rental agency within the company. This activity is structured in four main phases, as shown in Figure 14.2.

The *planning phase* is concerned with short-term planning and budgeting processes and culminates in the determination of the hourly charge to be used for each piece of equipment. Management personnel should be committed to this charge rate structure. For this purpose individual target rates are developed for each equipment unit. Each of the target rates is based on records that contain the information needed to estimate the hourly utilization and charges for each unit. The information used for establishing the hourly rate should be based on an estimate of the level of utilization of each piece of equipment in hours. A budgeted amount of fixed (i.e., ownership) expenses for the period and a budgeted amount of variable (i.e., operational) expenses related to the estimated level of utilization should also be established. In addition, a share of the overhead expense of the equipment division should also be prorated to each unit. These elements constitute the target rate charged for each piece of equipment.

The *expense collection phase* of the equipment control activity is concerned with the accumulation of actual expenses incurred for each equipment unit. For this purpose

**FIGURE 14.1  Recordkeeping Functions of the Company**

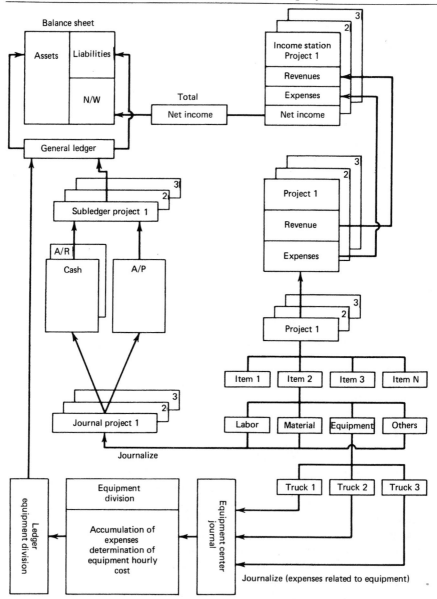

**FIGURE 14.2  Conceptual Framework of Equipment Cost Control System**

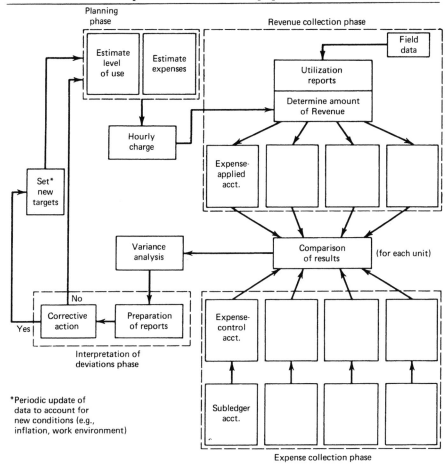

individual *control accounts* should be used. In addition, these control accounts should be supported by subledger accounts that provide a more detailed breakdown of the type of expenses that are accumulated.

The interface between these two parts of the management system is provided by a comparison of the control accounts and *expense-applied accounts* balances at the end of the control period. The expense-applied accounts are updated as part of the *revenue collection phase* shown in Figure 14.2. Amounts are posted to the expense-applied accounts based on the number of hours each piece of equipment is used times its charge rate. The integration of field data regarding the number of hours the equipment is used with the charge rate is accomplished by using a utilization report form.

The comparison of field utilization is reflected by the expense-applied accounts and the expense data logged in the control accounts is carried out in terms of *variance analysis.* This, the control aspect of the system, allows the manager to determine whether the original projections of cost are correct.

Variance analysis aids the manager in interpreting the data. The *interpretation phase* is key to the success of the system. In this phase, the manager makes corrections and revises targets as required. The interpretation phase implements one of the main purposes of management control, allowing the manager to evaluate performance and diagnose deviations from projected cost. This comparative process provides a vehicle for systematically learning how to adjust to the specific working environment by attempting to determine why plans and budgets were not fulfilled. Furthermore, from this comparative analysis an important understanding of cost factors for future projects can be gained.

## 14.2   PLANNING PHASE

In order to implement the control system, the first step is to obtain an estimated hourly charge for each equipment unit, so the categories of cost referred to in Section 13.2 should be considered.

To establish the charge rate it is necessary to set up records for each equipment unit. These records contain the estimated amount of operating hours, ownership cost, and operating cost. In addition, a share of the overhead expenses of the equipment division should be prorated to each equipment unit. A format for this record is presented in Figure 14.3.

If historical records are available, they should be considered in making these estimates. However, historical records should be analyzed cautiously before they are taken into consideration. Conditions may have changed between the periods under study in a way that invalidates their utilization. Another problem with some historical records is that because they are not properly formatted and organized, the data they contain may not be usable for estimating future costs.

The important point is that predetermined standard rates and budgets are the basis against which actual performance is compared. Therefore, the validity of such standards depends largely on how much care went into their development.

## 14.3   DATA ACQUISITION PHASE—REVENUE AND UTILIZATION

The control part of the system requires the collection of field data. Equipment units are assigned to different projects. Projects are charged for the use of these units on the basis of the estimated charge rate and the amount of hours worked. All expenses related to equipment are transferred from the projects to the equipment control group. The concept is shown in Figure 14.4. In the equipment division, individual records for each unit are kept of the expenses actually incurred (expense collection) and the revenues actually earned (revenue collection). Revenues in this sense (i.e., rental) áre registered in the expense-applied accounts. The revenues generated by the equipment

# FIGURE 14.3 Budgeted Hourly Rate Record

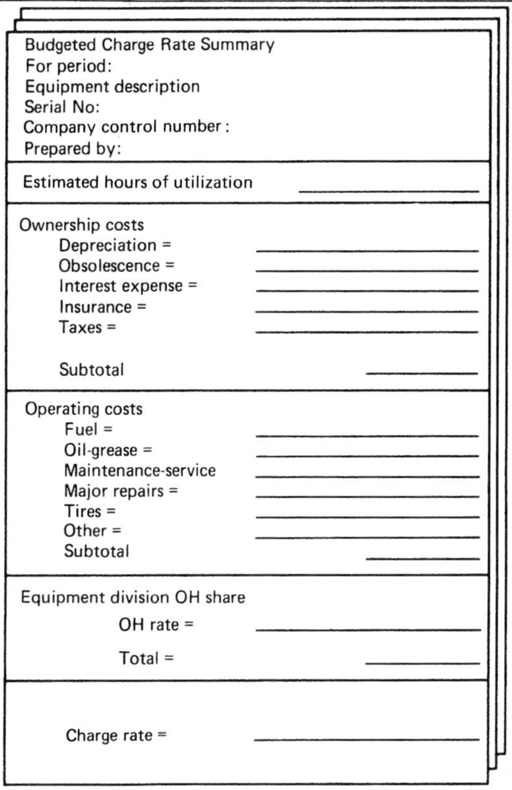

Budgeted Charge Rate Summary
For period:
Equipment description
Serial No:
Company control number:
Prepared by:

Estimated hours of utilization  _____

Ownership costs
    Depreciation =  _____
    Obsolescence =  _____
    Interest expense =  _____
    Insurance =  _____
    Taxes =  _____

    Subtotal  _____

Operating costs
    Fuel =  _____
    Oil-grease =  _____
    Maintenance-service  _____
    Major repairs =  _____
    Tires =  _____
    Other =  _____
    Subtotal  _____

Equipment division OH share

    OH rate =  _____

    Total =  _____

    Charge rate =  _____

# FIGURE 14.4 Data Flow from Projects to Equipment Control Group

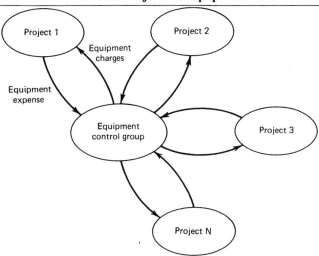

from a utilization point of view are called "expense-applied," since they represent the amount of expenses being recovered by applying a predetermined rate. The amount of revenue or expense applied for each unit is obtained by multiplying the number of hours worked by the standard charge rate.

The amount of hours worked is collected at the project level. It is recommended that these data be collected on a daily basis in order to avoid omissions. Due to the reluctance of field personnel to do paper work, it may be necessary to have the task of gathering these data performed by a full-time timekeeper. The accuracy of these data is most important, because the usefulness of the proposed control system is directly related to the accuracy of the data collected. A report form such as the one shown in Figure 13.1 should be used for this data collection.

Downtime for repairs should be included in the daily usage reports. Appreciable repair time can indicate inadequate equipment maintenance, fatigued units, difficult operating conditions, or operator abuse. If data related to repair time are not included,

**FIGURE 14.5  Weekly Equipment Report**

Weekly equipment summary

Prepared by: _____

For week:

| Equip-ment No. | Equipment Description | Number of hours by Project | | | | Total Hours | Hourly Rate | Total Cost |
|---|---|---|---|---|---|---|---|---|
| | | Project 1 | Project 2 | Project 3 | Project n | | | |
| | | | | | | | | |

it will be difficult for management to spot frequent downtime due to repairs. The dollar figures associated with actual repair expenses do not make such downtime obvious.

   In order not to overload the accounting and recordkeeping functions, the data contained in the daily reports are summarized on a weekly basis. In this way the data can be posted to the respective equipment expense-applied accounts once a week. Of course, this is not a rigid schedule, and different companies can adopt different posting intervals, according to their needs, capabilities, and internal policies. A sample format for a weekly consolidation report is given in Figure 14.5.

## 14.4   ACTUAL EXPENSE DATA COLLECTION

In addition to recording the amount of revenue generated by each unit, it is also necessary to record the amount of expenses incurred by each unit. In doing this it is important to record these expenses at the same level of detail and in terms of the units used for estimating them. This methodology is significant, since excessive costs can be corrected only if the exact cause can be determined.

   Therefore, in comprehensive equipment control systems the expense control account of each unit should be supported by subsidiary ledgers. Each subsidiary ledger contains one category of cost. Subledgers are shown schematically in Figure 14.6.

   Repair and overhaul expenses represent the single most significant variable cost item for equipment. Therefore, a subsidiary ledger for this cost source should be maintained. At this increased level of detail, repair expenses can be grouped according

**FIGURE 14.6   Subledgers Used to Support Expense Control Accounts**

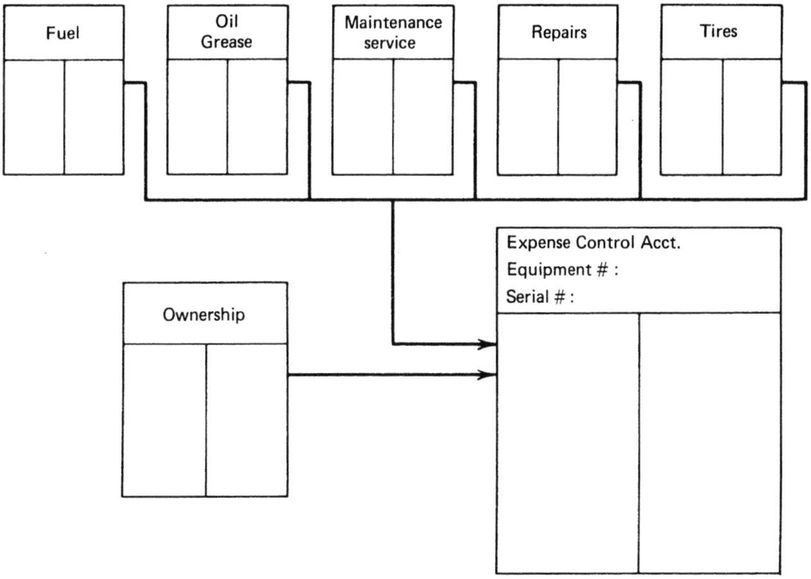

to the type of repair performed. For example, data can be separated by hydraulic system, engine, transmission, cooling system, etc.

The dollar amount related to a given type of repair is recorded in the appropriate subledger account. However, it is mandatory to create a special service record for each unit in order to maintain in detail the major repairs performed. This information should include the date of the repair and a brief description of the repair work and parts that were replaced, in addition to the dollar amount. These repair records are completely independent of the accounting system and should be used to provide to management greater insight as to the kind of expenses incurred. They are feeder documents to the equipment expense control accounts.

For the recordkeeping of the actual ownership costs, it is usually not necessary to use separate subledgers for each category of costs. The ownership costs are relatively stable and predictable. Therefore, a single subledger account for these fixed costs is used.

Due to the fact that all expenses related to equipment are paid by the equipment division, there should be a person on each project in charge of collecting all invoices related to equipment expense. For this purpose, it is convenient to prepare a folder for each unit in which all of the expense invoices are accumulated. Invoices in this folder are sent weekly to accounting for posting to the respective expense-control accounts for each unit.

## 14.5 COMPARISON OF RESULTS WITH STANDARDS

The total control process is shown in Figure 14.7. In the expense-applied accounts (shown above the equipment utilization report), the amount of revenue generated by each equipment unit is accumulated. The amount of revenue generated by each unit is based on the numbers of hours worked and the charge rate estimated at the beginning of the control period.

The hourly rate amount and the information used to estimate the hourly rate are contained in the records located in the bottom row of the figure. The numbers of hours worked for each unit are collected at the project level and summarized on a weekly basis in the equipment utilization report. In the utilization report the amount of revenue generated by each unit is calculated.

From the equipment utilization report the revenue amount is credited to the respective expense-applied accounts of a given unit and debited to the cash or A/R account of the equipment division. The cash account of the equipment division can be considered a subaccount of the company cash account. The bottom half of the figure is concerned with the revenue collection part of the control system.

Conversely, the upper half is concerned with the collection of actual expenses incurred for each unit. For this purpose expense control accounts for each unit are implemented. As can be seen in the figure, the expense-control accounts are supported by subledger accounts. These subledger accounts are used to initially collect the equipment expenses. The process followed in recording a given expense amount is to credit the cash or A/P account and debit the subsidiary expense account for the amount

**FIGURE 14.7   Detailed Equipment Control System**

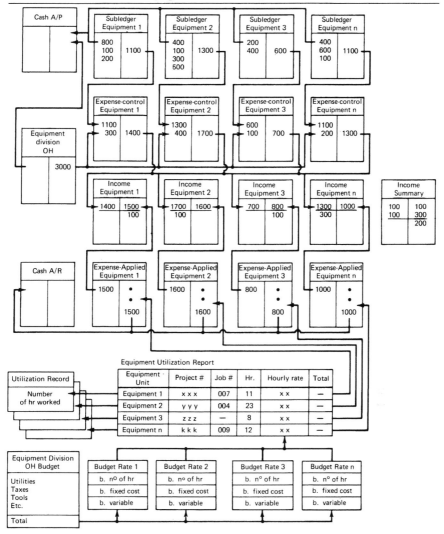

of the expense. On a periodic basis the expenses collected in the subledger accounts of each unit are closed to the expense-control account.

At the end of the control period the expense-applied and expense-control accounts are closed to an income account. The income accounts for each unit are shown in Figure 14.7 between the expense-control and expense-applied accounts. This part of the system can be viewed as the income recognition part of the control process.

Theoretically, the income obtained should be zero because the equipment division is considered a cost center and should operate at the breakeven point. Therefore, if there is any difference between the two accounts mentioned before, it will be reflected as a variance in the income account.

## 14.6   EQUIPMENT VARIANCE ANALYSIS

Comparative variance analysis is performed periodically, and is designed to direct management's attention to conditions that are not what they should be. Results of this analysis on an equipment-by-equipment basis can be summarized in a format similar to that shown in Figure 14.8. The use of this approach as a basis for comparison is shown schematically in Figure 14.9. If expenses applied exceed expenses incurred, there is a positive variance, and expenses are said to have been *overapplied*. If expenses

**FIGURE 14.8   Equipment Control Report**

Equipment control report

Control period:

Prepared by: _____

| Equip-ment No. | Equipment Description | Estimated Hours | Estimated Expenses | Actual Hours | Actual Expense | Standard Rate | Variance | | Eff | | Spen | |
|---|---|---|---|---|---|---|---|---|---|---|---|---|
| | | | | | | | Under | Over | F | U | F | U |
| | | | | | | | | | | | | |

**FIGURE 14.9   Schematic of Variance Analysis**

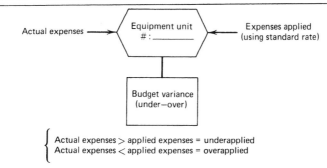

incurred exceed expenses applied, the amount charged is less than the amount needed to cover costs. This is a negative variance, and expenses applied are said to be *underapplied*. Variance analysis is discussed in detail in Section 16.2.

From a management viewpoint it is not enough to know that there is a variance. In most cases, in order to take corrective action, it is necessary to isolate the causes producing variances.

When variances for a given unit are caused by unrealistic estimates, the charge rate must be adjusted appropriately. However, if the variances are caused by operating inefficiencies, adjusting the charge rate is not recommended. In this case, transfer of the company's inefficiency to the customer can affect the competitive position of the company for bidding purposes. If the company is not in a position to efficiently manage a given type of equipment, the alternative of renting should be investigated. It is important to compare the company internal charge rates with market rental rates for similar units. Doing so will help better determine how efficiently the equipment fleet is being managed. The comparison of actual expenses, budgeted expenses, and rental rates should enable management to better understand how equipment expenses behave and to decide on the alternative best suited to a given project and company operations in general.

## REVIEW QUESTIONS

**14.1**   How does depreciation enter into profit and loss calculations?
**14.2**   How could an equipment cost-control system be used for equipment replacement decisions?
**14.3**   What objectives must be considered when setting up an equipment control system? What data should be used for estimating?
**14.4**   Who should be able to update the equipment-cost data and how often should it be done?

# Chapter 15

# JOB AND FIXED OVERHEAD COSTS

## 15.1 CHARGES FOR INDIRECT AND OVERHEAD EXPENSE

Contractor-incurred expenses associated with the construction of a given facility relate to:

1. Direct cost consumed in the realization of a physical subelement of the project (e.g., labor and material costs involved in pouring footer).
2. Production support costs incurred by project-related support resources or required by the contract (e.g., superintendent's salary, site office costs, builder's risk insurance).
3. Costs associated with the operation and management of the company as a viable business entity (e.g., home office overhead, such as the costs associated with preparation of payroll in the home office, preparation of the estimate, marketing, salaries of company officers).

The production support costs are normally referred to as *project indirect costs*. The home office charges are normally referred to as *home office overhead*. All of these costs must be recovered before income to the firm is generated. As was discussed in Chapter 2 (Section 2.8), the home office overhead, or general and administrative expense, can be treated as a period cost and charged separately from the project (direct costing). On the other hand, they may be prorated to the job and charged to the job cost overhead accounts and the work in progress general ledger accounts (absorption costing). If accounted for as period costs, these overhead expenses are reflected in appropriate G/L accounts but not in the job cost ledgers.

## 15.2 PROJECT INDIRECT COSTS

Job-related indirect costs such as those listed in the labor cost report of Figure 12.6 are typically incurred as part of the on-site related cost associated with realizing the project. As such, they are charged to appropriate accounts within the job cost system. The level and amount of these costs should be projected during the estimating phase and included in the bid as individual estimate line items. Although it is recommended that job indirects be precisely defined during estimate development, many contractors prefer to handle these charges by adding a flat rate amount to cover them. Under this approach, the contractor calculates the direct costs (as defined above) and multiplies these charges by a percentage factor to cover both project indirects and home office fixed overhead. To illustrate, assume that the direct costs for a given project are determined to be $200,000. If the contractor applies a fixed factor of 20 percent to cover field indirects and home office overhead, the required flat charge would be $40,000. If he adds 10 percent for profit, his total bid amount would be $264,000.

The estimate summary shown in Figure 10.5 establishes line items for indirect charges and calculates them on an item-by-item basis (rather than applying a flat rate). Typical items of job-related indirect cost that should be estimated for recovery in the bid are those listed in Figure 2.1 as project overhead accounts (.700 to .999). This is the recommended procedure, since it is felt that sufficient information is available to the contractor at the time of bid to allow relatively precise definition of these indirect costs. The R. S. Means method of developing overhead and profit (illustrated in Figure 10.11) represents a percentage rate approach that incorporates a charge into the estimate to cover overhead on a line item-by-item basis. This is essentially a variation of the flat rate application described above.

## 15.3  FIXED OVERHEAD

Whereas the project indirect charges are unique to the job and should be estimated on a job-by-job basis, home office overhead is a more or less fixed expense that maintains a constant level not directly tied to individual projects. In this case, the application of a percentage rate to prorate or allocate home office expense to each project is accepted practice, since it is not reasonable to try to estimate the precise allocation of home office to a given project. Rather, a percentage prorate or allocation factor is used to incorporate support of home office charges into the bid.

The calculation of this home office overhead allocation factor is based on:

1.  The G&A (home office) expenses incurred in the past year
2.  The estimated sales (contract) volume for the coming year
3.  The estimated gross margin (i.e., markup) for the coming year

This procedure is illustrated by Adrian in the following example:*

---

* James J. Adrian, *Construction Accounting* (Reston, Va.: 1979), p. 263.

| **Step** | **Estimate of Annual Overhead (General and** |  |
|---|---|---|
| **1** | **Administrative Expense)** |  |

| Last year's G & A | $270,000 |
| 10% inflation | 27,000 |
| Firm growth | 23,000 |
| Estimated G & A | $320,000 |

**Step 2**   **Estimate of $ of Cost Basis for Allocation**

| Estimated volume | $4,000,000 |
| Gross margin | 20% = $800,000 |
| Labor and material | $3,200,000 |

**Step 3**

$$\frac{\text{Overhead costs estimated (G \& A)}}{\text{Labor and material estimate}} = \frac{320,000}{3,200,000} = 10\%$$

**Step 4**   **Cost to Apply to a Specific Project**

| Estimated labor and material costs | 500,000 |
| Overhead to apply | 50,000 |
| Labor and material costs 500,000 times allocation rate 10% | |
| | $550,000 |

In the example, the anticipated volume for the coming year is $4,000,000. The G&A expense for home office operation in the previous year was $270,000. This value is adjusted for inflation effects and expected expansion of home office operations. The assumption is that the overhead allocation factor will be applied to the direct labor and materials costs. These direct costs are calculated by factoring out the 20-percent gross margin.

Direct costs amount to $3,200,000. The $320,000 in G&A costs to be recovered indicate a 10 percent prorate to be applied against the $3,200,000 of direct costs. This means that an overhead amount of $50,000 would be added to a contract bid based on $500,000 of direct cost to provide for G&A cost recovery. The profit would be added to the $550,000 base recovery amount.

## 15.4   UNALLOCATED OVERHEAD ACCOUNTS

Many firms establish catch-all financial accounts to simplify accounting procedures and to track overhead variation on a month-by-month basis. Accounts established for this purpose are typically called *unallocated overhead accounts*. Such accounts are used to record differences between the actual recorded fixed overhead costs and the estimated fixed overhead costs for each month. Accounts of this type are used by construction and construction-related firms that are marketing man-hours of service at a price that is developed on the basis of estimated overhead costs.

The overhead factor used in pricing jobs and planning for yearly revenues is derived generally from the method described in Section 15.3. Consider the following example.

## *Overhead Factor Calculation for a Typical Construction Management Firm*

| | |
|---|---|
| 19X1 Overhead costs | $200,000 |
| Known increase in overhead costs for 19X2 | 40,000 |
| Total overhead estimate 19X2 | $240,000 |
| | |
| Estimated 19X2 Project man-hours | 20,000 |
| Charged cost per man-hour | × $20 |
| Estimated 19X2 man-hour charges | $400,000 |

Overhead factor = $240,000 ÷ $400,000

19X2 overhead factor = 0.6

The overhead factor of 0.6 is an estimate based on man-hour projections and presently known fixed overhead costs. The overhead costs for the coming year (19X2) should be 60 percent of the project man-hours charged. When trying to estimate yearly revenues by using the formula:

19X2 revenues = Total man-hour cost × overhead factor × profit factor

the overhead factor used is 1.6.

At the end of each month, the actual overhead costs for the month are totaled and compared to the overhead costs estimated for the month using the overhead factor times project hours sales. There will always be some differences between the estimate and the actual costs. Most differences will be due to monthly fluctuations in revenue volumes and unanticipated overhead costs for the period. Fluctuations in volume will cause the actual monthly overhead factor to fluctuate because overhead costs are fixed. These costs do not generally fluctuate with changes in the revenue volume. The actual overhead percentage of total sales is an inverse function of volume.

The monthly differences between actual and estimated overhead are recorded in the unallocated overhead account. Since the account is an expense account (actual factor is greater than 1.6), when actual costs are less than the planned factor the difference is recorded as a credit. As costs change from month to month with debits and credits entered into the account, it is anticipated that by the end of the year the debits and credits will balance out to zero. Any remaining balance at the end of the year is closed out to zero by transfering the balance to the income summary account. For example:

19X2 overhead factor = 1.6

19X2 estimated project man-hour costs = $400,000

Therefore, 19X2 total revenues (less profit and growth factors) equals $640,000 (1.6 × $400,000), which equates to approximately $53,333 per month. Consider the following actual situation as of 31 January 19X1:

$$\text{Actual man-hour sales} = \$30,500$$

$$\text{Actual overhead costs} = \$13,050$$

Based on overhead factor of 0.6, the allocated overhead costs are $11,437.50.* The difference between the actual overhead costs of $13,050 and the estimated overhead costs of $11,437.50 is $1612.50, which is entered in the unallocated overhead account as a debit.

## Journal

| Account | Debit | Credit |
|---|---|---|
| Overhead expenses | $11,437.50 | |
| Unallocated overhead | 1,612.50 | |
| Cash | | $13,050 |

The situation changes as follows: on 28 February 19X2

| | | |
|---|---|---|
| Actual man-hour sales | = | $36,221 |
| Actual overhead costs | = | 12,089 |
| Estimated overhead 0.6 factor | = | 13,582.88 |
| Overhead difference | = | 1,493.88 |

## Journal

| Account | Debit | Credit |
|---|---|---|
| Overhead expenses | $13,582.88 | |
| Cash | | $12,089 |
| Unallocated overhead | | 1,493.88 |

In order to close out the unallocated overhead account so that it has a zero balance for the next year, the year-end balance must be transferred to the income summary account.

The unallocated overhead account is a good method for a company to use to track the accuracy of the estimate of overhead for the year without having to review several accounts and compile data to see how close the estimate has been to reality. A similar account could be established for any other expense that is estimated at the beginning of the year and is subject to fluctuate because of unknown factors. If at the end of the year the account balance is not within reason, then the next year's factor should be developed with the observed variation in mind.

---

* The portion of sales allocated to cover overhead is calculated as $\dfrac{0.6}{1.6} \times \$30,500$.

## 15.5   CONSIDERATIONS IN ESTABLISHING FIXED OVERHEAD

In considering costs from a business point of view, it is common to categorize them either as variable costs or fixed costs. Variable costs are costs directly associated with the production process. In construction they are the direct costs for labor, machines, and materials as well as the field indirect costs (i.e., production support costs). These costs are considered variable, since they vary as a function of the volume of work underway. Fixed costs are incurred at a more or less constant rate independent of the volume of work in progress. In order to be in business, a certain minimum of staff in the home office, space for home office operations, telephones, supplies, and the like must be maintained, and costs for these items are incurred. These central administrative costs are generally constant over a given range of sales/construction volume. If volume expands drastically, home office support may have to be expanded also. For purposes of analysis, however, these costs are considered fixed or constant over the year. Fixed costs are essentially the general and administrative costs referred to elsewhere in this text.

As described in Section 15.3, the level of G&A (fixed) costs can be estimated by referring to the actual costs incurred during the previous year's operation. The method of projecting fixed overhead as a percentage of the estimated total direct costs projected for the coming year is widely used. Since the fixed overhead incurred in the previous year is typically available as a percentage of the previous year's total sales volume, a simple conversion must be made to reflect it as a percentage of the total direct cost. The formula for this conversion is as follows:

$$P_c = \frac{P_s}{100 - P_s}$$

where $P_c$ = the percentage applied to project's total direct cost for the coming year

$P_s$ = the percentage of total volume in the reference year incurred as fixed or G&A expense

If, for instance, $800,000 is incurred as home office G&A expense in a reference year in which the total volume billed was $4,000,000, the $P_s$ value would be 20 percent ($800,000/$4,000,000 × 100). The calculated percentage to be added to direct cost estimates for the coming year to cover G&A fixed overhead would be

$$P_c = \frac{20}{100 - 20} = 25 \text{ percent}$$

If the direct-cost estimate (e.g., labor, materials, equipment and field indirects) for a job is $1,000,000, $250,000 would be added to cover fixed overhead. Profit would be added to the total of field direct and indirects plus fixed overhead. The field (variable) costs plus the fixed overhead (G&A) charge plus profit yield the bid price. In this example, if profit is included at 10 percent, the total bid would be $1,375,000. It is obvious that coverage of the fixed overhead is dependent on generating enough billings to offset both fixed and variable costs. This is discussed in the next section in terms of breakeven analysis.

Certain companies prefer to include a charge for fixed overhead that is more responsive to the source of overhead support. The assumption here is that home office support for management of certain resources is greater or smaller, and this effect should be included in charging for overhead. For instance, the cost of preparing payroll and support for labor in the field may be considerably higher than the support needed in administering materials procurement and subcontracts. Therefore, a 25 percent rate for fixed overhead is applied to labor and equipment direct cost, while a 15 percent rate on materials and subcontract costs is used. If differing fixed overhead rates are used on various subcomponents of the field (variable) costs in the bid, the fixed overhead charge will reflect the mix of resources used. This is shown in Table 15.1, in which a fixed rate of 20 percent on the total direct costs for three jobs is compared to the use of a 25 percent rate on labor and equipment and a 15 percent rate on materials and subcontracts.

It can be seen that the fixed overhead amounts using the 25/15 percent approach are smaller on Jobs 101 and 102 than the flat 20 percent rate. This reflects the fact that the amount of labor and equipment direct cost on these projects is smaller than the materials and subcontract costs. The assumption is that support requirements on labor and equipment will also be proportionately smaller. On Job 102, for instance, it appears that most of the job is subcontracted with only $200,000 of labor and equipment in house. Therefore, the support costs for labor and equipment will be minimal, and the bulk of the support cost will relate to management of materials procurement and subcontract administration. This leads to a significant difference in fixed overhead charge when the 20 percent flat rate is used, as opposed to the 25/15 percent modified rates (i.e., $440,000 versus $350,000).

On Job 103, the fixed overhead charge is the same with either of the rate structures, since the amount of labor and equipment cost is the same as the amount of the materials and subcontract cost.

**TABLE 15.1   Comparison of Fixed Overhead Rate Structures**

|  |  |  | *20% on Total Direct* | *25% on Labor and Equipment 15% on Material and Subcontracts* |
|---|---|---|---|---|
| Job 101 | Labor and equipment | $ 800,000 | $160,000 | $200,000 |
|  | Materials and subcontracts | 1,200,000 | 240,000 | 180,000 |
|  |  |  | $400,000 | $380,000 |
| Job 102 | Labor and equipment | 200,000 | $ 40,000 | $ 50,000 |
|  | Materials and subcontracts | 2,000,000 | 400,000 | 300,000 |
|  |  |  | $440,000 | $350,000 |
| Job 103 | Labor and equipment | 700,000 | $140,000 | $175,000 |
|  | Materials and subcontracts | 700,000 | 140,000 | 105,000 |
|  |  |  | $280,000 | $280,000 |

It should be obvious that in tight bidding situations use of the stylized rate system, which attempts to better link overhead costs to the types of support required, might give the bidder an edge in reducing his bid. Of course, in the example given (i.e., the 25/15 percent rate versus 20 percent) the 20 percent flat rate would yield a lower overall charge for fixed overhead on labor- and equipment-intensive jobs. The main point is that the charge for fixed overhead should be reflective of the support required. Because the multiple rate structure tends to reflect this better, many firms now arrive at fixed overhead charges by using this approach rather than the flat rate applied to total direct cost.

## 15.6    BREAK-EVEN ANALYSIS

Both variable and fixed costs (as described in the previous section) must be recovered by billings to the client. A certain percentage of each billed dollar must cover the variable costs incurred, while another portion of the billed dollar must cover the fixed overhead costs (G&A expenses). This is shown conceptually in Figure 15.1. If billings exceed a certain amount (a certain level of volume is achieved), profit will be generated by the billings. However, up to a certain level of billings, all of the dollars billed are essentially covering the combined fixed and variable costs. The point at which billed volume is equal to the fixed and variable cost incurred to generate it is called the *break-even point*.

To better understand this concept, consider the portion of each dollar billed that is used to cover variable costs. The amount left when this portion is paid out is available to be applied to fixed overhead. As described in the previous sections, the amount of fixed overhead (G&A) expense is normally projected from the previous year's operations. Assume that the amount of fixed overhead corrected for inflation and expanded operations for the present year is $400,000. Assume further that the percentage of each billed dollar required to cover operational costs is 80 percent. This means that 80 cents of each billed dollar is expended on variable costs. Twenty cents is left to

**FIGURE 15.1    Distribution of Billed Dollars to Cover Costs Incurred**

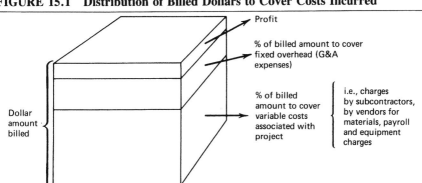

cover fixed overhead. How many dollars must be billed to cover the $400,000 of projected fixed overhead? Based on 20 cents per dollar for fixed overhead, the minimum amount of billings required to cover the projected G&A expense would be $400,000 ÷ 0.20, or $2 million. That is, until $2 million has been billed to clients, the fixed overhead has not been recovered. If more than $2 million is billed over the year, profit will be generated at the rate of 20 cents on the dollar for every dollar in excess of $2 million. On the other hand, if the volume of billings falls below $2 million, the fixed overhead will not be fully covered, and a loss will occur. A graphical representation of this situation is given in Figure 15.2. The point at which billings just cover fixed and variable costs is called the *break-even point*. The chart shown in the figure is called a *break-even chart*. The slope of the variable cost line is $0.80/billed dollar. The variable cost line starts on the $Y$ axis at the level of projected fixed cost (i.e., projected G&A costs for the year). The dotted line shows the billings amount profile. As the amount billed increases, the variable costs also increase. At $2 million billed, the billings amount curve intersects the total cost line (fixed + variable), indicating that billings equal fixed and variable costs. The chart is based on a few simple equations. At the break-even point, billings volume must equal fixed costs plus variable costs.

**FIGURE 15.2   Break-Even Analysis Chart**

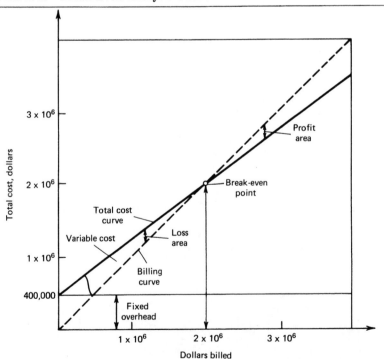

$$BV = FC + VC$$

Variable costs are $X$ percent of the billed dollar.

$$\text{therefore BV} = FC + \left(\frac{X}{100}\right)(BV)$$

$$\text{or BV} - \left(\frac{X}{100}\right)BV = FC$$

Solving for BV, we obtain

$$BV = \frac{FC}{1 - (X/100)}$$

In the example given, the break-even point is calculated as

$$BV = \frac{\$400,000}{1 - (80/100)} = \frac{\$400,000}{0.20} = \$2,000,000$$

## 15.7  BASIC RELATIONSHIPS GOVERNING THE BREAK-EVEN POINT

Several relationships are clear from a consideration of the break-even chart and the supporting equations.

1. Lowering the fixed overhead amount will move the break-even point to the left (reduce the volume required) if the variable cost percentage is constant. If the fixed overhead amount in the example is reduced to $300,000, the new break-even point will be $300,000 ÷ 0.20 = $1,500,000.
2. If the percentage of each billed dollar required to cover variable costs is reduced, the break-even point volume is reduced. This is obvious, since the total cost curve slope is reduced while the billings curve slope is constant (i.e., always 45°). If in the example the fixed overhead cost is $400,000 as originally stated, but the variable cost percentage is 75 percent, the new break-even point volume will be $1,600,000.
3. Below the break-even point, the loss associated with a given construction volume is the difference between the total cost curve and the billings curve. Above the break-even point, this difference represents the profit associated with a given volume. These relationships are illustrated in Figure 15.3.

**FIGURE 15.3a   Variation in Break-Even Point Based on Reducing Fixed Overhead**

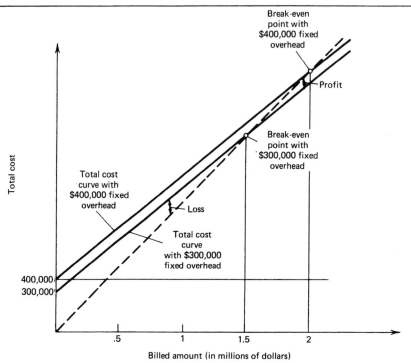

Billed amount (in millions of dollars)

Fixed overhead can be considered to be a steplike function of volume, moving up to provide higher levels of support across certain ranges of volume. For example, the following ranges of fixed overhead might be defined.

| Volume Range | Fixed Overhead |
|---|---|
| (1)  0–$1,000,000 | $300,000 |
| (2)  1,000,000–$1,800,000 | $350,000 |
| (3)  1,800,000–$2,500,000 | $400,000 |

If the original percentage coverage of variable costs (80 percent) is maintained, the break-even characteristics of each range can be studied. These are shown in Figure 15.4. As can be seen from the figure, the break-even point is not reached in range one, since the slope of the total cost curve stays above the billings curve through this range. The total cost curve is increased by $50,000 (slope remains $.80/dollar billed) in range two, due to the increase in fixed overhead across this range. The break-even point occurs at a volume of $1,750,000. In the last range the revised break-even point due to increasing fixed overhead to $400,000 is $2 million, as originally calculated.

**FIGURE 15.3*b*** **Variation in Break-Even Point Based on Reducing Variable Cost Recovery Rate**

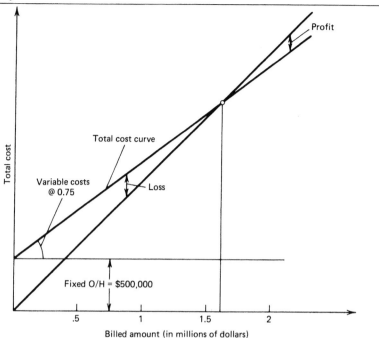

The number of total billed dollars (volume) required to cover one dollar of fixed overhead at the balance point for varying markups can be calculated by considering the ratio of volume to fixed overhead (BV/FC)

$$BV = FC + VC$$

If *M* = markup,

$$\frac{M}{100}(VC) = FC \text{ at the break-even point}$$

or

$$VC = \frac{100 \; FC}{M}$$

$$BV = FC + \frac{100 \; FC}{M}$$

**FIGURE 15.4   Step Variation of Fixed Overhead**

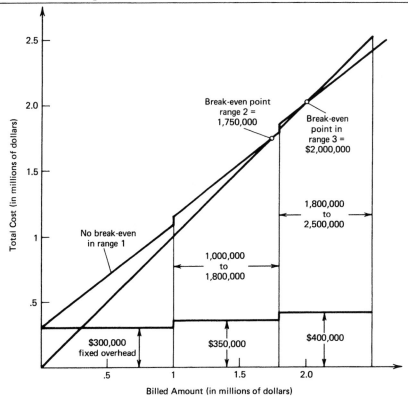

**TABLE 15.2   Billings Volume Required to Cover $1 of Fixed Overhead at Varying Levels of Markup**

| Percent Markup on Direct Costs | Volume Required per Dollar of Fixed Overhead |
|:---:|:---:|
| 1 | $101.00 |
| 2 | 51.00 |
| 3 | 34.33 |
| 4 | 26.00 |
| 5 | 21.00 |
| 10 | 11.00 |
| 15 | 7.67 |
| 20 | 6.00 |
| 25 | 5.00 |
| 30 | 4.33 |
| 35 | 3.86 |
| 40 | 3.50 |
| 45 | 3.22 |
| 50 | 3.00 |

**FIGURE 15.5   Plot of Markup Versus Volume/Fixed Cost**

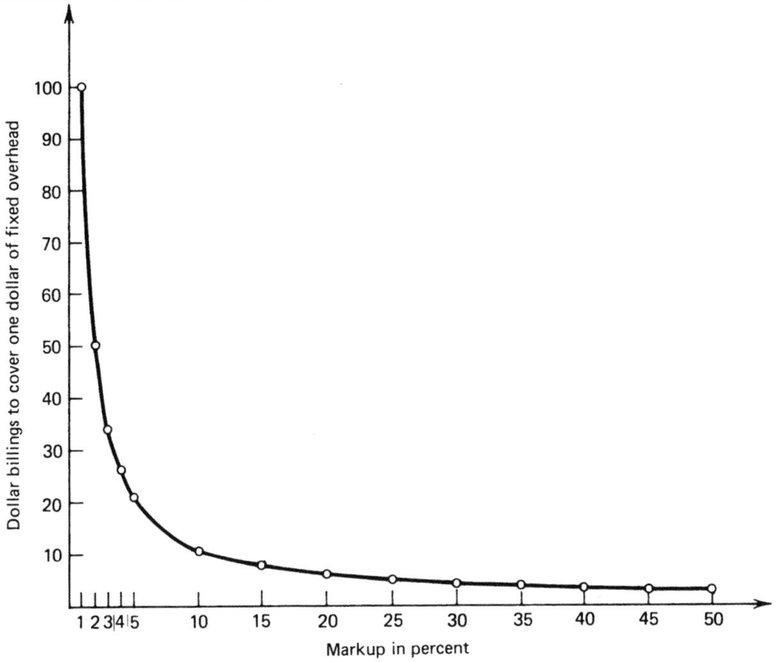

Dividing both sides by FC, we obtain

$$\frac{BV}{FC} = 1 + \left(\frac{100}{M}\right)$$

If the markup is 20 percent, the ratio of billing volume to fixed costs at the break-even point is $(1 + {}^{100}/_{20})$, or 6. This means that $6 of billed volume will cover $1 of fixed overhead. As the markup increases, the coverage ratio becomes stronger. At low markups, a large billing volume is required to cover fixed overhead. This relationship is shown in Table 15.2 and Figure 15.5.

## 15.8   REPORTING OF INDIRECT AND OVERHEAD CHARGES

The previous sections emphasized the need to include amounts in the bid price to recover expenses associated with project indirect costs and fixed overhead. Specific charges for indirect and overhead items during the operational year are captured in the payroll and accounts payable journals. These transactions are routed to the appropriate accounts in general ledger and job cost systems by the posting activity. Those charges associated with a particular project overhead activity (e.g., storage area rental,

FIGURE 15.6  Monthly Status Report Indirect/Overhead Cost

Use for Job Overhead
& Corporate Overhead

**MONTHLY STATUS REPORT**
**INDIRECT/OVERHEAD COST**

Page ——— of ———

Company ———

Month Ending ——— 19 ———

| Cost Code | | Description | Total Estimate | Job Status to Date | | | | % of Actual to Estimated Costs | Job Cost Trends | |
| Job or Corporate Entity | Account | | | Estimate | Actual | Variance | Unused Estimate | | % Unused Estimate | % Unused Time |
|---|---|---|---|---|---|---|---|---|---|---|
| | | | | | | | | | | |
| | | | | | | | | | | |
| | | | | | | | | | | |
| | | | | | | | | | | |
| | | | | | | | | | | |

*List active indirect cost account items for corporate, division, subsidiary and jobs. This report can also be used to summarize indirect costs for all jobs and corporate entities.

308

temporary fences, drinking water, and the like) are recorded in the appropriate job cost ledgers as well as in the project expense or work in progress accounts in the general ledger. General and administrative costs associated with the operation of the home and regional offices are recorded in the appropriate general ledger accounts and are reflected at the end of the accounting period as support expenses in the income statement.

A form for reporting pertinent data as they relate to indirect and overhead charges is shown in Figure 15.6. This form allows for account designation to include job or corporate subentity as well as account number. The expense description, amount, and status versus the estimate are reflected, as well as the variance from the estimate. Percentage amounts are also calculated to reflect performance versus projection.

As described in Section 15.4, specific rates are applied and then compared with actual costs incurred. Variance analysis is performed periodically. Corrections in the overhead rates are made based on this analysis to insure that incurred costs are being recovered.

## REVIEW QUESTIONS AND EXERCISES

**15.1** Based on the definition of fixed cost, list costs in the main office that are not really fixed, and field office costs that are fixed.

**15.2** Categorize the following costs as (a) direct, (b) project indirect, or (c) fixed overhead.
Labor
Materials
Main office rental
Tools and minor equipment
Field office
Performance bond
Sales tax
Main office utilities
Salaries of managers, clerical personnel, and estimators

**15.3** The following data are available on Del Fabbro International, Inc. The fixed (home office) overhead for the past year was $365,200. Total volume was $5,400,000. Del Fabbro uses a profit markup of 10 percent. The estimating department has indicated that the direct and field indirects for a renovation job will be $800,000. What bid price should be submitted to insure proper coverage of fixed overhead? Assume a 5 percent inflation factor and a 12 percent growth factor in the calculation.

**15.4** Using the data in Problem 15.3 calculate the break-even point for Del Fabbro International in the present year.

# Chapter 16

# CONCEPTS FOR CONTROLLING COST

## 16.1 MANAGEMENT BY EXCEPTION

The primary function of collecting data on costs and resource usage is to provide timely decision-making information to management. Management needs to be continuously updated with regard to field performance so that actual and budgeted values as contained in the control budget can be compared. The detection of deviations from programmed or budgeted cost and resource usage, highlights potential problem areas and triggers decisions on the part of management to correct overruns and monitor underruns. This approach to detecting and correcting deviations from planned progress is referred to as *management by exception*.

The action of controlling cost is based upon comparisons of actual performance to that originally projected in the *control budget*. In order to provide a structure or framework within which this comparison can take place, the control budget should contain the following elements:

1. Identification of each work item (by code and/or description)
2. Units of measure and expected quantity of each work item
3. Expected cost per unit and total cost for each work item
4. Provision for recording actual total and unit cost to date for each work item
5. Provision for projecting totals at completion of each work item on the basis of current knowledge
6. A way of expressing variances so that exceptions and large deviations stand out

The variance can be shown in percentage or in absolute amount. Small percentage variations on large work items can be more important than large deviations on small items. If aluminum hardware on a given project has an estimated cost of $600 and experiences a 50 percent overrun, this results in a deviation of $300. On a million-dollar project this is not significant. On the other hand, if the estimated cost of structural steel is $600,000, a 5 percent overrun results in a $30,000 deviation from expected cost. This deviation can certainly be significant. Therefore, it is recommended that both percentage and absolute amount of variance be reported. No matter how simple or complex the control budget is, it should be structured so as to provide the framework for comparison implied by the six items listed above.

## 16.2  VARIANCE ANALYSIS

Variance analysis is the most obvious and commonly used method of evaluating actual cost performance versus that budgeted. In its simplified form, it is a straightforward comparison of the deviation in cost between actual and estimated amounts. To illustrate, consider the labor cost report of Figure 12.6. The cost item "hand backfill" (Code 312) has a negative to-date variance of $1047 and a negative end-of-project variance of $209. This indicates an underrun on both to-date and end-of-project projected amounts. The to-date variance is based on 5000 of 6000 cubic yards having been placed to date. This is 83 percent of the total amount. The actual cost to date is running $1.29 per cubic yard versus an original estimate of $1.50 per yard.

The simple comparison of unit and total costs is referred to as a one-factor model. In the one-factor model,

$$V_i = AC_i - BC_i$$

where

$$V_i = \text{the variance for the } i\text{th item}$$

$$AC_i = \text{the actual cost of the } i\text{th item}$$

$$BC_i = \text{the budgeted cost of the } i\text{th item}$$

This model is limited in application, since it gives little more than the total variance associated with a given cost item.* In the case of the hand backfill, the simple one factor masks the fact that although the cost per unit is 21 cents lower than the budgeted amount (a cost saving per unit), the quantity will exceed the original estimate of 6000 cubic yards by 815 cubic yards. The projected end of project underrun is $209, yielding a total cost of $8791. At $1.29 per unit (the present project unit price), the total quantity will be 6815 yards versus the original estimate of 6000 cubic yards. This indicates a favorable *cost per unit* variance, which will be offset in part due to an unfavorable *quantity* variance. The overall projected variance of $209 will be, however, favorable to the contractor.

The variance of both quantity and cost per unit can be detected by using the so-called two-factor model. The structure of the two-factor model is defined as follows:

$$V_{ip} = AC_i - (AQ_i \times SP)$$

$$V_{iQ} = (AQ_i \times SP_i) - (SQ_i \times SP_i)$$

---

* Nicholas Dopuch, Jacob Birnberg, Joel Demski, *Cost Accounting,* 1974.

where

$V_{ip}$ is the price-related deviation for the $i$th item

$V_{iQ}$ is the quantity-related deviation for the $i$th item

$AC_i$ is the actual cost of the $i$th item

$AQ_i$ is the actual quantity of the $i$th item

$SP_i$ is the standard or estimated price per unit of the $i$th item

$SQ_i$ is the standard or estimated quantity required on the $i$th item to achieve the level of activity

This model presents a better means of analysis. Here variations associated with changes in the unit price of the resources and actual quantities of work items can be evaluated in terms of their effect upon the total variance. In the context of the hand backfill example the two-factor model can be illustrated schematically as shown in Figure 16.1. The same approach can be used to evaluate labor costs in which the standard or estimated unit of measure is a time unit, such as a man-hour. To illustrate the use of the two-factor model for labor comparisons, consider the following example. A contractor has completed a portion of a contract requiring 100,000 square feet glass store front to be placed on a site for a large shopping center. In 17.25 shifts, the supervisor and the crew of nine craftspeople and one laborer have placed 50,000 square feet of glazing at a total labor cost of $11,040 (excluding the supervisor's salary). In estimating this project, the contractor's estimators used a productivity rate of 2820 square feet per eight-hour shift standard with a total labor rate of $7.50 per man-hour. If it is assumed that wage rates were set by top management on this job, variance analysis can be used to evaluate the supervisor's performance.

To calculate the variance based on performance to date, the estimated number of man-hours for the work completed to date must be calculated. The percentage of work completed is 50 percent. The estimated hours required would be 50,000 divided by the shift rate times 80 man-hours per shift (based on the ten-person crew). This yields an estimated man-hour figure for the 50,000 square feet of 1419. The actual number of hours worked to date is 80 times 17.25 shifts, or 1380 man-hours. The two-factor variance is summarized in Figure 16.2.

**FIGURE 16.1   Material Cost Variance Using the Two-Factor Model**

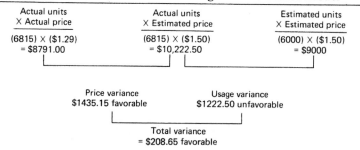

| Actual units X Actual price | Actual units X Estimated price | Estimated units X Estimated price |
|---|---|---|
| (6815) X ($1.29) = $8791.00 | (6815) X ($1.50) = $10,222.50 | (6000) X ($1.50) = $9000 |

Price variance
$1435.15 favorable

Usage variance
$1222.50 unfavorable

Total variance
= $208.65 favorable

**FIGURE 16.2   Labor Cost Variance Using the Two-Factor Model for Work to Date**

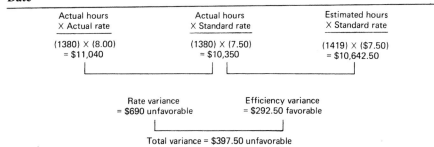

If the one-factor model were used, the supervisor would quite likely be given the responsibility for the unfavorable total labor variance of $397.50. When the two-factor analysis is used, the supervisor's performance is viewed more favorably, and the responsibility for the wage rate variance is placed upon whoever established the wage rate estimate.

The projected end-of-project variances are shown in Figure 16.3. The actual hours are projected as 2760 versus an initial estimate of 2937. This indicates that if the productivity rate and the labor price remain constant to the end of the glass installation, the improved productivity will offset the price escalation, reducing the unfavorable variance to $803.40 for the work item.

When we apply variance analysis to analyze job indirect costs, the procedure is essentially the same. Consider the trash-hauling cost item (Code 111) in Figure 12.6. The one-factor variance of this item to date is $3646 favorable. The end-of-project variance is shown in Figure 16.4. In this case, a projected end-of-job total favorable variance is due primarily to a favorable spending variance, even though the efficiency variance is unfavorable. That is, the costs are below projection, even though production is not as high as originally estimated.

The one- and two-factor variance models provide a simple and easily understood method of analyzing cost reports. The two-factor model also aids in detecting the reason for a favorable or unfavorable variance in terms of pricing and quantity variation.

**FIGURE 16.3   End-of-Project Variance**

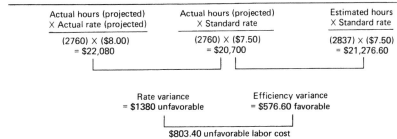

**FIGURE 16.4   Overhead Cost Variance for End of Project**

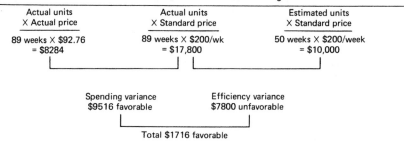

## 16.3   COST TREND ANALYSIS

Cost reports in the formats described in Chapters 12 and 13 are feedback mechanisms that provide the manager with vital information regarding job status. They provide the basis for applying variance analysis to determine the reasons for departure from estimated performance. They are limited, however, in that they provide only piecemeal information regarding slippage in original project duration and total project cost overruns.

> Many . . . managers try to keep on top of project cost by merely reviewing cost reports as they become available. This procedure is forever lagging behind what is actually occurring. What is needed is a method for forecasting costs and completion dates so that efforts to keep costs in line can be launched before things get out of hand.*

The use of trend analysis can help the manager to forecast cost and time performance and thereby provide a red flag indicating the need for management actions. The projection of total project cost is based on the prediction of principal project parameters. Davis maintains that the project completion date and the predicted cost at completion are the key parameters that must be monitored to maintain control of the project.* These principal parameters are continuously reviewed and updated by using graphical charts that are supported by data worksheets. The required data are developed from periodic cost reports. The following sections describe the development of the *completion date trend chart* and the *project cost trend chart*.

## 16.4   COMPLETION DATE TREND CHART*

A completion date trend chart is shown in Figure 16.5. The completion date trend chart is set up with one calendar date scale on the bottom axis and another on the left-

---

* J. Gordon Davis, "Keeping Project Costs in Line," *Machine Design*, 1976.

* Material in this and the following section is taken from an article by Davis printed in *Machine Design*.

**FIGURE 16.5** Completion Date Trend Chart

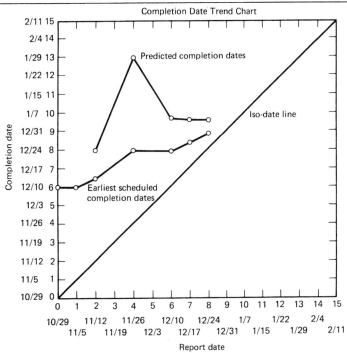

hand axis of a daily or weekly grid. The bottom calendar should start with the date of the project. This scale should be labeled Update Date, Report Date, or Effective Date of Schedule Information. The vertical calendar should start six to eight weeks ahead of the completion date initially scheduled for project completion or any other milestone that is to be tracked. This scale should be labeled Completion Date. An iso-date line is then drawn. This is the straight line passing through all points for which both calendar scales give the same date. This 45-degree line is the target line in the sense that trend lines will be projected to their intersection with the iso-date line. The points to be plotted on this chart are the earliest scheduled completion dates for the milestone activities being tracked. Each entry is made by finding on the bottom axis the date on which the latest schedule update was accomplished, then moving vertically to a point that shows the completion date resulting from the latest update.

If the initial schedule never had to be revised—that is, actual progress exactly matched the initial schedule—the completion date trend line would develop as a horizontal line. This line would intersect the iso-date line on the day that the milestone activity was completed. If progress toward that milestone was slower than scheduled, the line would have a positive slope. Conversely, better-than-scheduled performance would result in a negative slope to the trend line. Regardless of the level of performance, the trend line can be projected beyond the last plotted point to an intersection with

**FIGURE 16.6   Completion Date Trend Chart Worksheet**

| Y Orig. = 6 | 1 | 2 | 3 | 4 | 5 | 6 |
|---|---|---|---|---|---|---|
| | Report Date | Previous Report Date | Update Interval | Earliest Scheduled Completion | Slippage to Date | 2(3 × 5) |
| | | | 1–2 | | 4–Y orig. | |
| Original schedule | 0 | | | 6.0 | | |
| | 1 | 0 | 1 | 6.0 | 0 | 0 |
| | 2 | 1 | 1 | 6.5 | 0.5 | 1.0 |
| | 4 | 2 | 2 | 8.0 | 2.0 | 8.0 |
| | 6 | 4 | 2 | 8.0 | 2.0 | 8.0 |
| | | 6 | 1 | 8.5 | 2.5 | 5.0 |
| | | 7 | 1 | 9.0 | 3.0 | 6.0 |

the iso-date line. The point of intersection has a value on the completion date axis that may be interpreted as the predicted completion date.

Assume first that the performance level, whether high or low relative to scheduled performance, is a constant. This means that the trend line is straight and that a projection can be made by simply extending the trend line. Thus, a project on which time estimates were too low will show a trend line with a positive slope, and the projection will result in a predicted completion date greater than the currently scheduled earliest completion date. Similarly, a project on which time estimates were too high will develop a trend line with a negative slope, and its predicted completion date will be earlier than that resulting from the most recent update.

The visual projection of the trend line formed by the earliest-scheduled-completion-date points is subject to a great deal of variation. To standardize this projection, the following procedure should be used.

1. Assign week numbers starting with zero to each end-of-week point on each axis. Let $X$ equal the week number on the report date axis and $Y$ equal the week number on the completion date axis.
2. Set up the completion date trend chart worksheet as shown (see Fig. 16.6).
3. After each update, start a new worksheet row by filling in columns 1 and 4.
4. Calculate the entries for each remaining column in numerical order. Column 12 is the projected completion time expressed in work weeks from the start of the job.
5. The number in column 12 may be plotted on the completion date trend chart to assist in its interpretation.

## 16.5   COST TREND CHARTS

A project cost trend chart is shown in Figure 16.7. The project cost trend chart is set up with a calendar date scale on the bottom axis and a cost scale on the left axis of a rectangular grid. The calendar scale should start with the starting date of the project.

| 7 | 8 | 9 | 10 | 11 | 12 | 13 |
|---|---|---|---|---|---|---|
| | | | | | Predicted Completion Week | Predicted Completion Date |
| Σ6 | 7 ÷ 1² | 1 × 8 | 4 − 9 | 1 − 8 | | |
| | | | | | 10 ÷ 11 | 12 conv. |
| 0 | 0 | 0 | 6.0 | 1.0 | 6.0 | |
| 1.0 | 0.25 | 0.50 | 6.0 | 0.75 | 8.0 | |
| 9.0 | 0.56 | 2.24 | 5.76 | 0.44 | 13.09 | |
| 17.0 | 0.47 | 2.82 | 5.18 | 0.53 | 9.77 | |
| 22.0 | 0.44 | 3.08 | 5.42 | 0.56 | 9.67 | |
| 28.0 | 0.43 | 3.44 | 5.56 | 0.57 | 9.75 | |

This scale should be labeled Cost Report Date. The cost scale should start approximately 10 percent below the initial estimated project cost. This scale should be labeled Total Project Cost.

The points to be plotted on this chart are the sum of actual costs to date plus committed costs to date plus estimated additional costs to complete. Refer to these points as estimated costs at completion. These data are taken directly from the job cost status report.

The resulting adjacent points may be joined by straight lines to make it easier to visualize the trend of these points. If the original estimate and the project execution were both perfect, this line would be horizontal. However, many factors tend to cause

**FIGURE 16.7   Project Cost Trend Chart**

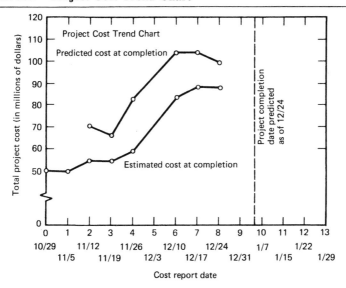

**FIGURE 16.8   Cost Trend Chart Worksheet**

| C orig. = $5.0 | 1 | 2 | 3 | 4 | 5 | 6 |
|---|---|---|---|---|---|---|
| | Report Date | Previous Report Date | Update Interval | Projected Completion Date | Remaining Time to Complete | Estimated Cost at Completion |
| | (input) | | 1 − 2 | (input) | 4−1 | (input) |
| Original | 0 | | | 6.0 | | 5.0 |
| | 1 | 0 | 1 | 6.0 | 5.0 | 5.0 |
| | 2 | 1 | 1 | 8.0 | 6.0 | 5.5 |
| | 3 | 2 | 1 | (8.0) | 5.0 | 5.5 |
| | 4 | 3 | 1 | 13.09 | 9.09 | 6.0 |
| | 6 | 4 | 2 | 9.77 | 3.77 | 8.5 |
| | 7 | 6 | 1 | 9.67 | 2.67 | 9.0 |
| | 8 | 7 | 1 | 9.75 | 1.75 | 9.0 |

the estimated cost at completion to rise from one cost report to the next. Projection of this trend to the anticipated project completion date will give a predicted cost at completion. The anticipated project completion date should be the predicted completion date from the completion date trend chart, rather than the earliest scheduled completion date, which could occur only if all factors causing project slippage were suddenly eliminated.

The visual projection of the cost trend line is subject to a great deal of variation. This projection can be made mathematically by use of the accompanying cost trend chart worksheet. The following procedure should be used:

1. On the cost trend chart, assign week numbers starting with zero to each end-of-week point on the report date axis. Let $X$ equal the week number on this axis. Let $C$ equal the cost figure on the total project cost axis.
2. After each update on the job cost status report, start a new worksheet row by filling in columns 1, 4, and 6. Column 12 of the completion date trend chart worksheet (see Figure 16.8) is the source of column 4 of the cost trend chart worksheet. If a cost figure is generated before a date on which no schedule update has been made, the most recent predicted completion date should be the column 4 entry on the cost trend chart worksheet.
3. Calculate the entries for each remaining column in numerical order. Column 12 is the predicted project cost at completion. This figure may be plotted on the cost trend chart to assist in its interpretation.

## 16.6   COST ANALYSIS IN A GAMING ENVIRONMENT

The time and cost estimates required to apply the concepts of variance and trend analysis are available from schedule and cost reports. In order to introduce the utilization of these techniques to managers, gaming techniques are useful in developing the required data in a pseudorealistic environment. The idea of gaming both as an educational as well as an analytical method of dealing with management topics is not new. One of the first historically recorded uses of games to model a real-world situation

| 7 | 8 | 9 | 10 | 11 | 12 |
|---|---|---|---|---|---|
| *Estimated Cost Overrun* | *2(3 λ 7)* | *Σ8* | *Cost Slope* | *Predicted Additional Overrun* | *Predicted Cost at Completion* |
| 6–C orig. | | | 9 ÷ 1² | 5 × 10 | 6 ÷ 11 |
| 0.0 | 0.0 | 0.0 | 0.0 | 0.0 | 5.0 |
| 0.5 | 1.0 | 1.0 | 0.25 | 1.5 | 7.0 |
| 0.5 | 1.0 | 2.0 | 0.22 | 1.1 | 6.6 |
| 1.0 | 2.0 | 4.0 | 0.25 | 2.27 | 8.27 |
| 3.5 | 14.0 | 18.0 | 0.50 | 1.89 | 10.39 |
| 4.0 | 8.0 | 26.0 | 0.53 | 1.42 | 10.42 |
| 4.0 | 8.0 | 34.0 | 0.53 | 0.93 | 9.93 |

dates back to the eighteenth century and the development of "war chess." Kibbee, ⟩ Craft, and Nanus in their book, *Management Games,* note that:

> Chess was early used as a form of war game, and many varieties of "war chess" have been created, the game being adapted to changing military concepts. Helwig, Master of Pages at the court of Duke of Brunswick, developed a form of war chess in the 18th century which used a board made up of 1,666 squares, and had pieces representing battalions of fusileers, squadrons of dragoons, batteries of siege guns, and so forth.*

Much of the literature concerning heuristically or intuitively formulated management games can be traced back to the late 1950s. In 1957, the American Management Association Academy at Saranac Lake, N.Y., introduced a computerized game designed to assist in the training of top management executives. Since the late nineteen-fifties, the proliferation of games for training purposes has been remarkable. The *Business Games Handbook* gives a selected list of over 180 general-purpose and functional management games in areas ranging from hotel management to egg production and from advertising to computer programming supervision.† The areas to which the gaming concept can be applied appear to be bounded only by the designer's imagination.

## 16.7  PROJECT-ORIENTED GAME

The project-oriented construction management game CONSTRUCTO has been developed to give the manager an environment in which he can experience to some degree the dynamics of project management. The game is designed to present the manager with realistic data projections that form the basis of decision making to control cost and time. This is accomplished by confronting him with situations in the game

---

* J. E. Kibbee, et al., *Management Games* (New York: Reinhold 1961).

† R. G. Graham, and C. F. Gray, *Business Games Handbook,* New York: American Management Association, Inc., 1969.

## FIGURE 16.9  CONSTRUCTO Flow Diagram

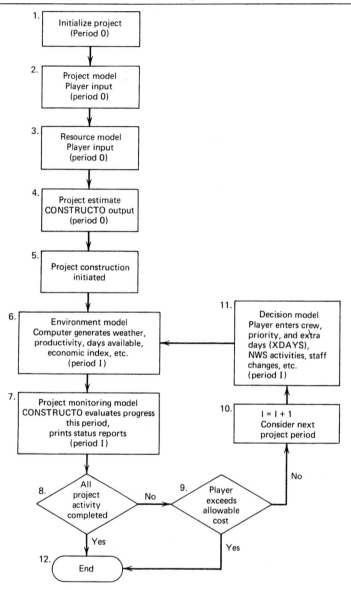

environment similar to those actually encountered on the job site. Admittedly, there is a degree of artificiality separating the real world and the game environment. However, the basic difficulty facing the real-world manager and the player are the same. That difficulty is one of sensing in which direction the situation will move and then making the correct decision. Correct decisions must be based upon an evaluation of the potential response of the "system" (i.e., the project) to the decision options available. A feeling for this dynamic response can be developed within the framework of a project-oriented game.

CONSTRUCTO confronts the player with a simulated situation described in terms of environmental and economic parameters and places him in the position of being in charge of a construction project. This is achieved by using the concept of man-machine interaction. The player, representing the manager, interacts with the computer, which generates the construction environment in terms of weather, economic situation, and work productivity. The model used to structure this interaction between player and machines is the well-known network or critical path representation of the project of interest. The interaction is represented in the flow diagram in Figure 16.9.

The player begins by selecting a project and breaking it into its component operations. This step can be bypassed if a library of projects developed for the game is available. In this case, a preselected project can be called to initiate the game. Crews for the operations within the project are selected by the player from a pool of 11 available crafts, and estimates of material costs for each operation are made. The player feeds this information to the computer in a free-form problem-oriented language (POL).

Following input of activities and crews, the computer generates an estimate of the construction progress (in terms of percentage complete) and the financial expenditures period by period throughout the life of the project. This estimate is designed to approximate the information available to the project manager from the company's estimating section at the beginning of the project. It assumes 20 days available for construction on each activity each month as an average. The CONSTRUCTO game is described in detail in *CONSTRUCTO—A Heuristic Game for Construction* by Halpin and Woodhead.

## 16.8    A PROJECT COST ANALYSIS

In order to better understand the application of the techniques discussed in this chapter, a reference project will be used. Figure 16.10 shows a 7200-ft$^2$ floor plan and approximate cross sections for a three-story office building. The project is conceived to provide office space with corresponding floor rest rooms and accesses. An underground parking and basement area is also provided with convenient mechanical rooms for the ventilation. Air conditioning, heating, and water supply are shown as well. Two staircases and one elevator allow movement through the building.

The first floor is assumed to be at the ground level, so general excavation to the basement level is required for the basement structure (retaining walls) and the building foundations. Beyond this level a minimum depth of additional excavation will provide room for footings and column pedestals. Reinforced concrete is the main structural element both for the building frame and for floor and basement slabs.

**FIGURE 16.10   Office Building Details**

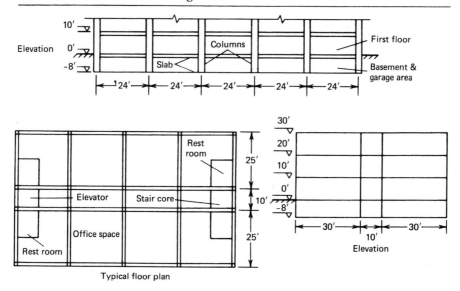

*Source:* D. W. Halpin and R. W. Woodhead, *CONSTRUCTO—A Heuristic Game for Construction Management,* copyright © 1973 by the Board of Trustees of the University of Illinois. Reprinted by permission of the University of Illinois Press.

The floors are sectioned off by drywall elements. Brick will be used for the exterior walls, and metal windows will be placed in the front face of the building. Acoustical ceiling with artificial lighting incorporated as an integral part will provide noise insulation as well as decorative effect. Floor access doors will be metal office doors, and closets and shelves will be wooden. Rest areas will be conveniently located on each floor. The input data for this project are given in Appendix F.

Relevant information for this project as generated by the CONSTRUCTO game is as follows:

```
PROJECTED EXPECTED PROFIT  =  154955.00
PROJECTED PROJECT DURATION  =  193
THE CONTRACT CALLS FOR THE WORK TO BE COMPLETED IN 193
WORKING DAYS.
THE WORK MUST BE COMPLETED WITHIN 10 MONTHS (IN
DECEMBER)
THE HIGH BID ON THE PROJECT WAS              1849864.00
THE AVERAGE BID ON THE PROJECT WAS           1733206.00
THE LOW BID ON THE PROJECT WAS               1666545.00
YOURS WAS THE LOW BID
YOU WERE LOWER THAN THE NEXT HIGHER BID BY     24998.17
YOUR EXPECTED COST FOR THIS PROJECT IS       1511590.00
                    GOOD LUCK
```

**FIGURE 16.11  Portion of Estimate/Budget for Small Office Building**

*Cumulative Progress Report Through March (Projected from Start)*

| Activity Number | Percent Complete | Labor Units to date | Labor Cost to Date | Total Cost to Date | Days to Date | Expected Total Duration | Days Remaining | Critical Index |
|---|---|---|---|---|---|---|---|---|
| * 1 | 100.0 | 453.83 | 2380.80 | 2880.80 | 4 | 4 | 0 | 1.000 |
| * 2 | 100.0 | 94.17 | 484.80 | 484.80 | 2 | 2 | 0 | 1.000 |
| * 3 | 100.0 | 70.21 | 368.80 | 848.80 | 2 | 2 | 0 | 1.000 |
| * 4 | 100.0 | 1530.01 | 8019.20 | 22499.20 | 7 | 7 | 0 | 1.000 |
| * 5 | 61.67 | 2784.32 | 13921.59 | 39823.38 | 7 | 12 | 5 | 1.000 |
| | 3.87 | 4932.53 | 30479.18 | 71840.88 | | | | |

. . .    . . .    . . .

(Reports for April to October have been deleted for compactness of presentation.)

*Cumulative Progress Report Through November (Projected from Start)*

| Activity Number | Percent Complete | Labor Units to date | Labor Cost to Date | Total Cost to Date | Days to Date | Expected Total Duration | Days Remaining | Critical Index |
|---|---|---|---|---|---|---|---|---|
| 31 | 100.00 | 6500.20 | 33139.16 | 45639.15 | 24 | 24 | 0 | 0.863 |
| * 40 | 100.00 | 695.34 | 3636.00 | 6036.00 | 5 | 5 | 0 | 1.000 |
| 41 | 92.20 | 5425.04 | 28720.78 | 37018.47 | 18 | 19 | 1 | 0.996 |
| 43 | 100.00 | 5013.07 | 26531.97 | 56531.97 | 15 | 15 | 0 | 0.991 |
| * 44 | 100.00 | 5179.05 | 26531.97 | 56531.97 | 15 | 15 | 0 | 1.000 |
| 46 | 100.00 | 4539.38 | 24295.98 | 45295.98 | 10 | 10 | 0 | 0.996 |
| * 47 | 21.94 | 971.84 | 4859.20 | 9466.73 | 2 | 10 | 8 | 1.000 |
| 48 | 100.00 | 124.80 | 646.40 | 646.40 | 2 | 2 | 0 | 0.958 |
| | 96.28 | 122845.94 | 691896.44 | 1463652.00 | | | | |

(cont.)

323

**FIGURE 16.11** (cont.)

*Cumulative Progress Report Through December (Projected from Start)*

| Activity Number | Percent Complete | Labor Units to Date | Labor Cost to Date | Total Cost to Date | Days to Date | Expected Total Duration | Days Remaining | Critical Index |
|---|---|---|---|---|---|---|---|---|
| * 1 | 100.0 | 453.83 | 2380.80 | 2880.80 | 4 | 4 | 0 | 1.000 |
| * 2 | 100.00 | 94.17 | 484.80 | 484.80 | 2 | 2 | 0 | 1.000 |
| * 3 | 100.00 | 70.21 | 368.80 | 848.80 | 2 | 2 | 0 | 1.000 |
| * 4 | 100.00 | 1530.01 | 8019.20 | 22499.20 | 7 | 7 | 0 | 1.000 |
| * 5 | 100.00 | 4514.80 | 23865.59 | 65865.56 | 12 | 12 | 0 | 1.000 |
| * 6 | 100.00 | 1797.86 | 9475.19 | 18475.19 | 4 | 4 | 0 | 1.000 |
| * 7 | 100.00 | 7377.66 | 37409.96 | 77409.94 | 15 | 15 | 0 | 1.000 |
| * 8 | 100.00 | 311.69 | 1590.40 | 6390.40 | 2 | 2 | 0 | 0.948 |
| * 9 | 100.00 | 4296.44 | 22852.77 | 28852.77 | 12 | 12 | 0 | 1.000 |
| 10 | 100.00 | 188.71 | 969.60 | 2769.60 | 2 | 2 | 0 | 0.948 |
| * 11 | 100.00 | 1078.93 | 5971.20 | 11971.20 | 4 | 4 | 0 | 1.000 |
| * 12 | 100.00 | 3648.69 | 19043.98 | 33043.98 | 10 | 10 | 0 | 1.000 |
| 13 | 100.00 | 308.79 | 1590.40 | 6390.40 | 2 | 2 | 0 | 0.948 |
| * 14 | 100.00 | 4317.61 | 22852.77 | 34852.77 | 12 | 12 | 0 | 1.000 |
| 15 | 100.00 | 185.54 | 969.60 | 2769.60 | 2 | 2 | 0 | 0.948 |
| * 16 | 100.00 | 1175.61 | 5971.20 | 11971.20 | 4 | 4 | 0 | 1.000 |
| 17 | 100.00 | 2655.96 | 13634.39 | 73634.38 | 19 | 19 | 0 | 0.484 |
| 18 | 100.00 | 1918.79 | 9665.59 | 59665.59 | 7 | 7 | 0 | 0.813 |
| 19 | 100.00 | 1057.05 | 5817.60 | 21817.60 | 6 | 6 | 0 | 0.780 |
| * 20 | 100.00 | 3463.50 | 19043.98 | 33043.98 | 10 | 10 | 0 | 1.000 |
| 21 | 100.00 | 294.66 | 1590.40 | 6390.40 | 2 | 2 | 0 | 0.947 |
| * 22 | 100.00 | 3530.82 | 17771.59 | 29771.59 | 11 | 11 | 0 | 1.000 |
| 23 | 100.00 | 191.55 | 969.60 | 2769.60 | 2 | 2 | 0 | 0.948 |

| | | | | | | | | | |
|---|---|---|---|---|---|---|---|---|---|
| | 24 | 100.00 | 1264.73 | 6574.39 | 12574.39 | 4 | 4 | 0 | 1.000 |
| * | 25 | 100.00 | 6765.35 | 35159.96 | 47659.96 | 25 | 25 | 0 | 0.726 |
| | 26 | 100.00 | 3971.19 | 20139.99 | 34139.99 | 10 | 10 | 0 | 1.000 |
| * | 27 | 100.00 | 460.31 | 2385.60 | 7185.60 | 3 | 3 | 0 | 0.953 |
| | 28 | 100.00 | 4210.66 | 22852.77 | 34852.77 | 12 | 12 | 0 | 1.000 |
| * | 29 | 100.00 | 183.29 | 969.60 | 2569.60 | 2 | 2 | 0 | 0.948 |
| | 30 | 100.00 | 1178.93 | 5971.20 | 11971.20 | 4 | 4 | 0 | 1.000 |
| * | 31 | 100.00 | 6500.20 | 33139.16 | 45639.16 | 24 | 24 | 0 | 0.863 |
| | 32 | 100.00 | 1685.46 | 8726.39 | 23726.39 | 9 | 9 | 0 | 0.838 |
| | 33 | 100.00 | 3084.04 | 16569.59 | 66569.56 | 12 | 12 | 0 | 0.991 |
| | 34 | 100.00 | 3042.84 | 16569.59 | 66569.56 | 12 | 12 | 0 | 0.985 |
| * | 35 | 100.00 | 3042.19 | 16569.59 | 66569.56 | 12 | 12 | 0 | 1.000 |
| | 36 | 100.00 | 1482.79 | 7756.80 | 23756.80 | 8 | 8 | 0 | 0.993 |
| | 37 | 100.00 | 1475.99 | 7756.80 | 23756.80 | 8 | 8 | 0 | 0.997 |
| * | 38 | 100.00 | 1448.32 | 7756.80 | 23756.80 | 8 | 8 | 0 | 1.000 |
| | 39 | 100.00 | 2532.82 | 12927.99 | 17727.99 | 10 | 10 | 0 | 0.978 |
| * | 40 | 100.00 | 695.34 | 3636.00 | 6036.00 | 5 | 5 | 0 | 1.000 |
| | 41 | 100.00 | 5884.21 | 30316.39 | 39316.39 | 19 | 19 | 0 | 0.996 |
| | 42 | 100.00 | 5254.62 | 26531.97 | 56531.97 | 15 | 15 | 0 | 0.994 |
| | 43 | 100.00 | 5013.07 | 26531.97 | 56531.97 | 15 | 15 | 0 | 0.991 |
| * | 44 | 100.00 | 5179.05 | 26531.97 | 56531.97 | 15 | 15 | 0 | 1.000 |
| | 45 | 100.00 | 4486.21 | 24295.98 | 45295.98 | 10 | 10 | 0 | 0.939 |
| | 46 | 100.00 | 4539.38 | 24295.98 | 45295.98 | 10 | 10 | 0 | 0.996 |
| * | 47 | 100.00 | 4429.40 | 24295.98 | 45295.98 | 10 | 10 | 0 | 1.000 |
| | 48 | 100.00 | 124.80 | 646.40 | 646.40 | 2 | 2 | 0 | 0.958 |
| | 49 | 100.00 | 4365.18 | 23975.98 | 38975.98 | 10 | 10 | 0 | 0.835 |
| * | 50 | 100.00 | 836.81 | 4507.99 | 4507.99 | 5 | 5 | 0 | 1.000 |
| | | 100.00 | 127599.50 | 722740.75 | 1511590.00 | | | | |

# FIGURE 16.12 September Progress Report

*Construct Small Office Building Period Report for Work Done in September*

| Mean Temp | Mean Rain | Economic Index |
|---|---|---|
| 79.29 | 1.16 | 1.05 |

**Craft Productivity**

| Carpenter | Ironworker | Mason | Finisher | Painter | Equipment Operator | Oiler | Forman | Labor | Electrician | Mechanic |
|---|---|---|---|---|---|---|---|---|---|---|
| 1.01 | 1.02 | 0.91 | 1.08 | 0.91 | 1.08 | 1.03 | 0.94 | 0.99 | 0.96 | 0.97 |

| Activity Number | Regular Days | Holidays | Period Labor Units | Activity Prod Ratio | Percent This Period | Period Labor Cost | Overtime Cost | Period Material Cost | Total Period Cost | Crash Rate |
|---|---|---|---|---|---|---|---|---|---|---|
| 26 | 1 | 0 | 123.60 | 0.94 | 3.44 | 2,014.00 | 0.0 | 560.98 | 2,574.98 | |
| 27 | 3 | 0 | 423.20 | 0.94 | 100.00 | 2,385.60 | 0.0 | 5,191.91 | 7,577.51 | |
| 28 | 12 | 0 | 4,157.46 | 0.94 | 100.00 | 22,852.77 | 0.0 | 11,961.27 | 34,814.04 | |
| 29 | 2 | 0 | 178.63 | 0.94 | 100.00 | 969.60 | 0.0 | 1,677.13 | 2,646.73 | |
| 30 | 5 | 0 | 1,155.25 | 0.94 | 100.00 | 7,463.99 | 0.0 | 5,974.95 | 13,438.94 | |
| 31 | 1 | 0 | 190.19 | 0.94 | 3.04 | 1,380.80 | 0.0 | 380.12 | 1,760.92 | |
| 32 | 1 | 0 | 133.55 | 0.94 | 8.29 | 969.60 | 0.0 | 1,246.71 | 2,216.31 | |
| 33 | 3 | 0 | 578.53 | 0.94 | 19.56 | 4,142.39 | 0.0 | 10,708.18 | 14,850.57 | |
| 34 | 12 | 0 | 2,956.18 | 0.94 | 100.00 | 16,569.59 | 0.0 | 53,496.79 | 70,066.38 | |
| 35 | 1 | 0 | 190.19 | 0.94 | 6.40 | 1,380.80 | 0.0 | 3,200.67 | 4,581.46 | |
| 36 | 8 | 0 | 1,389.25 | 0.94 | 100.00 | 7,756.80 | 0.0 | 17,126.17 | 24,882.97 | |
| 37 | 4 | 0 | 680.86 | 0.94 | 46.30 | 3,878.40 | 0.0 | 7,635.43 | 11,513.83 | |
| NWS 42 | 9 | 0 | 2,998.92 | 0.99 | 61.43 | 15,919.19 | 0.0 | 18,583.71 | 34,502.91 | |
| STAFF | 20 | 0 | 1,060.80 | | | 5,304.00 | | | 5,304.00 | |
| | | | 15,155.81 | | 12.67 | 92,987.38 | 0.0 | 137,743.81 | 230,731.38 | |

*Cumulative Progress Report Through December (Projected from September)*

| Activity Number | Percent Complete | Labor Units to Date | Labor Cost to Date | Total Cost to Date | Days to Date | Expected Total Duration | Days Remaining | Critical Index |
|---|---|---|---|---|---|---|---|---|
| 23 | 100.00 | 178.92 | 1454.40 | 3327.43 | 3 | 3 | 0 | |
| * 24 | 100.00 | 1218.59 | 7401.59 | 13893.04 | 3 | 3 | 0 | |

| | | | | | | | | |
|---|---|---|---|---|---|---|---|---|
| | 25 | 100.00 | 6609.79 | 39379.17 | 52748.13 | 28 | 28 | 0 |
| * | 26 | 100.00 | 3594.10 | 22153.98 | 37292.14 | 11 | 11 | 0 | 0.468 |
| | 27 | 100.00 | 423.20 | 2385.60 | 7577.51 | 3 | 3 | 0 | 0.392 |
| * | 28 | 100.00 | 4157.46 | 22852.77 | 34814.03 | 12 | 12 | 0 |
| | 29 | 100.00 | 178.63 | 969.60 | 2646.73 | 2 | 2 | 0 |
| * | 30 | 100.00 | 1155.25 | 7463.99 | 13438.94 | 5 | 5 | 0 |
| | 31 | 100.00 | 6259.85 | 38662.36 | 51173.36 | 28 | 28 | 0 |
| | 32 | 100.00 | 1610.89 | 9695.99 | 24733.39 | 10 | 10 | 0 |
| | 33 | 100.00 | 2957.91 | 17950.38 | 68920.63 | 13 | 13 | 0 |
| | 34 | 100.00 | 2956.18 | 16569.59 | 70066.31 | 12 | 12 | 0 |
| * | 35 | 100.00 | 2972.14 | 19331.18 | 69347.63 | 14 | 14 | 0 | 0.921 |
| | 36 | 100.00 | 1389.25 | 7756.80 | 24882.96 | 8 | 8 | 0 |
| | 37 | 100.00 | 1470.43 | 8726.39 | 25216.47 | 9 | 9 | 0 | 0.852 |
| * | 38 | 100.00 | 1448.32 | 7756.80 | 23756.80 | 8 | 8 | 0 | 0.991 |
| | 39 | 100.00 | 2532.82 | 12927.99 | 17727.99 | 10 | 10 | 0 | 0.858 |
| * | 40 | 100.00 | 695.34 | 3636.00 | 6036.00 | 5 | 5 | 0 | 0.996 |
| | 41 | 100.00 | 5884.21 | 30316.39 | 39316.39 | 19 | 19 | 0 | 0.886 |
| | 42 | 100.00 | 4881.58 | 26531.98 | 56782.19 | 15 | 15 | 0 | 0.873 |
| | 43 | 100.00 | 5013.07 | 26531.97 | 56531.97 | 15 | 15 | 0 | 0.865 |
| * | 44 | 100.00 | 5179.05 | 26531.97 | 56531.97 | 15 | 15 | 0 | 0.994 |
| | 45 | 100.00 | 4486.21 | 24295.98 | 45295.98 | 10 | 10 | 0 | 0.692 |
| | 46 | 100.00 | 4539.38 | 24295.98 | 45295.98 | 10 | 10 | 0 | 0.781 |
| * | 47 | 100.00 | 4429.40 | 24295.98 | 45295.98 | 10 | 10 | 0 | 0.985 |
| | 48 | 100.00 | 124.80 | 646.40 | 646.80 | 2 | 2 | 0 | 0.860 |
| | 49 | 100.00 | 4365.18 | 23975.98 | 38975.98 | 10 | 10 | 0 | 0.437 |
| * | 50 | 100.00 | 836.81 | 4507.99 | 4507.99 | 5 | 5 | 0 | 0.994 |
| | | 100.00 | 123377.38 | 778092.69 | 1578433.00 | | | |

Projected expected profit = 88,112.00
Projected project duration = 197
Warning: This expected duration exceeds the contract limit by 4 days.

327

The project budget as generated is shown in part in Figure 16.11.

Based on the budget, the maximum cost (bid price) is $1,666,545 with an expected cost of $1,511,590. The project is estimated to last approximately ten months with a total duration of 193 days. Hence, the contract allows 193 days for completion.

The actual progress report for the month of September is given in Figure 16.12. The report indicates that activities 26 through 37 were worked on during the month. The relevant portion of the scheduling network is shown in Figure 16.13. The network is shown in precedence notation. Activity 26 was completed during the period. Data required for variance analysis are available in the original estimate (Figure 16.11) and the September progress report (Figure 16.13). The analysis is carried out based on labor units and labor cost. The relevant data are shown below.

A two-factor analysis of these data is shown in Figure 16.14.

**FIGURE 16.13   Network Segment as of September**

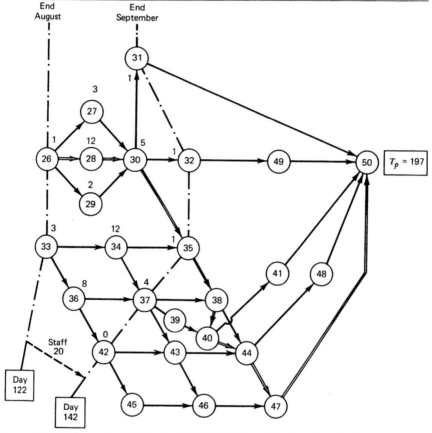

**FIGURE 16.14   Two-Factor Analysis of Activity 26**

|  | Labor units | Labor cost |
|---|---|---|
| Budget | 3971.19 | $20,139.99 |
| Completed amount<br>(September report) | 3594.10 | $22,153.98 |

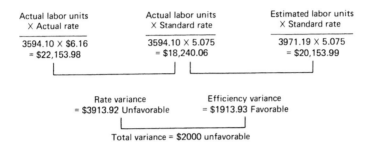

The labor unit as used in the CONSTRUCTO game is defined as follows:

$$LU \text{ (craft K)} = \frac{\text{wage rate (craft K)}}{\text{wage rate (laborer)}}$$

The crew for activity 26 is constituted as follows:

|  | Wage Rate for Craft | (Hourly) LU Value |
|---|---|---|
| 3 foremen | $10.10 | 1.47 |
| 3 carpenters | 8.75 | 1.28 |
| 6 iron workers | 9.35 | 1.36 |
| 8 laborers | 6.85 | 1.00 |
| 1 equipment operator | 9.35 | 1.36 |

The hourly LU value for this crew is given as

$$\text{Hourly LUs} = \sum_{k=1}^{N} \{N(K) \times LU(K)\}$$

where $N(K)$ is the number of crew members of craft K

$LU(K)$ is the LU value of craft K as defined above

For this crew the hourly value is

$$\text{Hourly LUs} = 3(1.47) + 3(1.28) + 6(1.36)$$
$$+ 8(1.00) + 1(1.36) = 25.77 \text{ LUs/hour}$$

**TABLE 16.1 Completion Date Trend Chart Worksheet (Working Days)**

| | 1 | 2 | 3 | 4 | 5 | 6 | 7 | 8 | 9 | 10 | 11 | 12 |
|---|---|---|---|---|---|---|---|---|---|---|---|---|
| | Report Date | Previous Report Date | Update Interval | Earliest Scheduled Completion | Slippage to Date | $2(3 \times 5)$ | $\Sigma 6$ | $7 \div 11^2$ | $1 \times 8$ | $4$–$9$ | $1$–$8$ | $10 \div 11$ |
| Original schedule | 0 | | | 193 | | | | | | | | |
| | 20 | 0 | 20 | 202 | 9 | 360 | 360 | 0.9 | 18 | 184 | 0.1 | 1840 |
| | 40 | 20 | 20 | 210 | 17 | 680 | 1040 | 0.65 | 26 | 184 | 0.35 | 527.7 |
| | 60 | 40 | 20 | 203 | 10 | 400 | 1440 | 0.4 | 24 | 179 | 0.6 | 298 |
| | 80 | 60 | 20 | 197 | 4 | 160 | 1600 | 0.25 | 20 | 177 | 0.75 | 236 |
| | 100 | 80 | 20 | 196 | 3 | 120 | 1720 | 0.172 | 17.2 | 178.8 | 0.828 | 216 |
| | 120 | 100 | 20 | 194 | 1 | 40 | 1760 | 0.122 | 14.6 | 179.4 | 0.878 | 204 |
| | 140 | 120 | 20 | 197 | 4 | 160 | 1920 | 0.098 | 13.7 | 183.3 | 0.902 | 203 |

**TABLE 16.2  Cost Trend Chart Worksheet**

| | 1 | 2 | 3 | 4 | 5 | 6 | 7 | 8 | 9 | 10 | 11 | 12 |
|---|---|---|---|---|---|---|---|---|---|---|---|---|
| | Report Date | Previous Report Date | Update Interval | Projected Completion Date | Remaining Time to Complete | Estimated Cost to Complete | Estimated Overrun | $2(3 \times 7)$ | $\Sigma 8$ | Cost Slope $9 \div (1)^2$ | Predicted Additional Overrun $5 \times 10$ | Predicted Cost at Completion $6 \div 11$ |
| Original cost estimate | 0 | | | 193 | | 1,511,590 | | | | | | |
| | 20 | 0 | 20 | 1840 | 1820 | 1,528,099 | 16,909 | 676,360 | 676,360 | 1691 | | |
| | 40 | 20 | 20 | 526 | 486 | 1,530,000 | 18,810 | 752,400 | 1,428,760 | 893 | 433,998 | 1,963,999 |
| | 60 | 40 | 20 | 298 | 238 | 1,580,000 | 68,810 | 2,752,400 | 4,181,160 | 1161 | 276,318 | 1,856,318 |
| | 80 | 60 | 20 | 236 | 156 | 1,570,000 | 58,810 | 2,352,400 | 6,533,560 | 1021 | 159,276 | 1,729,276 |
| | 100 | 80 | 20 | 216 | 116 | 1,560,000 | 48,410 | 1,936,400 | 8,469,960 | 847 | 98,252 | 1,658,252 |
| | 120 | 100 | 20 | 204 | 84 | 1,556,527 | 44,937 | 1,797,480 | 10,267,440 | 713 | 59,892 | 1,616,419 |
| | 140 | 120 | 20 | 203 | 63 | 1,578,433 | 66,843 | 2,673,720 | 12,941,160 | 660 | 41,580 | 1,620,013 |

Based on performance through September, the costs to date exceed those originally projected by:

$$\text{Overrun} = \begin{pmatrix} \text{total cost as} \\ \text{of September} \end{pmatrix} - \begin{pmatrix} \text{original budget} \\ \text{total cost} \end{pmatrix}$$

$$\text{Cost} \quad = (\$1,578,433) \quad - (\$1,511,590) = \$66,843$$

This is a projected overrun based on actual performance to date. By using cost trend analysis the forecasted overrun as of the end of the project can be calculated. This projection will take into account the expected time overrun (if any) as well as cost trends as they have developed to date. This analysis is based on data available from the CONSTRUCTO game output up through and including the month of September. The relevant data and supporting calculations are shown in Tables 16.1 (completion date trend chart worksheet) and 16.2 (cost trend chart worksheet). The graphical plots of the relevant completion date and project cost trends are shown in Figures 16.15 and 16.16.

**FIGURE 16.15   Completion Date Trend Chart for Small Office Building**

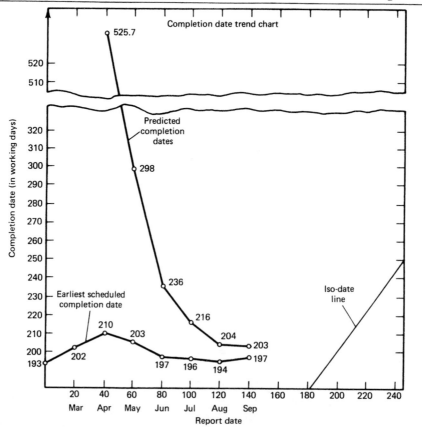

**FIGURE 16.16   Project Cost Trend Chart for Small Office Building**

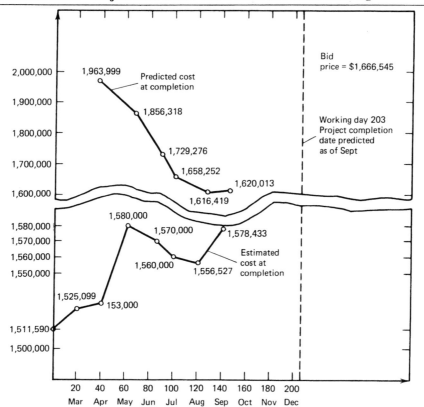

The completion date trend chart and supporting worksheet indicate that the earliest completion date reported in the September progress report is 197 working days (40 weeks). By using the methods of Section 16.4, the predicted completion date is forecast as 203 working days. The completion date trend data are used as input to column 4 of the cost trend chart worksheet. The cost trend chart indicates that the predicted cost at completion based on production through September is $1,620,013. This forecast indicates that the actual overrun will be greater than projected in the September program report. The cost trend analysis indicates an overrun:

$$\text{Cost overrun} = \left(\begin{matrix}\text{predicted cost}\\\text{at completion}\end{matrix}\right) - \left(\begin{matrix}\text{original budget}\\\text{total cost}\end{matrix}\right)$$

$$= (\$1,620,013) - (\$1,511,590)$$

$$= \$108,423$$

The profit in the job based on the original estimate and progress to date as well as forecast is summarized below:

1.  Original profit                                    $154,955
2.  Profit based on September progress report           88,112
3.  Predicted profit based on cost trend                46,532


## 16.9    SUMMARY

Cost reports provide the manager with the feedback information needed to properly control a project. The interpretation and analysis of this information is key to detecting trouble spots and taking appropriate management action. Appropriate corrective action may entail reallocating resources such as labor and machines from one activity to another. The proper analysis of cost reports should provide the manager with early warning regarding cost overruns and time losses that will reduce profit.

Two methods have been discussed in this chapter. The variance analysis method uses project status information to detect deviation from expected cost and production performance. The two-factor variance model provides the best insight into the reasons for cost over- or underrun.

The cost trend analysis technique described in this chapter is oriented toward forecasting deviation in both time and cost from originally budgeted figures. Whereas variance analysis focuses primarily on interpretation of cost item variation, the trend analysis forecasts project level cost and time deviations. Both methods should be used. Successfully applied, they provide the manager with a method of identifying and heading off problems before they get out of control.


## REVIEW QUESTIONS AND EXERCISES

**16.1**   For a particular job, one cost account was estimated to require 500 hours of work time at $12.50 per hour. The work was completed in 522 hours at a cost of $10.48 per hour. The original duration of the activity was projected as 10 working days. At the end of five days, the number of hours expended was 210 and the hourly cost was $9.55 per hour.
After the first five days, what are the

(a) Labor rate variance
(b) Labor efficiency variance
(c) Total labor variance
at five days and projected to the end of the activity. Designate whether favorable or unfavorable.

**16.2**   The estimated number of man-hours required for trench excavation on the Camp Creek Pumping Station (Div. 3) was 180. The actual number of hours required was 143. Eighty of these hours were straight-time hours at $9.64 per hour. The remaining hours were premium hours at time and a half. The job was bid and based on straight time only. Calculate the rate variance, the efficiency variance, and the total variance on this account.

**16.3**   Calculate price variance and efficiency variance for Problem 12.6. Suggest actions to be taken to correct deviations.

**16.4**   Using the CONSTRUCTO output provided for the Coffer Dam project of Figure P16.4 to include the original estimate cash flows for months March, April, and June 19X1, draw the S curves over the life of the project for:

**FIGURE P16.4**

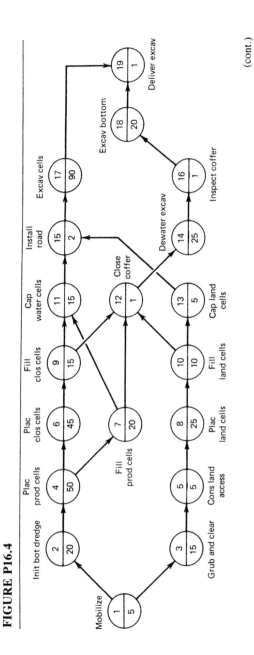

(cont.)

335

## FIGURE P16.4  (cont.)

Coffer Dam Project Data

ESTIMATED PROGRESS REPORT THROUGH MARCH 19X1

|   | ACTIVITY NUMBER | PERCENT COMPLETE | LABOR UNITS TO DATE | LABOR COST TO DATE | TOTAL COST TO DATE |
|---|---|---|---|---|---|
| * | 1 | 100.00 | 383.04 | 2740.00 | 2740.00 |
| * | 2 | 75.54 | 1036.20 | 7098.00 | 7098.00 |
|   | 3 | 100.00 | 394.59 | 2766.00 | 2766.00 |
|   |   | -------- | -------- | --------- | --------- |
|   |   | 7.73 | 1813.84 | 18308.00 | 18308.00 |

ESTIMATED PROGRESS REPORT THROUGH APRIL 19X1

|   | ACTIVITY NUMBER | PERCENT COMPLETE | LABOR UNITS TO DATE | LABOR COST TO DATE | TOTAL COST TO DATE |
|---|---|---|---|---|---|
| * | 2 | 100.00 | 1371.66 | 9464.00 | 9464.00 |
|   | 3 | 100.00 | 394.59 | 2766.00 | 2766.00 |
| * | 4 | 32.23 | 2173.14 | 14886.00 | 18109.12 |
|   | 5 | 100.00 | 240.75 | 1744.00 | 2744.00 |
|   | 8 | 64.41 | 643.80 | 4410.00 | 8274.58 |
|   |   | -------- | -------- | --------- | --------- |
|   |   | 22.17 | 5206.98 | 47418.00 | 55505.71 |

ESTIMATED PROGRESS REPORT THROUGH MAY 19X1

|   | ACTIVITY NUMBER | PERCENT COMPLETE | LABOR UNITS TO DATE | LABOR COST TO DATE | TOTAL COST TO DATE |
|---|---|---|---|---|---|
| * | 4 | 75.21 | 5070.66 | 35726.40 | 43247.02 |
|   | 8 | 100.00 | 999.53 | 7056.00 | 13056.00 |
|   | 10 | 100.00 | 780.14 | 5692.00 | 7692.00 |
|   |   | -------- | -------- | --------- | --------- |
|   |   | 39.34 | 9240.37 | 82300.40 | 98821.02 |

ESTIMATED PROGRESS REPORT THROUGH JUNE 19X1

|   | ACTIVITY NUMBER | PERCENT COMPLETE | LABOR UNITS TO DATE | LABOR COST TO DATE | TOTAL COST TO DATE |
|---|---|---|---|---|---|
| * | 4 | 100.00 | 6742.34 | 47635.20 | 57635.20 |
| * | 6 | 12.30 | 628.91 | 4308.00 | 5292.27 |
|   | 7 | 26.92 | 95.47 | 654.00 | 1730.78 |
|   | 10 | 100.00 | 780.14 | 5692.00 | 7692.00 |
|   | 13 | 100.00 | 397.98 | 2846.00 | 3346.00 |
|   |   | -------- | -------- | --------- | --------- |
|   |   | 51.23 | 12034.40 | 107721.20 | 129282.25 |

ESTIMATED PROGRESS REPORT THROUGH JULY 19X1

|   | ACTIVITY NUMBER | PERCENT COMPLETE | LABOR UNITS TO DATE | LABOR COST TO DATE | TOTAL COST TO DATE |
|---|---|---|---|---|---|
| * | 6 | 61.52 | 3144.53 | 22401.60 | 27322.95 |
|   | 7 | 100.00 | 354.67 | 2485.20 | 6485.20 |
|   |   | -------- | -------- | --------- | --------- |
|   |   | 63.04 | 14809.22 | 133350.00 | 161771.35 |

ESTIMATED PROGRESS REPORT THROUGH AUGUST 19X1

| ACTIVITY NUMBER | PERCENT COMPLETE | LABOR UNITS TO DATE | LABOR COST TO DATE | TOTAL COST TO DATE |
|---|---|---|---|---|
| *    6 | 100.00 | 5111.65 | 37048.80 | 45048.80 |
| | -------- | -------- | --------- | --------- |
| | 71.42 | 16776.34 | 153701.20 | 185201.20 |

ESTIMATED PROGRESS REPORT THROUGH SEPTEMBER 19X1

| ACTIVITY NUMBER | PERCENT COMPLETE | LABOR UNITS TO DATE | LABOR COST TO DATE | TOTAL COST TO DATE |
|---|---|---|---|---|
| *    6 | 100.00 | 5111.65 | 37048.80 | 45048.80 |
| *    9 | 100.00 | 1826.27 | 12924.00 | 18924.00 |
| 12 | 100.00 | 81.56 | 569.20 | 569.20 |
| 14 | 17.55 | 92.50 | 633.60 | 703.79 |
| | -------- | -------- | --------- | --------- |
| | 79.93 | 18776.67 | 173532.00 | 211102.19 |

ESTIMATED PROGRESS REPORT THROUGH OCTOBER 19X1

| ACTIVITY NUMBER | PERCENT COMPLETE | LABOR UNITS TO DATE | LABOR COST TO DATE | TOTAL COST TO DATE |
|---|---|---|---|---|
| *   11 | 100.00 | 0.00 | 0.00 | 1000.00 |
| 14 | 100.00 | 527.15 | 3801.60 | 4201.60 |
| *   15 | 100.00 | 79.73 | 588.00 | 1338.00 |
| *   17 | 9.76 | 262.07 | 1795.20 | 2478.64 |
| | -------- | -------- | --------- | --------- |
| | 83.24 | 19553.12 | 184787.20 | 225120.64 |

ESTIMATED PROGRESS REPORT THROUGH NOVEMBER 19X1

| ACTIVITY NUMBER | PERCENT COMPLETE | LABOR UNITS TO DATE | LABOR COST TO DATE | TOTAL COST TO DATE |
|---|---|---|---|---|
| 14 | 100.00 | 527.15 | 3801.60 | 4201.60 |
| 16 | 100.00 | 7.69 | 54.80 | 54.80 |
| *   17 | 34.17 | 917.26 | 6507.60 | 8899.62 |
| 18 | 90.31 | 1348.55 | 9237.60 | 11043.77 |
| | -------- | -------- | --------- | --------- |
| | 91.80 | 22364.46 | 215938.80 | 261689.41 |

ESTIMATED PROGRESS REPORT THROUGH DECEMBER 19X1

| ACTIVITY NUMBER | PERCENT COMPLETE | LABOR UNITS TO DATE | LABOR COST TO DATE | TOTAL COST TO DATE |
|---|---|---|---|---|
| *   17 | 58.58 | 1572.44 | 11220.00 | 15320.61 |
| 18 | 100.00 | 1493.28 | 10264.00 | 12264.00 |
| | -------- | -------- | --------- | --------- |
| | 95.20 | 22364.46 | 215938.80 | 261689.41 |

ESTIMATED PROGRESS REPORT THROUGH JANUARY 19X2

| ACTIVITY NUMBER | PERCENT COMPLETE | LABOR UNITS TO DATE | LABOR COST TO DATE | TOTAL COST TO DATE |
|---|---|---|---|---|
| *   17 | 82.99 | 2227.62 | 15932.40 | 21741.60 |
| | -------- | -------- | --------- | --------- |
| | 97.99 | 23019.64 | 226355.20 | 273814.40 |

# FIGURE P16.4 (cont.)

ESTIMATED PROGRESS REPORT THROUGH FEBRUARY 19X2

| ACTIVITY NUMBER | PERCENT COMPLETE | LABOR UNITS TO DATE | LABOR COST TO DATE | TOTAL COST TO DATE |
|---|---|---|---|---|
| * 17 | 100.00 | 2684.25 | 19298.40 | 26298.40 |
| | --------- | --------- | --------- | --------- |
| | 99.94 | 23476.27 | 235425.20 | 284075.20 |

ESTIMATED PROGRESS REPORT THROUGH MARCH 19X2

| ACTIVITY NUMBER | PERCENT COMPLETE | LABOR UNITS TO DATE | LABOR COST TO DATE | TOTAL COST TO DATE |
|---|---|---|---|---|
| * 1 | 100.00 | 383.04 | 2740.00 | 2740.00 |
| * 2 | 100.00 | 1371.66 | 9464.00 | 9464.00 |
| 3 | 100.00 | 394.59 | 2766.00 | 2766.00 |
| * 4 | 100.00 | 6742.34 | 47635.20 | 57635.20 |
| 5 | 100.00 | 240.75 | 1744.00 | 2744.00 |
| * 6 | 100.00 | 354.67 | 2485.20 | 6485.20 |
| 8 | 100.00 | 999.53 | 7056.00 | 13056.00 |
| * 9 | 100.00 | 1826.27 | 12924.00 | 18924.00 |
| 10 | 100.00 | 780.14 | 5692.00 | 7692.00 |
| * 11 | 100.00 | 0.00 | 0.00 | 1000.00 |
| 12 | 100.00 | 81.56 | 569.20 | 569.20 |
| 13 | 100.00 | 397.98 | 2846.00 | 3346.00 |
| 14 | 100.00 | 527.15 | 3801.60 | 4201.60 |
| * 15 | 100.00 | 79.73 | 588.00 | 1338.00 |
| 16 | 100.00 | 7.69 | 54.80 | 54.80 |
| * 17 | 100.00 | 2684.25 | 19298.40 | 26298.40 |
| 18 | 100.00 | 1493.28 | 10264.00 | 12264.00 |
| * 19 | 100.00 | 15.93 | 109.60 | 109.60 |
| | --------- | --------- | --------- | --------- |
| | 100.00 | 23492.20 | 241238.80 | 289888.80 |

PROJECTED EXPECTED PROFIT = 29363.32
PROJECTED PROJECT DURATION = 243.

THE CONTRACT CALLS FOR THE FOR TO BE COMPLETED IN 243. WORKING DAYS
THE WORK MUST BE COMPLETED WITHIN 13 MONTHS (IN MARCH )

| | |
|---|---|
| THE HIGH BID ON THE PROJECT WAS | 354369.85 |
| THE AVERAGE BID ON THE PROJECT WAS | 332022.20 |
| THE LOW BID ON THE PROJECT WAS | 319252.12 |
| YOURS WAS THE LOW BID | |
| YOU WERE LOWER THAN THE NEXT HIGHER BID BY | 4788.78 |
| YOUR EXPECTED COST FOR THIS PROJECT IS | 289888.80 |

GOOD LUCK

ACTUAL PROGRESS REPORT THROUGH MARCH 19X1

| ACTIVITY NUMBER | PERCENT COMPLETE | LABOR UNITS TO DATE | LABOR COST TO DATE | TOTAL COST TO DATE |
|---|---|---|---|---|
| 1 | 100.00 | 349.84 | 4384.00 | 4384.00 |
| * 2 | 38.18 | 517.98 | 6151.60 | 6151.60 |
| 3 | 53.13 | 201.85 | 2397.20 | 2397.20 |
| | --------- | --------- | --------- | --------- |
| | 4.57 | 1069.67 | 18922.00 | 18922.00 |

TOTAL PROJECTED PROJECT COST = $297,840.80

ACTUAL PROGRESS REPORT THROUGH APRIL 19X2

| | ACTIVITY NUMBER | PERCENT COMPLETE | LABOR UNITS TO DATE | LABOR COST TO DATE | TOTAL COST TO DATE |
|---|---|---|---|---|---|
| | 1 | 100.00 | 349.84 | 4384.00 | 4384.00 |
| * | 2 | 100.00 | 100.00 | 1356.65 | 14196.00 |
| | 3 | 100.00 | 379.91 | 4056.80 | 4056.80 |
| | 5 | 100.00 | 240.75 | 1744.00 | 2744.00 |
| | 8 | 15.17 | 151.59 | 1176.00 | 2085.98 |
| | | 10.58 | 2478.74 | 37250.00 | 39159.98 |

TOTAL PROJECTED PROJECT COST = $309,547.70

ACTUAL PROGRESS REPORT THROUGH JUNE 19X2

| | ACTIVITY NUMBER | PERCENT COMPLETE | LABOR UNITS TO DATE | LABOR COST TO DATE | TOTAL COST TO DATE |
|---|---|---|---|---|---|
| | 1 | 100.00 | 349.84 | 4384.00 | 4384.00 |
| | 2 | 100.00 | 1356.65 | 15615.60 | 15615.60 |
| | 3 | 100.00 | 379.91 | 4425.60 | 4425.60 |
| * | 4 | 87.52 | 5836.63 | 43665.60 | 52008.13 |
| | 5 | 100.00 | 231.79 | 2790.40 | 3740.66 |
| | 8 | 100.00 | 993.75 | 7644.00 | 13632.70 |
| | 10 | 100.00 | 771.60 | 5692.00 | 7623.80 |
| | 13 | 100.00 | 383.17 | 2846.00 | 3321.13 |
| | | 44.19 | 10303.35 | 111305.20 | 128993.62 |

PROJECTED PROGRESS REPORT THROUGH MARCH 19X2 BASED ON STATUS END OF JUNE 19X1

| | ACTIVITY NUMBER | PERCENT COMPLETE | LABOR UNITS TO DATE | LABOR COST TO DATE | TOTAL COST TO DATE |
|---|---|---|---|---|---|
| | 1 | 100.00 | 349.84 | 4384.00 | 4384.00 |
| | 2 | 100.00 | 1356.65 | 15615.60 | 15615.60 |
| | 3 | 100.00 | 379.91 | 4425.60 | 4425.60 |
| * | 4 | 100.00 | 6668.58 | 49620.00 | 59151.67 |
| | 5 | 100.00 | 231.79 | 2790.40 | 3740.66 |
| * | 6 | 100.00 | 5111.65 | 37048.80 | 45048.80 |
| | 7 | 100.00 | 354.67 | 2485.20 | 6485.20 |
| | 8 | 100.00 | 993.75 | 7644.00 | 13632.70 |
| * | 9 | 100.00 | 1826.27 | 12924.00 | 18924.00 |
| | 10 | 100.00 | 771.60 | 5692.00 | 7623.80 |
| * | 11 | 100.00 | 0.00 | 0.00 | 1000.00 |
| | 12 | 100.00 | 81.56 | 569.20 | 569.20 |
| | 13 | 100.00 | 383.17 | 2846.00 | 3321.13 |
| | 14 | 100.00 | 527.15 | 3801.60 | 4201.60 |
| * | 15 | 100.00 | 79.73 | 588.00 | 1338.00 |
| | 16 | 100.00 | 7.69 | 54.80 | 54.80 |
| * | 17 | 100.00 | 2684.25 | 19298.40 | 26298.40 |
| | 18 | 100.00 | 1493.28 | 10264.00 | 12264.00 |
| * | 19 | 100.00 | 15.93 | 109.60 | 109.60 |
| | | 100.00 | 23317.47 | 255739.20 | 303766.76 |

*339*

**FIGURE P16.4   (cont.)**

---

PROJECTED EXPECTED PROFIT = 15485.36

PROJECTED PROJECT DURATION = 259.

WARNING THIS EXPECTED DURATION EXCEEDS THE CONTRACT
LIMIT BY 16. DAYS

THE PROJECTED PROJECT DURATIONS AT THE ENDS OF MARCH
AND APRIL ARE 260 DAYS RESPECTIVELY

(a) Total estimated dollars
(b) Total labor units (LUs) expended
(c) Actual expenditure of dollars and LUs through the end of June 19X1
Do a cash-flow analysis based on the initial estimate information.
Assume 10-percent retainage through the project.
Assume markup as indicated by the data provided.
Progress payment requests (billings) are submitted at the end of one month and received one month later.
Draw the overdraft profile.

**16.5** Using the data provided for the Coffer Dam project of problem 16.4, prepare a completion date trend chart and a project cost trend chart based on project status as of the end of June 19X1.

**16.6** Using the data for the Coffer Dam project of problem 16.4, calculate the efficiency variance, rate variance, and total variance for activities 2, 10, and 13 as of the end of June 19X1.

# Appendix A

# CHART OF ACCOUNTS

## GENERAL LEDGER ACCOUNTS

### *Assets*

| | |
|---|---|
| 10. | Petty cash |
| 11. | Bank deposits |
| .1 | General bank account |
| .2 | Payroll bank account |
| .3 | Project bank accounts |
| .4 | |
| 12. | Accounts receivable |
| .1 | |
| .2 | Parent, associated, or affiliated companies |
| .3 | Notes receivable |
| .4 | Employees' accounts |
| .5 | Sundry debtors |
| .6 | |
| 13. | Deferred receivables |
| | All construction contracts are charged to this account, being diminished by progress payments as received. This account is offset by Account 48.0, Deferred Income. |
| 14. | Property, plant, and equipment |

### *Property and General Plant*

| | |
|---|---|
| .100 | Real estate and improvements |
| .200 | Leasehold improvements |
| .300 | Shops and yards |

### *Mobile Equipment*

| | |
|---|---|
| .400 | Motor vehicles |
| .500 | Tractors |

.01  Repairs, parts, and labor
.05  Outside service
.12  Tire replacement
.15  Tire repair
.20  Fuel
.25  Oil, lubricants, filters
.30  Licenses, permits
.35  Depreciation
.40  Insurance
.45  Taxes
.510  Power shovels
.520  Bottom dumps
.525

## Stationary Equipment

.530  Concrete mixing plant
.540  Concrete pavers
.550  Air compressors
.560

## Small Power Tools and Portable Equipment

.600  Welders
.610  Concrete power buggies
.620  Electric drills
.630

## Marine Equipment

.700

## Miscellaneous Construction Equipment

.800  Scaffolding
.810  Concrete forms
.820  Wheelbarrows
.830

## Office and Engineering Equipment

.900  Office equipment
.910  Office furniture
.920  Engineering instruments
.930
15.  Reserve for depreciation
16.  Amortization for leasehold
17.  Inventory of materials and supplies

.1      Lumber
.2      Hand shovels
.3      Spare parts
.4

These accounts show the values of all expendable materials and supplies. Charges against these accounts are made by authenticated requisitions showing project where used.

18.     Returnable deposits
.1      Plan deposits
.2      Utilities
.3


## Liabilities

40.     Accounts payable
41.     Subcontracts payable
42.     Notes payable
43.     Interest payable
44.     Contracts payable
45.     Taxes payable
.1      Old-age, survivors, and disability insurance (withheld from employees' pay)
.2      Federal income taxes (withheld from employees' pay)
.3      State income taxes (withheld from employees' pay)
.4
46.     Accrued expenses
.1      Wages and salaries
.2      Old-age, survivors, and disability insurance (employer's portion)
.3      Federal unemployment tax
.4      State unemployment tax
.51     Payroll insurance (public liability and property damage)
.52     Payroll insurance (workmen's compensation)
.6      Interest
.7
47.     Payrolls payable
48.     Deferred income
49.     Advances by clients


## Net Worth

50.     Capital stock
51.     Earned surplus
52.     Paid-in-surplus
53.

## Income

| | |
|---|---|
| 70. | Income accounts |
| .101 | Project income |
| .102 | |
| .2 | Cash discount earned |
| .3 | Profit or loss from sale of capital assets |
| .4 | Equipment rental income |
| .5 | Interest income |
| .6 | Other income |

## Expense

| | |
|---|---|
| 80. | Project expense (Expenses directly chargeable to the projects. See Figure 2.2.) |
| .100 | Project work accounts |
| .700 | Project overhead accounts |
| | These are control accounts for the detail project cost accounts that are maintained in the detail cost ledgers. |
| 81. | Office expense |
| .10 | Officer salaries |
| .11 | Insurance on property and equipment |
| .20 | Donations |
| .21 | Utilities |
| .22 | Telephone and telegraph |
| .23 | Postage |
| .30 | Repairs and maintenance |
| 82. | Yard and warehouse expense (not assignable to a particular project) |
| .10 | Yard salaries |
| .11 | Yard supplies |
| 83. | Estimating department expense accounts |
| .10 | Estimating salaries |
| .11 | Estimating supplies |
| .12 | Estimating travel |
| 84. | Engineering department expense accounts |
| .10 | |
| 85. | Cost of equipment ownership |
| .1 | Depreciation |
| .2 | Interest |
| .3 | Taxes and licenses |
| .4 | Insurance |
| .5 | Storage |
| 86. | Loss on bad debts |
| 87. | Interest |
| 90. | Expense on office employees |

|       |                                            |
|-------|--------------------------------------------|
| .1    | Workmen's compensation insurance           |
| .2    | Old-age, survivors, and disability insurance |
| .3    | Employees' insurance                       |
| .4    | Other insurance                            |
| .5    | Federal and state unemployment taxes       |
| .6    |                                            |
| 91.   | Taxes and licenses                         |
| .1    | Sales taxes                                |
| .2    | Compensating taxes                         |
| .3    | State income taxes                         |
| .4    | Federal income taxes                       |

# Appendix B

# FURTHER ILLUSTRATIONS OF TRANSACTIONS

## Composite of Ledger Entries 1–20

| Cash | | Accounts Receivable | |
|---|---|---|---|
| DR + | CR − | DR + | CR − |
| $ 50,000(1) | $330,000(5) | | |
| 50,000(1) | 5,000(5) | $414,000(2) | |
| 360,000(4) | 85,000(5) | 112,500(2) | |
| 20,000(4) | 2,000(5) | | $360,000(4) |
| 90,000(4) | 11,000(9) | 27,000(6) | 90,000(4) |
| 6,000(4) | 2,250(10) | | 20,000(7) |
| 5,000(7) | 33,500(11) | | |
| 20,000(7) | 3,160(13) | | |
| 4,000(20) | | | |

| Retainage Receivable | | Accrued Interest Receivable | |
|---|---|---|---|
| DR + | CR − | DR + | CR − |
| $46,000(2) | | $4,000(19) | |
| 12,500(2) | | | $4,000(20) |
| | $20,000(4) | | |
| | 6,000(4) | | |
| 3,000(6) | 5,000(7) | | |

| Work in Progress Expense (Asset)[a] | | Prepaid Insurance | |
|---|---|---|---|
| DR + | CR − | DR + | CR − |
| $370,000(3) | | $7,200(14) | |
| 100,000(3) | | | |
| 11,000(8) | | | |
| 2,250(10) | | | |

[a]This expense account is a nominal account used to capture expenses on jobs in progress pending reconciliation at closing of accounts. These expenses are viewed as assets until closed to the real account, "contract expense." This occurs at the end of the accounting period.

| Fixed Assets ||
|---|---|
| DR + | CR − |
| $90,000(16) | |

| Accumulated Depreciation ||
|---|---|
| DR + | CR − |
| | $7,500(18) |

| Accounts Payable ||
|---|---|
| DR − | CR + |
| | $340,000(3) |
| | 92,000(3) |
| $330,000(5) | |
| 85,000(5) | |
| | 10,000(8) |
| 10,000(9) | 3,160(12) |
| 3,160(13) | 7,200(14) |

| Accrued Liability ||
|---|---|
| DR − | CR + |
| | $3000(15) |
| | 4800(17) |

| Retainage Payable ||
|---|---|
| DR − | CR + |
| | $30,000(3) |
| | 8,000(3) |
| $5,000(5) | |
| 2,000(5) | |
| 1,000(9) | 1,000(8) |

| Billings in Excess of Revenues ||
|---|---|
| DR − | CR + |
| | |

| Notes Payable—Short-term ||
|---|---|
| DR − | CR + |
| | $30,000(16) |

| Notes Payable—Long-Term ||
|---|---|
| DR − | CR + |
| | $60,000(16) |

| Advanced Billings ||
|---|---|
| DR − | CR + |
| | $460,000(2) |
| | 125,000(2) |
| | 30,000(6) |

| Retained Earnings ||
|---|---|
| DR − | CR + |
| | |

| Capital—Common Stock ||
|---|---|
| DR − | CR + |
| | $50,000(1) |
| | 50,000(1) |

| Project Revenue (Sales) ||
|---|---|
| DR − | CR + |
| | |

| Project Expense ||
|---|---|
| DR + | CR − |
| | |

| General Overhead ||
|---|---|
| DR + | CR − |
| $33,500(11) | |
| 3,160(12) | |
| 3,000(15) | |

| Interest Income | |
| --- | --- |
| *DR* – | *CR* + |
| | $4000(19) |

| Interest Expense | |
| --- | --- |
| *DR* + | *CR* – |
| $4800(17) | |

| Depreciation Expense | |
| --- | --- |
| *DR* + | *CR* – |
| $7500(18) | |

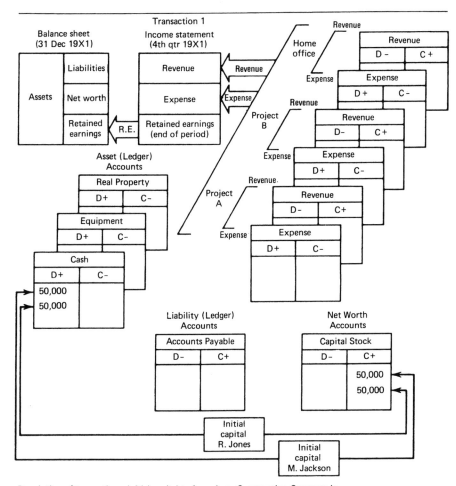

Description of transaction: Initial capital to form Apex Construction Company is contributed by R. Jones and M. Jackson. Each contributes $50,000. Stock in the amount of 10,000 shares is issued toeach.

| Journal entry: | Description | DR | CR |
|---|---|---|---|
| | Cash | 50,000 | |
| | Capital stock—R. Jones | | 50,000 |
| | Cash | 50,000 | |
| | Capital stock—M. Jackson | | 50,000 |

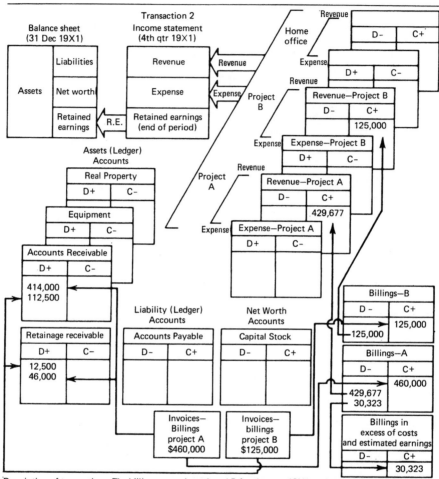

Description of transaction: The billings on projects A and B for the year 19X1 as shown in Table 3.3 will be posted based on the percentage-of-completion method. The actual billed amounts on A and B are $460,000 and $125,000, respectively. A retainage of 10 percent is held on each contract. Although many individual billings would be made over the year, they have been consolidated into single entries for the purpose of this illustration. The calculated revenues on A and B (with POC) are $429,677 and $125,000, respectively. The overbilling on A is accounted for in the liability account Billings in excess of costs and estimated earnings.

| Description | DR | CR | |
|---|---|---|---|
| Accounts receivable | 414,000 | | Accumulation of entries to billings through period |
| Retainage receivable | 46,000 | | |
| Billings—Project A | | 460,000 | |
| Accounts receivable | 112,500 | | |
| Retainage receivable | 12,500 | | |
| Billings—Project B | | 125,000 | |
| Billings—Project A | 460,000 | | Closing entries |
| Revenue—Project A | | 429,677 | |
| Billings in Excess of costs and estimated Earnings | | 30,323 | |
| Billings—Project B | 125,000 | | |
| Revenue—Project B | | 125,000 | |

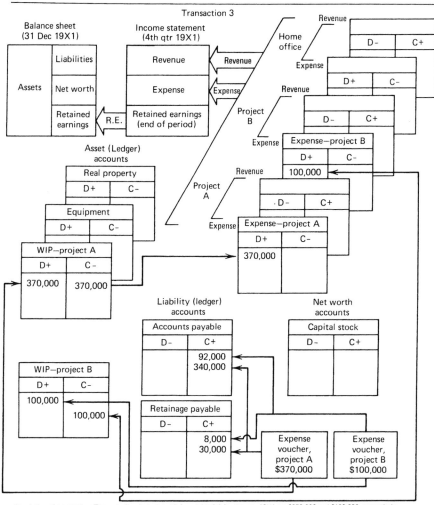

**Transaction 3**

Balance sheet (31 Dec 19X1)

| | |
|---|---|
| Assets | Liabilities |
| | Net worth |
| | Retained earnings |

Income statement (4th qtr 19X1)

| |
|---|
| Revenue |
| Expense |
| Retained earnings (end of period) |

R.E.

Home office

Revenue

| D – | C + |
|---|---|

Expense

| D + | C – |
|---|---|

Project B

Revenue

| D – | C + |
|---|---|

Expense–project B

| D + | C – |
|---|---|
| 100,000 | |

Project A

Revenue

| . D – | C + |
|---|---|

Expense–project A

| D + | C – |
|---|---|
| 370,000 | |

Asset (Ledger) accounts

Real property

| D + | C – |
|---|---|

Equipment

| D + | C – |
|---|---|

WIP–project A

| D + | C – |
|---|---|
| 370,000 | 370,000 |

Liability (ledger) accounts

Accounts payable

| D – | C + |
|---|---|
| | 92,000 |
| | 340,000 |

WIP–project B

| D + | C – |
|---|---|
| 100,000 | |
| | 100,000 |

Retainage payable

| D – | C + |
|---|---|
| | 8,000 |
| | 30,000 |

Net worth accounts

Capital stock

| D – | C + |
|---|---|

| Expense voucher, project A $370,000 | Expense voucher, project B $100,000 |
|---|---|

Description of transaction: The cost of projects A and B from table 3.3 for the year 19X1 are $370,000 and $100,000, respectively. Retainage amounts of $30,000 and $8,000 respectively, on A and B are held by Apex from subcontract billings. These costs are carried as expenses under the appropriate project expense accounts. Although the costs would accumulate for many individual transactions through the year, they are shown as single lump sum entries in this example.

Journal entry:

| Description | DR | CR | |
|---|---|---|---|
| WIP–expenses–project A | 370,000 | | |
| accounts payable | | 340,000 | Accumulation of |
| Retainage payable | | 30,000 | expense during |
| WIP–expenses–project B | 100,000 | | the period |
| Accounts payable | | 92,000 | |
| Retainage payable | | 8,000 | |
| Project expenses–project A | 370,000 | | |
| Project expenses–project B | 100,000 | | |
| WIP expense–project A | | 370,000 | Closing actions |
| WIP–expense–project B | | 100,000 | |

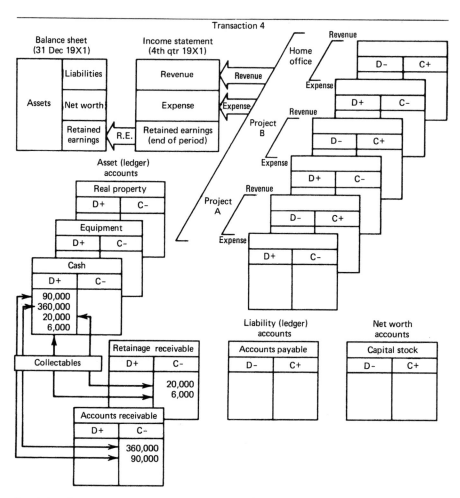

Transaction 4

Description of transaction: Payments received on project A from client amount to $360,000 in billings and $20,000 in retainage. Payments received on project B amount to $90,000 in billings and $6,000 in retainage. These are accumulations of individual collections during the year.

Journal entry:

| Description | DR | CR |
|---|---|---|
| Cash | 360,000 | |
| Accounts receivable | | 360,000 |
| Cash | 20,000 | |
| Retainage receivable | | 20,000 |
| Cash | 90,000 | |
| Accounts receivable | | 90,000 |
| Cash | 6,000 | |
| Retainage receivable | | 6,000 |

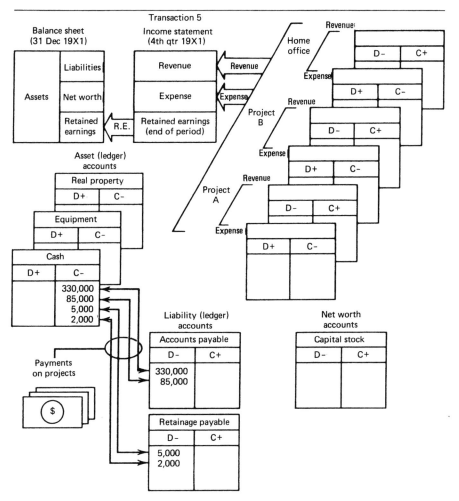

Description of transaction: Payments to vendors and subcontractors on project A amount to $330,000 in regular payments and $5,000 in retainage during period. Payments to vendors and subcontractors on Project B amount to $85,000 in regular payments and $2,000 in retainage during the period.

| Journal entry: | Description | DR | CR |
|---|---|---|---|
| | Accounts payable—A | 330,000 | |
| | Cash | | 330,000 |
| | Retainage payable—A | 5,000 | |
| | Cash | | 5,000 |
| | Accounts payable—B | 85,000 | |
| | Cash | | 85,000 |
| | Retainage payable—B | 2,000 | |
| | Cash | | 2,000 |

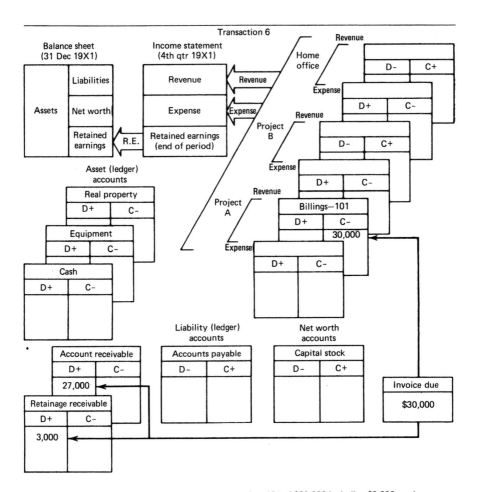

Description of transaction: Billing to a customer on project 101 of $30,000 including $3,000 retainage.
*Billing accounts are "work-in-progress—billings." Accounts during the period prior to clearing and closing.

| Journal entry: | Description | DR | CR |
|---|---|---|---|
| | Accounts receivable | 27,000 | |
| | Retainage receivables | 3,000 | |
| | Billings—project 101 | | 30,000 |

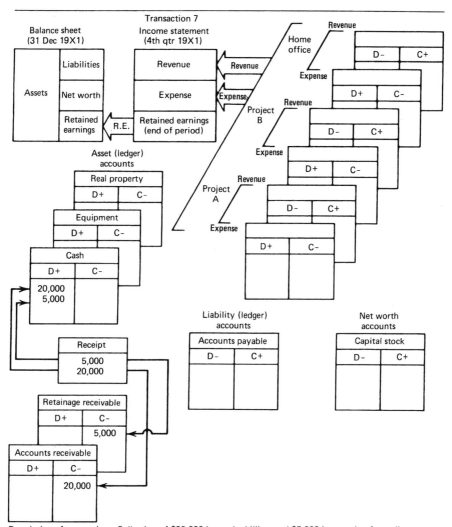

Description of transaction: Collection of $20,000 in regular billings and $5,000 in retention from client on project 101.

| Journal entry: | Description | DR | CR |
|---|---|---|---|
| | Cash | 5,000 | |
| | Cash | 20,000 | |
| | Retainage receivable | | 5,000 |
| | Accounts receivable | | 20,000 |

*355*

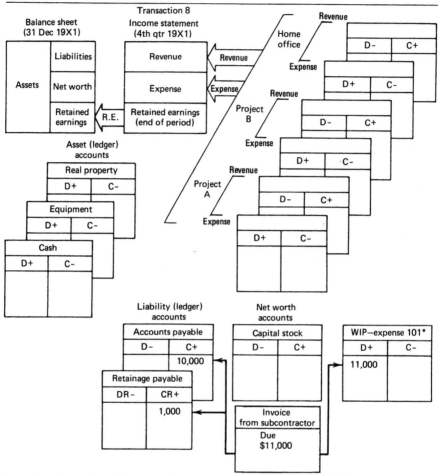

Description of transaction: Billing received from subcontractor in the amount of $10,000, including $1,000 retainage on project 101.

*Expenses are booked "work in progress—expense." Accounts prior to closing out of expense accounts at the end of the accounting period.

| Journal entry: | Description | DR | CR |
|---|---|---|---|
| | Work in progress— expense proj 101 (Jones Elec) | 11,000 | |
| | Accounts payable | | 10,000 |
| | Retainage payable | | 1,000 |

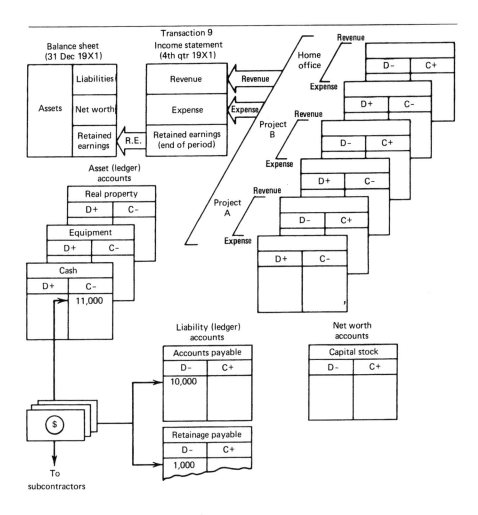

Transaction 9

Description of transaction: Payment of $10,000 in regular billings and $1,000 in retainage on project 101. Subcontractor billings received previously.

| Journal entry: | Description | DR | CR |
|---|---|---|---|
| | Accounts payable | 10,000 | |
| | Retainage payable | 1,000 | |
| | Cash | | 11,000 |

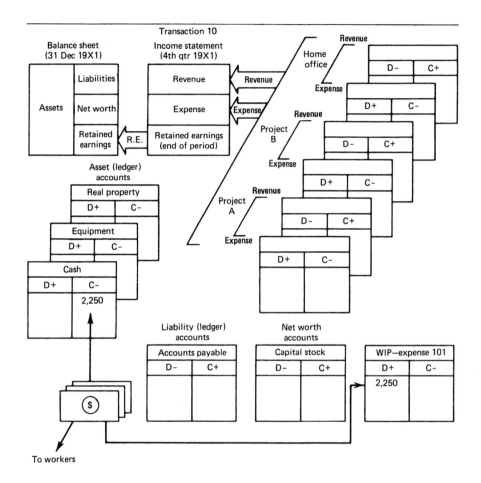

Transaction 10

Description of transaction: Payment of weekly salary for craft workers on project 101 in the amount of $2,250.

| | Description | DR | CR |
|---|---|---|---|
| Journal entry: | WIP—expense<br>   Job 101<br>   (payroll)<br>Cash | 2,250 | 2,250 |

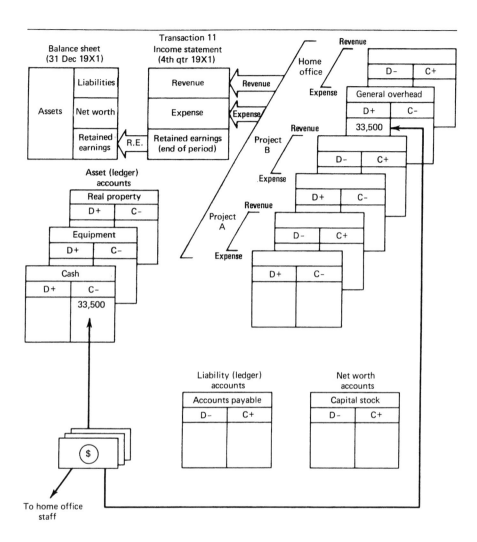

Balance sheet
(31 Dec 19X1)

Transaction 11
Income statement
(4th qtr 19X1)

Description of transaction: Payment of home office staff for month of June in the amount of $33,500.

| Journal entry: | Description | DR | CR |
|---|---|---|---|
| | General overhead (monthly home office salary) | 33,500 | |
| | Cash | | 33,500 |

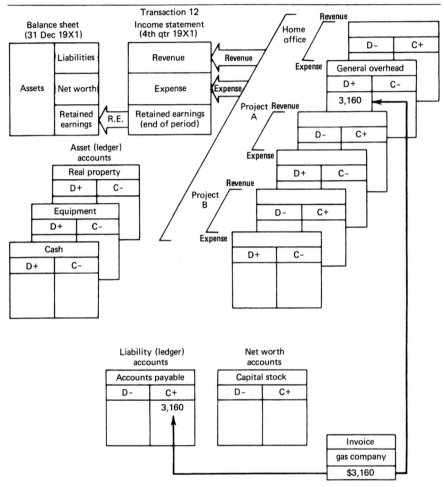

Description of transaction: Invoice for home office costs for heating headquarters building in the amount of $3,160.

| Journal entry: | Description | DR | CR |
|---|---|---|---|
| | General overhead (heating) | 3,160 | |
| | Accounts payable | | 3,160 |

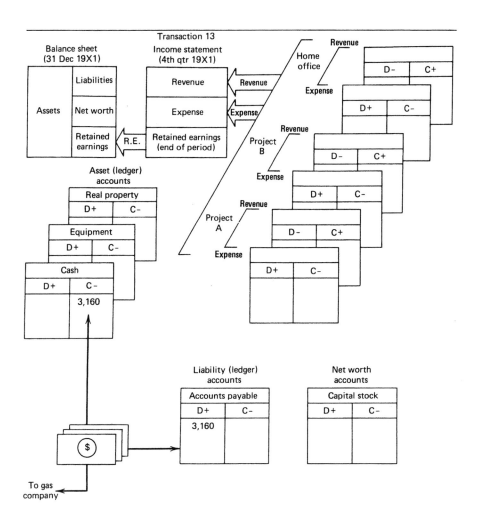

Transaction 13

Balance sheet (31 Dec 19X1)

| Assets | Liabilities |
| | Net worth |
| | Retained earnings |

Income statement (4th qtr 19X1)

| Revenue |
| Expense |
| Retained earnings (end of period) |

R.E.

Home office

Revenue

| D – | C + |

Expense

| D + | C – |

Project B

Revenue

| D – | C + |

Expense

| D + | C – |

Project A

Revenue

| D – | C + |

Expense

| D + | C – |

Asset (ledger) accounts

Real property

| D + | C – |

Equipment

| D + | C – |

Cash

| D + | C – |
| | 3,160 |

Liability (ledger) accounts

Accounts payable

| D + | C – |
| 3,160 | |

Net worth accounts

Capital stock

| D + | C – |
| | |

To gas company

Description of transaction: Payment of invoice in Transaction 12.

Journal entry:

| Description | DR | CR |
|---|---|---|
| Accounts payable (city gas companY) | 3,160 | |
| Cash | | 3,160 |

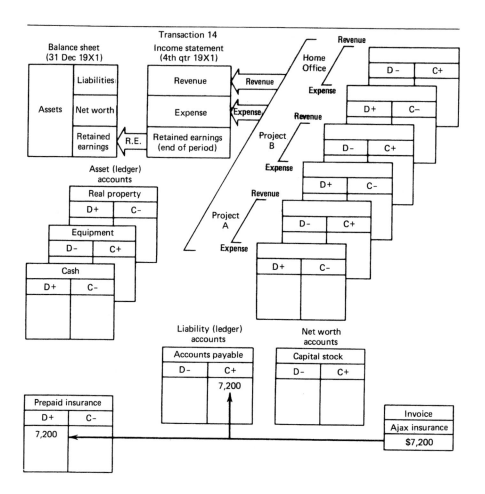

Description of transaction: An invoice for insurance for the coming year is received in the amount of $7,200. The payment represents a prepayment of insurance.

| Journal entry: | Description | DR | CR |
|---|---|---|---|
| | Prepaid insurance | 7,200 | |
| | Accounts payable | | 7,200 |

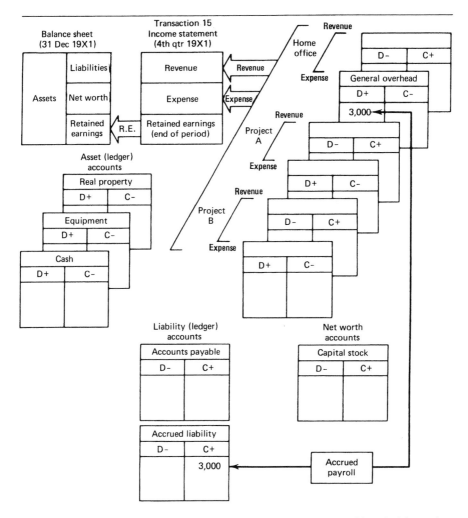

Description of transaction: Assume that company security personnel are paid on the fifteenth of the month. The books must be closed as of the end of December. Since 15 December, $3,000 in payroll to be paid (presumably on 15 January) has accrued. No specific invoice is received, but this obligation is incurred.

| Journal entry: | Description | DR | CR |
|---|---|---|---|
| | General overhead | 3,000 | |
| | Accrued liability | | 3,000 |

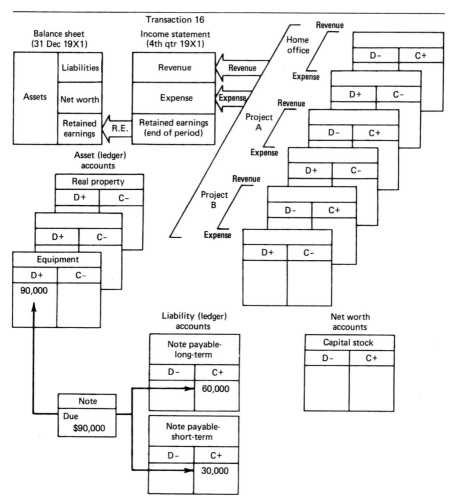

Transaction 16

**Balance sheet**
(31 Dec 19X1)

**Income statement**
(4th qtr 19X1)

| Assets | Liabilities |
|---|---|
| | Net worth |
| | Retained earnings |

Revenue

Expense

Retained earnings
(end of period)

R.E.

Home office

Revenue

Expense

Project A

Revenue

Expense

Project B

Revenue

Expense

| D– | C+ |
|---|---|

| D+ | C– |
|---|---|

| D– | C+ |
|---|---|

| D+ | C– |
|---|---|

| D– | C+ |
|---|---|

| D+ | C– |
|---|---|

Asset (ledger) accounts

**Real property**

| D+ | C– |
|---|---|

| D+ | C– |
|---|---|

**Equipment**

| D+ | C– |
|---|---|
| 90,000 | |

Liability (ledger) accounts

**Note payable-long-term**

| D– | C+ |
|---|---|
| | 60,000 |

**Note**

Due $90,000

**Note payable-short-term**

| D– | C+ |
|---|---|
| | 30,000 |

Net worth accounts

**Capital stock**

| D– | C+ |
|---|---|

Description of transaction: A ditching machine is purchased on 1 July at a cost of $90,000. This company borrows this amount on a note at the First National Bank. $30,000 is due within one year. The remainder is due in years two and three of the three-year note.

Journal entry:

| Description | DR | CR |
|---|---|---|
| Fixed assets—equipment | 90,000 | |
| Notes payable—short-term | | 30,000 |
| Notes payable—long-term | | 60,000 |

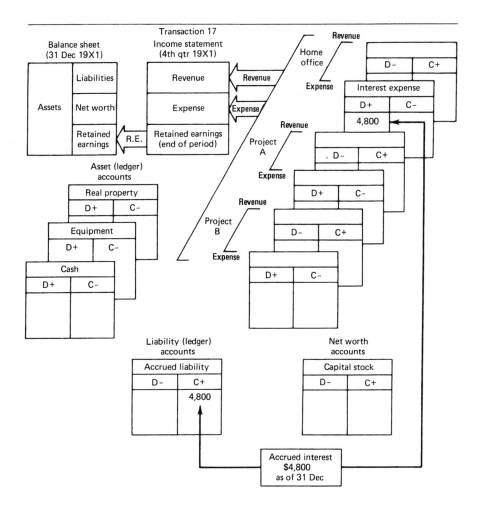

Description of transaction: Interest due on the note in Transaction 16 is computed to be $4,800 as of 31 December of the year of purchase. This is reflected as an accrued liability for statement preparation purposes.

| Journal entry: | Description | DR | CR |
|---|---|---|---|
| | Interest expense | 4,800 | |
| | Accrued liability | | 4,800 |

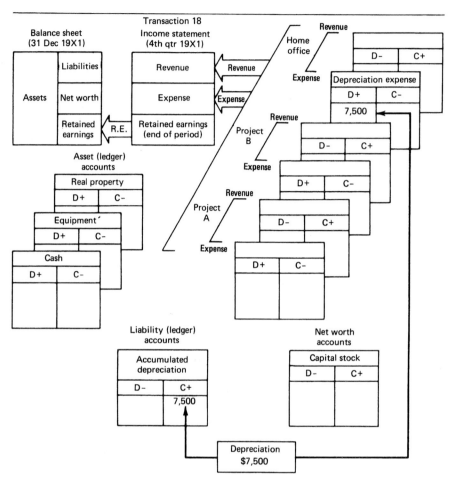

Description of transaction: Depreciation on the ditching machine of Transaction 16 as of 31 December is $7,500. An account is maintained to accumulate this expense. This type of account is called a contra account. It allows the original value of the asset to be maintained in the "fixed asset—equipment" account while placing the accumulated depreciation in the counter-balancing contra account. The contra account is considered a liability account.

| Journal entry: | Description | DR | CR |
|---|---|---|---|
| | Depreciation expense | 7,500 | |
| | Accumulated depreciation | | 7,500 |

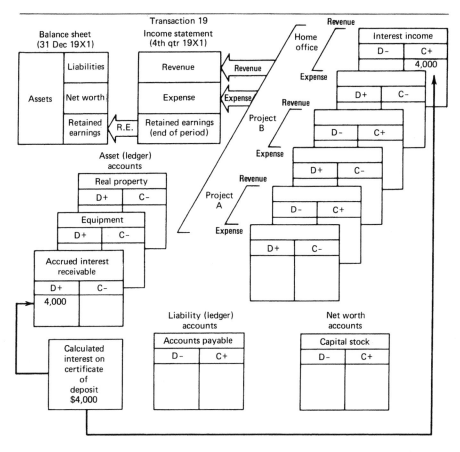

Description of transaction: Interest in the amount of $4,000 has accrued on a certificate of deposit (interest-producing-note) at the bank as of 31 December. This must be reflected in the year-end statements and is revenue to Apex Construction Company.

| Journal entry: | Description | DR | CR |
|---|---|---|---|
| | Accrued interest receivable | 4,000 | |
| | Interest income | | 4,000 |
| | | | |
| | | | |
| | | | |

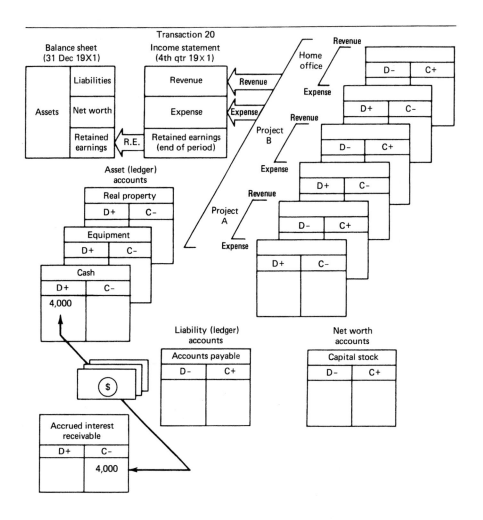

Description of transaction: Interest in Transaction 19 is paid into construction company account.

| Journal entry: | Description | DR | CR |
|---|---|---|---|
| | Cash | 4,000 | |
| | Accrued interest receivable | | 4,000 |
| | | | |
| | | | |
| | | | |

# Appendix C

# A COMPUTER-BASED CASH-FLOW MODEL FOR DETERMINING PROJECT MARKUP

## C.1 PROJECT MARKUP BASED ON CASH FLOW

By using the techniques described in Chapter 11 and rate-of-return methods of analysis from engineering economy, it is possible to examine the project cash-flow sequence (i.e., incomes and expenses) and to determine the expected rate of return for a given project. In this analysis the percentage markup, as well as the delays in payment and the percentage retainage, are known.* The reverse of this procedure is proposed as a means of determining the level of markup to be used on a project.† Rather than the *rate of return* being the unknown, it is assumed to be a factor decided by the contractor. The contractor decides what level of return he wants on the project and determines his level of markup accordingly. In this approach, the markup is the unknown that is to be calculated, and the rate of return is known. The rate of return desired by the contractor is assumed to be the minimum attractive rate of return (MARR). This is normally based on the return that the contractor could achieve by taking the monies he will utilize to finance the project construction and investing them in a secure certificate of deposit or similar instrument at the bank. That is, it is based on the rate he would receive at the bank rather than investing in the project.

---

* This analysis is described in D. W. Halpin and R. W. Woodhead, *Construction Management*, (New York: Wiley, 1980), Section 8.4.

† John W. Fondahl and Ricardo R. Bacarreza, "Construction Contract Markup Related to Forecasted Cash Flow," Tech Report No. 161, Dept. of Civil Engineering, Stanford University, Stanford, Calif., November 1972.

**FIGURE C.1   Project Net Cash Flow**

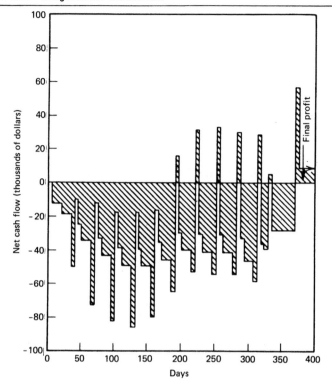

Expenses actually occur at discrete times, and revenues lag the billing by some interval. The effect of this lag on the cash flow profile is shown in Figure C.1.

This figure indicates the discrete nature of expenditures and revenues in the stair-step appearance, showing negative and positive cash flow at specific times. The discrete nature of these flows is accounted for in the report by Fondahl and Bacarreza by referencing them to the end of each progress payment period. The progress payment period is assumed to be a calendar month. The start date is referenced to the end of the first month to determine the length of the first progress payment period.

Rather than breaking the project into schedule activities (e.g., activity bars in the bar chart), the costs generated are related to six major subelements of the entire job. The six cost elements are as follows:

| Cost Type | Lag (days) |
|---|---|
| 1. Bare labor | −15 |
| 2. Labor burden | +15 |
| 3. Materials | +15 |
| 4. Subcontracts | +25 |
| 5. Field overhead | 0 |
| 6. Home office overhead | 0 |

These six cost items account for all expense on the project. The standard lags defining the points in time at which these expenditures occur are given above with the cost type. These are offsets in time referenced to the end of the payment period. Bare labor, for instance, for a given payment period is assumed to be a single expenditure occurring 15 days before the end of the payment period ($-15$ days). Labor burdens for a given period occur as a single expense 15 days after the end of the period ($+15$ days). Material costs for a given period occur as a single expense 15 days after the end of the period ($+15$ days). Subcontracts are paid as a single expenditure 25 days after the end of the period, and so on.

The amounts charged (expended) for each of the six cost elements is calculated from simple linear expense curves, as shown in Figure C.2. These curves are defined for each of the six cost elements. The charge for a given cost element at a point in the job is calculated by entering the appropriate cost curve on the $X$ axis at the relevant point in time (the percentage of total project duration represented by point in time) and reading out the percentage of the expense incurred as of that time on the $Y$ axis. This is a cumulative percentage of expense. By comparing it with the cumulative percentage last charged, the period percentage and expense can be calculated. This procedure is illustrated in Table C.1 for an example project. Five of the cost elements are shown. This project starts on 10 November. Therefore, the length of the first pay period is 21 days, as shown in column 2.

The Fondahl and Bacarreza approach allows for a variety of policies regarding billings to the client. In effect, the amount charged to the client can be different from the expense incurred in a given period (plus the markup). That is, the billings may reflect an under- or overbilling policy during various parts of the project. Some standard billings curves are shown in Figure C.3. The A100 curve at the top left indicates that the expenses will be billed at 100 percent of the amount incurred throughout the job. The C115 curve indicates an overbilling policy throughout the job. The B90 curve is used on the example project expenses (Table C.1) to calculate the billings for each period. These billings and associated revenues are shown in Table C.2.

The technique determines the level of markup by referring all of the expenditures

**FIGURE C.2  Three Types of Simple Cost Curves**

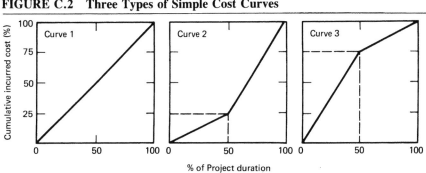

**TABLE C.1  Calculation of Cumulative Incurred Expenses by Cost Categories**

| 1 | 2 | 3 | 4 | 5 | 6 | 7 | 8 | 9 | 10 | 11 | 12 | 13 |
|---|---|---|---|---|---|---|---|---|---|---|---|---|
| | | | | | | | *Cumulative Expenses by Cost Components* | | | | | |
| | | | *Blab = $120,000* | | *Lbur = $40,000* | | *Mat = $96,000* | | *Sub = $480,000* | | *Foh = $64,000* | |
| *Prog. Pay No.* | *Cutoff Day* | *% Dur* | *Cum %* | *Cum $* | *Cum %* | *Cum $* | *Cum %* | *Cum $* | *Cum %* | *Cum $* | *Cum %* | *Cum $* |
| 1 | 21 | 6.77 | 10.16 | 12,192 | 10.16 | 4,064 | 10.16 | 9,754 | 3.38 | 16,224 | 6.77 | 4,333 |
| 2 | 51 | 16.45 | 24.68 | 9,872 | 24.68 | 9,872 | 24.68 | 23,693 | 8.22 | 39,456 | 16.45 | 10,528 |
| 3 | 82 | 26.45 | 39.68 | 47,616 | 39.68 | 15,872 | 39.68 | 38,093 | 13.22 | 63,456 | 26.45 | 16,928 |
| 4 | 113 | 36.45 | 54.68 | 65,616 | 54.68 | 21,872 | 54.68 | 52,493 | 18.22 | 87,456 | 36.45 | 23,328 |
| 5 | 141 | 45.48 | 68.22 | 81,864 | 68.22 | 27,288 | 68.22 | 65,491 | 22.74 | 109,152 | 45.48 | 29,107 |
| 6 | 172 | 55.48 | 77.74 | 93,288 | 77.74 | 31,096 | 77.74 | 74,630 | 33.22 | 159,456 | 55.48 | 35,507 |
| 7 | 202 | 65.16 | 82.58 | 99,096 | 82.58 | 33,032 | 82.58 | 79,277 | 47.74 | 229,152 | 65.16 | 41,702 |
| 8 | 233 | 75.16 | 87.58 | 105,096 | 87.58 | 35,032 | 87.58 | 84,077 | 62.74 | 301,152 | 75.16 | 48,102 |
| 9 | 263 | 84.84 | 92.42 | 110,904 | 92.42 | 36,968 | 92.42 | 88,723 | 77.26 | 370,848 | 84.84 | 54,298 |
| 10 | 294 | 94.84 | 97.42 | 116,904 | 97.42 | 38,968 | 97.42 | 93,523 | 92.26 | 442,848 | 94.84 | 60,698 |
| 11 | 310 | 100 | 100 | 120,000 | 100 | 40,000 | 100 | 96,000 | 100 | 480,000 | 100 | 64,000 |

**FIGURE C.3   Four Types of Billing Curves**

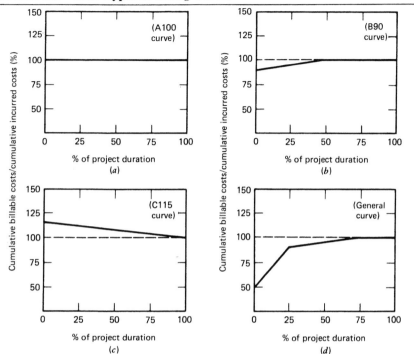

and revenues to present worth and solving for the markup. This is illustrated in Figure C.4. The MARR is the effective rate of return.

Let

$$\text{Present worth of disbursements} = C$$

$$\text{Present worth of receipts without markup} = R$$

$$\text{Markup in \%} = MK$$

then

$$\left( R\left[ 1 + \frac{MK}{100} \right] \right) = C \qquad \text{at } i = \text{MARR}$$

In the example project calculations shown in Figure C.4, the right side of the equation is $725,673.

**TABLE C.2**

| Prog. Pay No. | Cutoff Day | % Dur. | Cumulative Total Job Expenses | % of Billable Expenses | Cumulative Billing | Retention | Progress Payment | Day of Payment | PW Factor | Present Worth |
|---|---|---|---|---|---|---|---|---|---|---|
| 1 | 21 | 6.77 | 46,567 | 91.35 | 42,539 | 4,254 | 38,285 | 41 | .9706 | 37,159 |
| 2 | 51 | 16.45 | 113,165 | 93.29 | 105,572 | 10,557 | 56,730 | 71 | .9496 | 53,871 |
| 3 | 82 | 26.45 | 181,965 | 95.29 | 173,394 | 17,339 | 61,040 | 102 | .9284 | 56,670 |
| 4 | 113 | 36.45 | 250,765 | 97.29 | 243,969 | 24,397 | 63,517 | 133 | .9077 | 57,654 |
| 5 | 141 | 45.48 | 312,902 | 99.10 | 310,086 | 31,009 | 59,505 | 161 | .8894 | 52,924 |
| 6 | 172 | 55.48 | 393,977 | 100 | 393,977 | 39,398 | 75,502 | 192 | .8696 | 65,657 |
| 7 | 202 | 65.16 | 482,259 | 100 | 482,259 | 48,226 | 79,454 | 222 | .8508 | 67,599 |
| 8 | 233 | 75.16 | 573,459 | 100 | 573,459 | 57,346 | 82,080 | 253 | .8318 | 68,274 |
| 9 | 263 | 84.84 | 661,741 | 100 | 661,741 | 66,174 | 79,454 | 283 | .8138 | 64,660 |
| 10 | 294 | 94.84 | 752,941 | 100 | 752,941 | 75,294 | 82,080 | 314 | .7957 | 65,311 |
| 11 | 310 | 100 | 800,000 | 100 | 800,000 | 80,000 | 42,353 | 330 | .7865 | 33,311 |
| Final | 310 | | | | | | 800,000 | 370 | .7639 | 61,112 |

Total PW—All Receipts 684,202

374

**FIGURE C.4   Calculation of Markup and Contract Price**

| Item | Present Worth |
|---|---|
| Total BLAB | $109,263 |
| Total LBUR | 36,026 |
| Total MAT | 86,459 |
| Total SUB | 401,803 |
| Total FOH | 56,691 |
| Total HOH | 35,431 |
| Total Disbursements | $725,673 |
| Total Receipts | $684,202 |

$$\text{Markup} = \left(\frac{\text{P.W. Disbursements}}{\text{P.W. Receipts}} - 1\right) \times 100$$

$$= \left(\frac{\$725,673}{\$684,202} - 1\right) \times 100$$

$$= 6.06\%$$

$$\text{Contract Price} = \$800,000 + (0.0606) \times (\$800,000)$$

$$= \$848,480$$

The equation reduces to

$$\text{MK} = \left(\frac{C}{R} - 1\right) \times 100$$

$$= \left(\frac{725,673}{684,202} - 1\right) \times 100 = 6.06\%$$

This level of markup applied to the costs in Table C.1 yields the bid price below.

Contract price $= 1.0606(120,000 + 40,000 + 96,000 + 480,000 + 64,000)$

$$= 1.0606(800,000) = \$848,480$$

Home office costs are not assumed to be billable and are recovered through the markup. It will be noted that the $800,000 to which the markup is applied does not include home office overhead (i.e., G & A distributed to the job). This expense is recovered as a part of the markup.

## C.2   A CASH-FLOW MARKUP PROGRAM

In the report by Fondahl and Bacarreza, a computer program to determine the level of markup based on this technique is presented. This program in the BASIC programming language is available from the Construction Institute, Stanford University, Stanford, California. The primary input variables required in the program are briefly described below.

MARR. This is the minimum attractive rate of return that the contractor expects to achieve on the project. This reflects the level of return that the company might obtain from other investment of funds. The level of return may vary from division to division within a given company. Given similar jobs, however, it should remain relatively constant. It is not designed to reflect risk associated with a job.

RET%, REPOL. These are variables pertaining to the client's retainage policy applied to the job. RET% indicates the percentage amount retained on billings. REPOL is the variable that represents over what percentage of the job retainage is withheld. Only 50 or 100 percent are acceptable values. That is, monies are retained either on 50 percent or 100 percent of the bid price.

DUR. DUR is the estimated duration of the project in calendar days based on 30-day months.

The variables associated with the percentage of total contract amount allocated to various direct and indirect costs are as follows:

BLAB% = % of total contract amount allocated to direct labor wage (bare labor costs)

LBUR% = % for labor burden (FICA, unemployment, etc.)

MAT% = % for materials

SUB% = % for subcontracts

FOH% = % for field overhead

HOH% = % for home office overhead

EQP% = equipment percent

X1,X2 = % for other cost items or special sub-items

All percentages (excluding HOH%) must sum to 100. The HOH is considered non-billable and is handled separately as part of the markup.

Note that since the major cost components are defined in terms of their percentage of total contract amount, the actual amount of these costs and of the final bid price need not be known precisely. They must only be projected in terms of their percentage of the total contract amount. This means that the level of markup can be determined very early in the preplanning stage with relatively sketchy cost and estimate information. With this approach, jobs that obviously have cash flow characteristics that require noncompetitively high markups can be "weeded out."

## Cost Curves

The cost curves are defined as segmented linear plots as shown below in Figure C.5. The pertinent variables for each curve are

$CCNO$ = cost curve number

$X2, Y2$ = the coordinates of curve point 2 as shown in Figure C.5(a)

$X3, Y3$ = the coordinates of curve point 3 as shown in Figure C.5(a)

The origin of each curve is $(0, 0)$. The end point is $(100, 100)$. The $X$ and $Y$ values must be in the range 0 to 100. For curves in which $X2, X3, Y2, Y3$ are not defined, they are assumed to be zero, and the curve is as shown in Figure C.5(b).

**FIGURE C.5   Cost Curves**

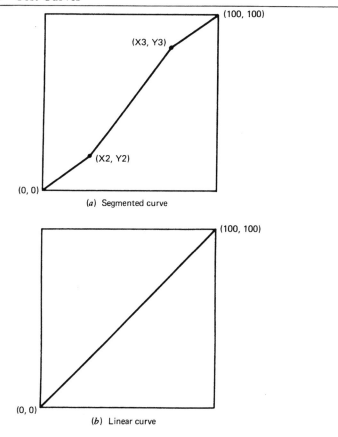

(*a*) Segmented curve

(*b*) Linear curve

## Cost Curve Designation

Designation of cost curves for each cost component is established by the input variables LCUR, MCUR, SCUR, FCUR, HCUR, ECUR, X1CUR, X2CUR. Integers indicating the CCNO of the distribution curves to be used for each cost component are linked to the appropriate cost center variable (e.g., LCUR, MCUR, etc.). The cost component curve variables are

LCUR   = bare labor and labor burden curve
MCUR   = material curve
SCUR   = sub curve
FCUR   = field overhead curve
HCUR   = home office overhead curve
ECUR   = equipment curve
X1CUR  = X1 cost curve
X2CUR  = X2 cost curve

If, for instance, LCUR is set to 1, this means the labor costs and labor burdens will be distributed throughout the job duration in accordance with cost curve 1.

PPAY, FINPAY. PPAY is the lag or delay in days between the end of the cutoff or pay period and the receipt of billing payment. FINPAY is the lag or delay in days from project completion until receipt of final payment to include retainage.

## Positioning of Expenditures

As noted above, the points at which the period expenses for each cost component occur is determined by lags tied to the end of the payment period. The lag variables are

LLAG  = lag for labor charge (default $-15$)
BLAG  = lag for labor burden (default $+15$)
MLAG  = lag for materials (default $+10$)
SLAG  = lag for subcontracts (default same as PPAY)
   FLAG, HLAG, ELAG, X1LAG, X2LAG (default for each = 0)

## Billing Curves

The definition of the billings curve follows a format similar to that used for defining the cost curves. The input parameters are $Y1$, $X2$, $Y2$, $X3$, as shown in Figure C.6. $X1$ is set to zero ($X1 = 0$). $Y3$ is set to 100 ($Y3 = 100$). This curve establishes the percentage of the cost incurred that will be billed as a function of the percentage of the project duration. If $(X2, Y2)$ is (30, 70), this means that 70 percent of incurred costs are billed when the job is 30 percent complete.

**FIGURE C.6 Billing Curve**

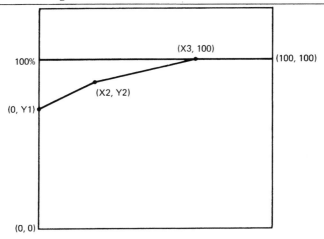

## Cost Variation

The program allows the user to check the sensitivity of the markup to variations in the major cost components. This feature is illustrated in Figure C.9. Various versions of the program allow the user to define policies regarding retainage on payment to subcontractors and other retainage factors (e.g., partial release of retainage at job completion and the like).

## C.3 SAMPLE INPUT TO MARKUP PROGRAM

The input sequence to the markup program is shown in Figure C.7. The first query from the program requests the PRE-TAX ROI in percentage. This is the MARR and in the example is given as 30. The retainage policy is that the client holds 10 percent over 100 percent of the project. The cost components percentage as given in the example are

Labor 10%
Burden 3%
Materials 35%
Subcontracts 40%
Equipment 2%
Field overhead 10%

The home office overhead is 8 percent. Only two cost curves are defined (see Figure C.8). The other cost curves (3 and above) default to the straight line of Figure C.5(b). The

## FIGURE C.7   Sample Input to Markup Program

```
--------------------------------------------
MARKUP % BASED ON FORECASTED CASH FLOW
--------------------------------------------

THIS PROGRAM CALCULATES THE MARKUP % NECESSARY TO RECOVER A STATED HOME OFFICE
OVERHEAD AND TO GENERATE A SPECIFIED RETURN ON INVESTMENT.

RESULTS ARE BASED ON A FORECASTED CASH FLOW FOR THE PROJECT.

THE FORECASTED CASH FLOW IS DETERMINED FROM DATA THAT YOU ESTIMATE AND TYPE IN
RESPONSE TO MY INPUT REQUESTS.
--------------------------------------------
INPUT EXPECTED PRE-TAX ROI IN %
?30
INPUT RET%, REPOL {100 OR 50}
?10,100
INPUT DUR - DAYS {1 MO. = 30 DAYS}
?360
INPUT BLAB%, LBUR%, MAT%, SUB%, EQP%, FOH%, X1%, X2%. {8 INPUTS REQ. EVEN IF ZERO.}
?10,3,35,40,2,10,0,0
INPUT HOH%
?8
TO DEFINE A COST CURVE, INPUT CCNO.; IF NO FURTHER COST CURVE DATA, TYPE 10
?1
INPUT X2,Y2,X3,Y3.
?30,30,70,70
CURVE 1 HAS BEEN STORED. CONTINUE.
TO DEFINE A COST CURVE, INPUT CCNO.; IF NO FURTHER COST CURVE DATA, TYPE 10
?2
INPUT X2,Y2,X3,Y3.
?30,20,70,80
CURVE 2 HAS BEEN STORED. CONTINUE.
TO DEFINE A COST CURVE, INPUT CCNO.; IF NO FURTHER COST CURVE DATA, TYPE 10
?10
INPUT LCUR, MCUR, SCUR, ECUR, FCUR, X1CUR, X2CUR, HCUR. {8 INPUTS, 1 TO 9, EVEN
IF COST COMPONENT UNUSED.}
?1,2,2,3,3,3,3,3
INPUT LAG VALUES FOR: PPAY, FINPAY.
?15,20
PRESET COST LAGS ARE: {1}LLAG = -15, {2}BLAG = 15, {3}MLAG = 10, {4}SLAG = PPAY,
{5} ELAG = {6}FLAG = {7}X1LAG = {8}X2LAG = {9}HLAG = 0.
TO MODIFY, TYPE LAG NO., OTHERWISE TYPE 10.
?10
TO DEFINE BILLING CURVE INPUT: Y1, X2, Y2, X3.

?100,50,100,70

--------------------------------

MARKUP = 9.49686131%

--------------------------------
```

**FIGURE C.8    Cost Curves for the Example Project**

Cost curve 1

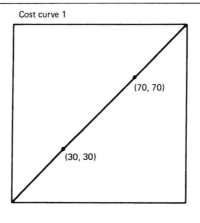

(70, 70)

(30, 30)

Cost curve 2

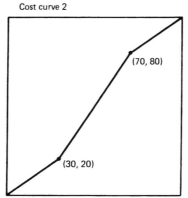

(70, 80)

(30, 20)

assignment of curves is as follows: LCUR = 1, MCUR = 2, SCUR = 2. All others are set to 3. The delay in receipt of payment following billing is 15 days. The delay in receipt of final payment is 20 days. The lags associated with each cost component are the default values as shown. The billings curve as defined establishes billing based on 100 percent of incurred cost throughout the job. The calculated level of markup based on these input parameters is 9.49 percent. From this input stream it is apparent that no dollar figures have been used. This emphasizes the utility of the program in evaluating a project based on relatively preliminary data.

The program is also well adapted for performing sensitivity analysis. The calculated markup for a range of MARR values between 20 and 45 is shown in Figure C.9. The lowest curve shows the response of the markup level to the mix of labor and subcontracts, as defined in Figure C.7. That is, this curve reflects 10 percent of the job performed with in-house labor and 40 percent with subcontracts. The range of markups is from 9.03 percent for 20 percent ROI to 10.10 percent for 45 percent ROI. Doing

**FIGURE C.9   Comparison of Factors Affecting Markup**

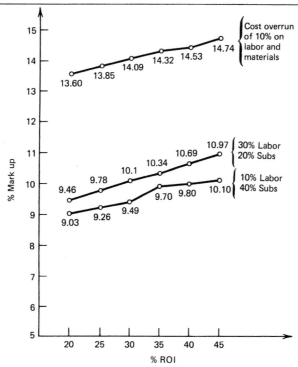

more of the work with in-house forces causes an increase in the level of markup, as shown in the middle curve. This curve indicates that the cash flow is more favorable to the contractor when he "subs out" a larger part of the work. This is reasonable, since the financing of a larger portion of the job is being done by the subcontractors. The top curve indicates the impact on markup of interjecting the possibility of as much as 10 percent overrun on the labor and materials cost components. A significant increase in markup is needed to protect the contractor from the negative cash flow aspects of such an overrun.

# Appendix D

# THE ASSOCIATED GENERAL CONTRACTORS OF AMERICA SUBCONTRACT FOR BUILDING CONSTRUCTION

# SUBCONTRACT FOR BUILDING CONSTRUCTION

## TABLE OF ARTICLES

This Agreement has important legal and insurance consequences. Consultation with an attorney and insurance consultant is encouraged with respect to its completion or modification and particularly when used with other than AIA A201 General Conditions of the Contract for Construction, August 1976 edition.

# TABLE OF CONTENTS

## ARTICLE 1

### AGREEMENT

This Agreement made this _____ day of _____ , 19___, and effective the _____ day of _____ ,
19___, by and between _____ ,
_____

hereinafter called the Contractor and _____
_____ ,

hereinafter called the Subcontractor, to perform part of the Work on the following Project:

PROJECT:

OWNER:

ARCHITECT:

CONTRACTOR:

SUBCONTRACTOR:

CONTRACT PRICE:

Notice to the parties shall be given at the above addresses.

## ARTICLE 2

### SCOPE OF WORK

**2.1 SUBCONTRACTOR'S WORK.** The Contractor employs the Subcontractor as an independent contractor, to perform the work described in Article 16. The Subcontractor shall perform such work (hereinafter called the "Subcontractor's Work") under the general direction of the Contractor and in accordance with this Agreement and the Contract Documents.

**2.2 CONTRACT DOCUMENTS.** The Contract Documents which are binding on the Subcontractor are as set forth in Article 16.5.

Upon the Subcontractor's request the Contractor shall furnish a copy of any part of these documents.

**2.3 CONFLICTS.** In the event of a conflict between this Agreement and the Contract Documents, this Agreement shall govern, except as follows:

## ARTICLE 3

### SCHEDULE OF WORK

**3.1 TIME IS OF ESSENCE.** Time is of the essence for both parties, and they mutually agree to see to the performance of their respective work and the work of their subcontractors so that the entire Project may be completed in accordance with the Contract Documents and the Schedule of Work. The Contractor shall prepare the Schedule of Work and revise such schedule as the Work progresses.

**3.2 DUTY TO BE BOUND.** Both the Contractor and the Subcontractor shall be bound by the Schedule of Work. The Subcontractor shall provide the Contractor with any requested scheduling information for the Subcontractor's Work. The Schedule of Work and all subsequent changes thereto shall be submitted to the Subcontractor in advance of the required performance.

**3.3 SCHEDULE CHANGES.** The Subcontractor recognizes that changes will be made in the Schedule of Work and agrees to comply with such changes subject to a reservation of rights arising hereunder.

**3.4 PRIORITY OF WORK.** The Contractor shall have the right to decide the time, order and priority in which the various portions of the Work shall be performed and all other matters relative to the timely and orderly conduct of the Subcontractor's Work.

The Subcontractor shall commence its work within _____ days of notice to proceed from the Contractor and if such work is interrupted for any reason the Subcontractor shall resume such work within two working days from the Contractor's notice to do so.

## ARTICLE 4

### CONTRACT PRICE

The Contractor agrees to pay to the Subcontractor for the satisfactory performance of the Subcontractor's Work the sum of _____

_____

Dollars ($ _____ ) in accordance with Article 5, subject to additions or deductions per Article 6.

## ARTICLE 5

### PAYMENT

**5.1 GENERAL PROVISIONS**

**5.1.1 SCHEDULE OF VALUES.** The Subcontractor shall provide a schedule of values satisfactory to the Contractor

and the Owner no more than fifteen (15) days from the date of execution of this Agreement.

**5.1.2 ARCHITECT VERIFICATION.** Upon request the Contractor shall give the Subcontractor written authorization to obtain directly from the Architect the percentage of completion certified for the Subcontractor's Work.

**5.1.3. PAYMENT USE RESTRICTION.** No payment received by the Subcontractor shall be used to satisfy or secure any indebtedness other than one owed by the Subcontractor to a person furnishing labor or materials for use in performing the Subcontractor's Work.

**5.1.4 PAYMENT USE VERIFICATION.** The Contractor shall have the right at all times to contact the Subcontractor's subcontractors and suppliers to ensure that the same are being paid by the Subcontractor for labor or materials furnished for use in performing the Subcontractor's Work.

**5.1.5 PARTIAL LIEN WAIVERS AND AFFIDAVITS.** When required by the Contractor, and as a prerequisite for payment, the Subcontractor shall provide, in a form satisfactory to the Owner and the Contractor, partial lien or claim waivers and affidavits from the Subcontractor, and its sub-subcontractors and suppliers for the completed Subcontractor's Work. Such waivers may be made conditional upon payment.

**5.1.6 SUBCONTRACTOR PAYMENT FAILURE.** In the event the Contractor has reason to believe that labor, material or other obligations incurred in the performance of the Subcontractor's Work are not being paid, the Contractor shall give written notice of such claim or lien to the Subcontractor and may take any steps deemed necessary to insure that any progress payment shall be utilized to pay such obligations.

If upon receipt of said notice, the Subcontractor does not:

(a) supply evidence to the satisfaction of the Contractor that the monies owing to the claimant have been paid; or

(b) post a bond indemnifying the Owner, the Contractor, the Contractor's surety, if any, and the premises from such claim or lien;

then the Contractor shall have the right to retain out of any payments due or to become due to the Subcontractor a reasonable amount to protect the Contractor from any and all loss, damage or expense including attorney's fees arising out of or relating to any such claim or lien until the claim or lien has been satisfied by the Subcontractor.

**5.1.7 PAYMENT NOT ACCEPTANCE.** Payment to the Subcontractor is specifically agreed not to constitute or imply acceptance by the Contractor or the Owner of any portion of the Subcontractor's Work.

## 5.2 PROGRESS PAYMENTS

**5.2.1 APPLICATION.** The Subcontractor's progress payment application for work performed in the preceding payment period shall be submitted to the Contractor per the terms of this Agreement and specifically Articles 5.1.1, 5.2.2, 5.2.3, and 5.2.4 for approval of the Contractor and _____

The Contractor shall forward, without delay, the approved value to the Owner for payment.

**5.2.2 RETAINAGE/SECURITY.** The rate of retainage shall not exceed the percentage retained from the Contractor's payment by the Owner for the Subcontractor's Work provided the Subcontractor furnishes a bond or other security to the satisfaction of the Contractor.

If the Subcontractor has furnished such bond or security; its work is satisfactory and the Contract Documents provide for reduction of retainage at a specified percentage of completion, the Subcontractor's retainage shall also be reduced when the Subcontractor's Work has attained the same percentage of completion and the Contractor's retainage for the Subcontractor's Work has been so reduced by the Owner.

However, if the Subcontractor does not provide such bond or security, the rate of retainage shall be _____%.

**5.2.3 TIME OF APPLICATION.** The Subcontractor shall submit progress payment applications to the Contractor no later than the _____ day of each payment period for work performed up to and including the _____ day of the payment period indicating work completed and, to the extent allowed under Article 5.2.4, materials suitably stored during the preceding payment period.

**5.2.4 STORED MATERIALS.** Unless otherwise provided in the Contract Documents, and if approved in advance by the Owner, applications for payment may include materials and equipment not incorporated in the Subcontractor's Work but delivered and suitably stored at the site or at some other location agreed upon in writing. Approval of payment application for such stored items on or off the site shall be conditioned upon submission by the Subcontractor of bills of sale and applicable insurance or such other procedures satisfactory to the Owner and Contractor to establish the Owner's title to such materials and equipment or otherwise protect the Owner's and Contractor's interests therein, including transportation to the site.

**5.2.5 TIME OF PAYMENT.** Progress payments to the Subcontractor for satisfactory performance of the Subcontractor's Work shall be made no later than seven (7) days after receipt by the Contractor of payment from the Owner for such Subcontractor's Work.

**5.2.6 PAYMENT DELAY.** If for any reason not the fault of the Subcontractor, the Subcontractor does not receive a progress payment from the Contractor within seven (7) days after the date such payment is due, as dc ˙ned in Article 5.2.5, then the Subcontractor, upon giving an additional seven (7) days written notice to the Contractor, and without prejudice to and in addition to any other legal remedies, may stop work until payment of the full amount owing to the Subcontractor has been received. To the extent obtained by the Contractor under the Contract Documents, the contract price shall be increased by the amount of the Subcontractor's reasonable costs of shut-down, delay, and start-up, which shall be effected by appropriate Change Order.

If the Subcontractor's Work has been stopped for thirty (30) days because the Subcontractor has not received progress payments as required hereunder, the Subcontractor may terminate this Agreement upon giving the Contractor an additional seven (7) days written notice.

## 5.3 FINAL PAYMENT

**5.3.1 APPLICATION.** Upon acceptance of the Subcontractor's Work by the Owner, the Contractor, and if necessary, the Architect, and upon the Subcontractor furnishing evidence of fulfillment of the Subcontractor's obligations in accordance with the Contract Documents and Article 5.3.2, the Contractor shall forward the Subcontractor's application for final payment without delay.

**5.3.2 REQUIREMENTS.** Before the Contractor shall be required to forward the Subcontractor's application for final payment to the Owner, the Subcontractor shall submit to the Contractor:

(a) an affidavit that all payrolls, bills for materials and equipment, and other indebtedness connected with the Subcontractor's Work for which the Owner or his property or the Contractor or the Contractor's surety might in any way be liable, have been paid or otherwise satisfied;

(b) consent of surety to final payment, if required;

(c) satisfaction of required closeout procedures; and

(d) other data if required by the Contractor or Owner, such as receipts, releases, and waivers of liens to the extent and in such form as may be designated by the Contractor or Owner.

Final payment shall constitute a waiver of all claims by the Subcontractor relating to the Subcontractor's Work, but shall in no way relieve the Subcontractor of liability for the obligations assumed under Article 9.10 hereof, or for faulty or defective work appearing after final payment.

**5.3.3 TIME OF PAYMENT.** Final payment of the balance due of the contract price shall be made to the Subcontractor:

(a) upon receipt of the Owner's waiver of all claims related to the Subcontractor's Work except for unsettled liens, unknown defective work, and non-compliance with the Contract Documents or warranties; and

(b) within seven (7) days after receipt by the Contractor of final payment from the Owner for such Subcontractor's Work.

**5.3.4 FINAL PAYMENT DELAY.** If the Owner or its designated agent does not issue a Certificate for Final Payment or the Contractor does not receive such payment for any cause which is not the fault of the Subcontractor, the Contractor shall promptly inform the Subcontractor in writing. The Contractor shall also diligently pursue, with the assistance of the Subcontractor, the prompt release by the Owner of the final payment due for the Subcontractor's Work. At the Subcontractor's request and joint expense, to the extent agreed upon in writing, the Contractor shall institute all reasonable legal remedies to mitigate the damages and pursue full payment of the Subcontractor's application for final payment including interest thereon.

**5.4 LATE PAYMENT INTEREST.** To the extent obtained by the Contractor under the Contract Documents, progress payments or final payment due and unpaid under this Agreement shall bear interest from the date payment is due at the rate provided in the Contract Documents, or, in the absence thereof, at the legal rate prevailing at the place of the Project.

## ARTICLE 6

### CHANGES, CLAIMS AND DELAYS

**6.1 CHANGES.** When the Contractor so orders in writing, the Subcontractor, without nullifying this Agreement, shall make any and all changes in the Work which are within the general scope of this Agreement.

Adjustments in the contract price or contract time, if any, resulting from such changes shall be set forth in a Subcontract Change Order pursuant to the Contract Documents.

No such adjustment shall be made for any such changes performed by the Subcontractor that have not been so ordered by the Contractor.

**6.2 CLAIMS RELATING TO OWNER.** The Subcontractor agrees to make all claims for which the Owner is or may be liable in the manner provided in the Contract Documents for like claims by the Contractor upon the Owner.

Notice of such claims shall be given by the Subcontractor to the Contractor within one (1) week prior to the beginning of the Subcontractor's Work or the event for which such claim is to be made, or immediately upon

the Subcontractor's first knowledge of the event, whichever shall first occur, otherwise, such claims shall be deemed waived.

The Contractor agrees to permit the Subcontractor to prosecute said claim, in the name of the Contractor, for the use and benefit of the Subcontractor in the manner provided in the Contract Documents for like claims by the Contractor upon the Owner.

**6.3 CLAIMS RELATING TO CONTRACTOR.** The Subcontractor shall give the Contractor written notice of all claims not included in Article 6.2 within five (5) days of the beginning of the event for which claim is made; otherwise, such claims shall be deemed waived.

All unresolved claims, disputes and other matters in question between the Contractor and the Subcontractor not relating to claims included in Article 6.2 shall be resolved in the manner provided in Article 14 herein.

**6.4 DELAY.** If the progress of the Subcontractor's Work is substantially delayed without the fault or responsibility of the Subcontractor, then the time for the Subcontractor's Work shall be extended by Change Order to the extent obtained by the Contractor under the Contract Documents and the Schedule of Work shall be revised accordingly.

The Contractor shall not be liable to the Subcontractor for any damages or additional compensation as a consequence of delays caused by any person not a party to this Agreement unless the Contractor has first recovered the same on behalf of the Subcontractor from said person, it being understood and agreed by the Subcontractor that, apart from recovery from said person, the Subcontractor's sole and exclusive remedy for delay shall be an extension in the time for performance of the Subcontractor's Work.

**6.5 LIQUIDATED DAMAGES.** If the Contract Documents provide for liquidated or other damages for delay beyond the completion date set forth in the Contract Documents, and are so assessed, then the Contractor may assess same against the Subcontractor in proportion to the Subcontractor's share of the responsibility for such delay. However the amount of such assessment shall not exceed the amount assessed against the Contractor.

## ARTICLE 7

### CONTRACTOR'S OBLIGATIONS

**7.1 OBLIGATIONS DERIVATIVE.** The Contractor binds itself to the Subcontractor under this Agreement in the same manner as the Owner is bound to the Contractor under the Contract Documents.

**7.2 AUTHORIZED REPRESENTATIVE.** The Contractor shall designate one or more persons who shall be the Contractor's authorized representative(s) a) on-site and b) off-site. Such authorized representative(s) shall be the only person(s) the Subcontractor shall look to for instructions, orders and/or directions, except in an emergency.

**7.3 STORAGE ALLOCATION.** The Contractor shall allocate adequate storage areas, if available, for the Subcontractor's materials and equipment during the course of the Subcontractor's Work.

**7.4 TIMELY COMMUNICATIONS.** The Contractor shall transmit, with reasonable promptness, all submittals, transmittals, and written approvals relating to the Subcontractor's Work.

**7.5 NON-CONTRACTED SERVICES.** The Contractor agrees, except as otherwise provided in this Agreement, that no claim for non-contracted construction services rendered or materials furnished shall be valid unless the Contractor provides the Subcontractor notice:

(a) prior to furnishing of the services or materials, except in an emergency affecting the safety of persons or property;

(b) in writing of such claim within three days of first furnishing such services or materials; and

(c) the written charges for such services or materials no later than the fifteenth (15th) day of the calendar month following that in which the claim originated.

### ARTICLE 8

### SUBCONTRACTOR'S OBLIGATIONS

**8.1 OBLIGATIONS DERIVATIVE.** The Subcontractor binds itself to the Contractor under this Agreement in the same manner as the Contractor is bound to the Owner under the Contract Documents.

**8.2 RESPONSIBILITIES.** The Subcontractor shall furnish all of the labor, materials, equipment, and services, including, but not limited to, competent supervision, shop drawings, samples, tools, and scaffolding as are necessary for the proper performance of the Subcontractor's Work.

The Subcontractor shall provide a list of proposed sub-subcontractors, and suppliers, be responsible for taking field dimensions, providing tests, ordering of materials and all other actions as required to meet the Schedule of Work.

**8.3 TEMPORARY SERVICES.** The Subcontractor shall furnish all temporary services and/or facilities necessary to perform its work, except as provided in Article 16. Said article also identifies those common temporary services (if any) which are to be furnished by this subcontractor.

**8.4 COORDINATION.** The Subcontractor shall:

(a) cooperate with the Contractor and all others whose work may interfere with the Subcontractor's Work;

(b) specifically note and immediately advise the Contractor of any such interference with the Subcontractor's Work; and

(c) participate in the preparation of coordination drawings and work schedules in areas of congestion.

**8.5 AUTHORIZED REPRESENTATIVE.** The Subcontractor shall designate one or more persons who shall be the authorized Subcontractor's representative(s) a) on-site and b) off-site. Such authorized representative(s) shall be the only person(s) to whom the Contractor shall issue instructions, orders or directions, except in an emergency.

**8.6 PROVISION FOR INSPECTION.** The Subcontractor shall notify the Contractor when portions of the Subcontractor's Work are ready for inspection. The Subcontractor shall at all times furnish the Contractor and its representatives adequate facilities for inspecting materials at the site or any place where materials under this Agreement may be in the course of preparation, process, manufacture or treatment.

The Subcontractor shall furnish to the Contractor in such detail and as often as required, full reports of the progress of the Subcontractor's Work irrespective of the location of such work.

**8.7 SAFETY AND CLEANUP.** The Subcontractor shall follow the Contractor's clean-up and safety directions, and

(a) at all times keep the building and premises free from debris and unsafe conditions resulting from the Subcontractor's Work; and

(b) broom clean each work area prior to discontinuing work in the same.

If the Subcontractor fails to immediately commence compliance with such safety duties or commence clean-up duties within 24 hours after receipt from the Contractor of written notice of noncompliance, the Contractor may implement such safety or cleanup measures without further notice and deduct the cost thereof from any amounts due or to become due the Subcontractor.

**8.8 PROTECTION OF THE WORK.** The Subcontractor shall take necessary precautions to properly protect the Subcontractor's Work and the work of others from damage caused by the Subcontractor's operations. Should the Subcontractor cause damage to the Work or property of the Owner, the Contractor or others, the Subcontractor shall promptly remedy such damage to the satisfaction of the Contractor, or the Contractor may so remedy and deduct the cost thereof from any amounts due or to become due the Subcontractor.

**8.9 PERMITS, FEES AND LICENSES.** The Subcontractor shall give adequate notices to authorities pertaining to the Subcontractor's Work and secure and pay for all permits, fees, licenses, assessments, inspections and taxes necessary to complete the Subcontractor's Work in accordance with the Contract Documents.

To the extent obtained by the Contractor under the Contract Documents, the Subcontractor shall be compensated for additional costs resulting from laws, ordinances, rules, regulations and taxes enacted after the date of the Agreement.

**8.10  ASSIGNMENT.** The Subcontractor shall not assign this Agreement nor its proceeds nor subcontract the whole nor any part of the Subcontractor's Work without prior written approval of the Contractor which shall not be unreasonably withheld. See Article 16.4 for sub-subcontractors and suppliers previously approved by the Contractor.

**8.11  NON-CONTRACTED SERVICES.** The Subcontractor agrees, except as otherwise provided in this Agreement, that no claim for non-contracted construction services rendered or materials furnished shall be valid unless the Subcontractor provides the Contractor notice:

(a)  prior to furnishing of the services or materials, except in an emergency affecting the safety of persons or property;

(b)  in writing of such claim within three days of first furnishing such services or materials; and

(c)  the written charge for such services or materials to the contractor no later than the fifteenth day (15th) of the calendar month following that in which the claim originated.

## ARTICLE 9

## SUBCONTRACT PROVISIONS

**9.1  LAYOUT RESPONSIBILITY AND LEVELS.** The Contractor shall establish principal axis lines of the building and site whereupon the Subcontractor shall lay out and be strictly responsible for the accuracy of the Subcontractor's Work and for any loss or damage to the Contractor or others by reason of the Subcontractor's failure to set out or perform its work correctly. The Subcontractor shall exercise prudence so that actual final conditions and details shall result in perfect alignment of finish surfaces.

**9.2  WORKMANSHIP.** Every part of the Subcontractor's Work shall be executed in strict accordance with the Contract Documents in the most sound, workmanlike, and substantial manner. All workmanship shall be of the best of its several kinds, and all materials used in the Subcontractor's Work shall be furnished in ample quantities to facilitate the proper and expeditious execution of the work, and shall be new except such materials as may

be expressly provided in the Contract Documents to be otherwise.

**9.3  MATERIALS FURNISHED BY OTHERS.** In the event the scope of the Subcontractor's Work includes installation of materials or equipment furnished by others, it shall be the responsibility of the Subcontractor to examine the items so provided and thereupon handle, store and install the items with such skill and care as to ensure a satisfactory and proper installation. Loss or damage due to acts of the Subcontractor shall be deducted from any amounts due or to become due the Subcontractor.

**9.4  SUBSTITUTIONS.** No substitutions shall be made in the Subcontractor's Work unless permitted in the Contract Documents and only then upon the Subcontractor first receiving all approvals required under the Contract Documents for substitutions. The Subcontractor shall indemnify the Contractor for any increased costs incurred by the Contractor as a result of such substitutions, whether or not the Subcontractor has obtained approval thereof.

**9.5  USE OF CONTRACTOR'S EQUIPMENT.** The Subcontractor, its agents, employees, subcontractors or suppliers shall not use the Contractor's equipment without the express written permission of the Contractor's designated representative.

If the Subcontractor or any of its agents, employees, suppliers or lower tier subcontractors utilize any machinery, equipment, tools, scaffolding, hoists, lifts or similar items owned, leased, or under the control of the Contractor, the Subcontractor shall be liable to the Contractor as provided in Article 12 for any loss or damage (including personal injury or death) which may arise from such use, except where such loss or damage shall be found to have been due solely to the negligence of the Contractor's employees operating such equipment.

**9.6  CONTRACT BOND REVIEW.** The Contractor's Payment Bond for the Project, if any, may be reviewed and copied by the Subcontractor.

**9.7  OWNER ABILITY TO PAY.** The Subcontractor shall have the right to receive from the Contractor information relative to the Owner's financial ability to pay for the Work.

**9.8  PRIVITY.** Until final completion of the Project, the Subcontractor agrees not to perform any work directly for the Owner or any tenants thereof, or deal directly with the Owner's representatives in connection with the Project, unless otherwise directed in writing by the Contractor. All work for this Project performed by the Subcontractor shall be processed and handled exclusively by the Contractor.

**9.9  SUBCONTRACT BOND.** If a Performance and Payment Bond is not required of the Subcontractor under

Article 16, then within the duration of this Agreement, the Contractor may require such bonds and the Subcontractor shall provide same.

Said bonds shall be in the full amount of this Agreement in a form and by a surety satisfactory to the Contractor.

The Subcontractor shall be reimbursed without retainage for cost of same simultaneously with the first progress payment hereunder.

The reimbursement amount for the bonds shall not exceed the manual rate for such subcontractor work.

Retainage reduction provisions of Article 5.2.2 shall not apply when bonds are furnished under the terms of this Article.

In the event the Subcontractor shall fail to promptly provide such requested bonds, the Contractor may terminate this Agreement and re-let the work to another Subcontractor and all Contractor costs and expenses incurred thereby shall be paid by the Subcontractor.

**9.10  WARRANTY.** The Subcontractor warrants its work against all deficiencies and defects in materials and/or workmanship and as called for in the Contract Documents.

The Subcontractor agrees to satisfy such warranty obligations which appear within the guarantee or warranty period established in the Contract Documents without cost to the Owner or the Contractor.

If no guarantee or warranty is required of the Contractor in the Contract Documents, then the Subcontractor shall guarantee or warranty its work as described above for the period of one year from the date(s) of substantial completion of all or a designated portion of the Subcontractor's Work or acceptance or use by the Contractor or Owner of designated equipment, whichever is sooner.

The Subcontractor further agrees to execute any special guarantees or warranties that shall be required for the Subcontractor's Work prior to final payment.

## ARTICLE 10

### RECOURSE BY CONTRACTOR

#### 10.1  FAILURE OF PERFORMANCE

**10.1.1  NOTICE TO CURE.** If the Subcontractor refuses or fails to supply enough properly skilled workers, proper materials, or maintain the Schedule of Work, or it fails to make prompt payment for its workers, sub-subcontractors or suppliers, disregards laws, ordinances, rules, regulations or orders of any public authority having jurisdiction, or otherwise is guilty of a material breach of a provision of this Agreement, and fails within three (3) working days after receipt of written notice to commence and continue satisfactory correction of such default with diligence and promptness, then the Contractor, without

prejudice to any rights or remedies, shall have the right to any or all of the following remedies:

(a) supply such number of workers and quantity of materials, equipment and other facilities as the Contractor deems necessary for the completion of the Subcontractor's Work, or any part thereof which the Subcontractor has failed to complete or perform after the aforesaid notice, and charge the cost thereof to the Subcontractor, who shall be liable for the payment of same including reasonable overhead, profit and attorney's fees;

(b) contract with one or more additional contractors to perform such part of the Subcontractor's Work as the Contractor shall determine will provide the most expeditious completion of the total Work and charge the cost thereof to the Subcontractor;

(c) withhold payment of any monies due the Subcontractor pending corrective action to the extent required by and to the satisfaction of the Contractor and _____; and

(d) in the event of an emergency affecting the safety of persons or property, the Contractor may proceed as above without notice.

**10.1.2  TERMINATION BY CONTRACTOR.** If the Subcontractor fails to commence and satisfactorily continue correction of a default within three (3) working days after receipt by the Subcontractor of the notice issued under Article 10.1.1, then the Contractor may, in lieu of or in addition to Article 10.1.1, issue a second written notice, by certified mail, to the Subcontractor and its surety, if any. Such notice shall state that if the Subcontractor fails to commence and continue correction of a default within seven (7) working days after receipt by the Subcontractor of the notice, the Contractor may terminate this Agreement and use any materials, implements, equipment, appliances or tools furnished by or belonging to the Subcontractor to complete the Subcontractor's Work. The Contractor also may furnish those materials, equipment and/or employ such workers or Subcontractors as the Contractor deems necessary to maintain the orderly progress of the Work.

All of the costs incurred by the Contractor in so performing the Subcontractor's Work, including reasonable overhead, profit and attorney's fees, shall be deducted from any monies due or to become due the Subcontractor. The Subcontractor shall be liable for the payment of any amount by which such expense may exceed the unpaid balance of the subcontract price.

**10.1.3  USE OF SUBCONTRACTOR'S EQUIPMENT.** If the Contractor performs work under this Article or sublets such work to be so performed, the Contractor and/or the persons to whom work has been sublet shall have the right to take and use any materials, implements, equipment, appliances or tools furnished by, belonging or delivered to the Subcontractor and located at the Project.

## 10.2 BANKRUPTCY

**10.2.1 TERMINATION ABSENT CURE.** Upon the appointment of a receiver for the Subcontractor or upon the Subcontractor making an assignment for the benefit of creditors, the Contractor may terminate this Agreement upon giving three (3) working days written notice, by certified mail, to the Subcontractor and its surety, if any. If an order for relief is entered under the bankruptcy code with respect to the Subcontractor, the Contractor may terminate this Agreement by giving three (3) working days written notice, by certified mail, to the Subcontractor, its trustee and its surety, if any, unless the Subcontractor, the surety, or the trustee:

(a) promptly cures all defaults;
(b) provides adequate assurances of future performance;
(c) compensates the Contractor for actual pecuniary loss resulting from such defaults; and
(d) assumes the obligations of the Subcontractor within the statutory time limits.

**10.2.2 INTERIM REMEDIES.** If the Subcontractor is not performing in accordance with the Schedule of Work at the time of entering an order for relief, or at any subsequent time, the Contractor, while awaiting the decision of the Subcontractor or its trustee to reject or to accept this Agreement and provide adequate assurance of its ability to perform hereunder, may avail itself of such remedies under this Article as are reasonably necessary to maintain the Schedule of Work.

The Contractor may offset against any sums due or to become due the Subcontractor all costs incurred in pursuing any of the remedies provided hereunder, including, but not limited to, reasonable overhead, profit and attorney's fees.

The Subcontractor shall be liable for the payment of any amount by which such expense may exceed the unpaid balance of the contract price.

**10.3 SUSPENSION BY OWNER.** Should the Owner suspend the Prime Contract or any part of the Prime Contract which includes the Subcontractor's Work, the Contractor shall so notify the Subcontractor in writing and upon receipt of said notice the Subcontractor shall immediately suspend the Subcontractor's Work.

In the event of such Owner suspension, the Contractor's liability to the Subcontractor is limited to the extent of the Contractor's recovery on the Subcontractor's behalf under the Contract Documents. The Contractor agrees to cooperate with the Subcontractor, at the Subcontractor's expense, in the prosecution of any Subcontractor claim arising out of an Owner suspension and to permit the Subcontractor to prosecute said claim, in the name of the Contractor, for the use and benefit of the Subcontractor.

**10.4 TERMINATION BY OWNER.** Should the Owner terminate the Prime Contract or any part of the Prime Contract which includes the Subcontractor's Work, the Contractor shall so notify the Subcontractor in writing and upon receipt of said notice, this Agreement shall also be terminated and the Subcontractor shall immediately stop the Subcontractor's Work.

In the event of such Owner termination, the Contractor's liability to the Subcontractor is limited to the extent of the Contractor's recovery on the Subcontractor's behalf under the Contract Documents.

The Contractor agrees to cooperate with the Subcontractor, at the Subcontractor's expense, in the prosecution of any Subcontractor claim arising out of the Owner termination and to permit the Subcontractor to prosecute said claim, in the name of the Contractor, for the use and benefit of the Subcontractor, or assign the claim to the Subcontractor.

**10.5 TERMINATION FOR CONVENIENCE.** The Contractor may order the Subcontractor in writing to suspend, delay, or interrupt all or any part of the Subcontractor's Work for such period of time as may be determined to be appropriate for the convenience of the Contractor.

The Subcontractor shall notify the Contractor in writing within ten (10) working days after receipt of the Contractor's order of the effect of such order upon the Subcontractor's Work, and the contract price or contract time shall be adjusted by Subcontract Change Order for any increase in the time or cost of performance of this Agreement caused by such suspension, delay, or interruption.

No claim under this Article shall be allowed for any costs incurred more than ten (10) working days prior to the Subcontractor's notice to the Contractor.

Neither the contract price nor the contract time shall be adjusted under this Article for any suspension, delay or interruption to the extent that performance would have been so suspended, delayed, or interrupted by the fault or negligence of the Subcontractor.

**10.6 WRONGFUL EXERCISE.** If the Contractor wrongfully exercises any option under this Article, the Contractor shall be liable to the Subcontractor solely for the reasonable value of work performed by the Subcontractor prior to the Contractor's wrongful action, including reasonable overhead and profit, less prior payments made, and attorney's fees.

## ARTICLE 11

### LABOR RELATIONS

(Insert here any conditions, obligations or requirements relative to labor relations and their effect on the project. Legal counsel is recommended.)

## ARTICLE 12

### INDEMNIFICATION

**12.1 SUBCONTRACTOR'S PERFORMANCE.** To the fullest extent permitted by law, the Subcontractor shall indemnify and hold harmless the Owner, the Architect, the Contractor (including its affiliates, parents and subsidiaries) and other contractors and subcontractors and all of their agents and employees from and against all claims, damages, loss and expenses, including but not limited to attorney's fees, arising out of or resulting from the performance of the Subcontractor's Work provided that

(a) any such claim, damage, loss, or expense is attributable to bodily injury, sickness, disease, or death, or to injury to or destruction of tangible property (other than the Subcontractor's Work itself) including the loss of use resulting therefrom, to the extent caused or alleged to be caused in whole or in any part by any negligent act or omission of the Subcontractor or anyone directly or indirectly employed by the Subcontractor or anyone for whose acts the Subcontractor may be liable, regardless of whether it is caused in part by a party indemnified hereunder.

(b) such obligation shall not be construed to negate, or abridge, or otherwise reduce any other right or obligation of indemnity which would otherwise exist as to any party or person described in this Article 12.

**12.2 NO LIMITATION UPON LIABILITY.** In any and all claims against the Owner, the Architect, the Contractor (including its affiliates, parents and subsidiaries) and other contractors or subcontractors, or any of their agents or employees, by any employee of the Subcontractor, anyone directly or indirectly employed by the Subcontractor or anyone for whose acts the Subcontractor may be liable, the indemnification obligation under this Article 12 shall not be limited in any way by any limitation on the amount or type of damages, compensation or benefits payable by or for the Subcontractor under worker's or workmen's compensation acts, disability benefit acts or other employee benefit acts.

**12.3 ARCHITECT EXCLUSION.** The obligations of the Subcontractor under this Article 12 shall not extend to the liability of the Architect, its agents or employees, arising out of (a) the preparation or approval of maps, drawings, opinions, reports, surveys, Change Orders, designs or specifications, or (b) the giving of or the failure to give directions or instructions by the Architect, its agents or employees provided such giving or failure to give is the primary cause of the injury or damage.

**12.4 COMPLIANCE WITH LAWS.** The Subcontractor agrees to be bound by, and at its own cost, comply with all federal, state and local laws, ordinances and regulations (hereinafter collectively referred to as "laws") applicable to the Subcontractor's Work including, but not limited to, equal employment opportunity, minority business enterprise, women's business enterprise, disadvantaged business enterprise, safety and all other laws with which the Contractor must comply according to the Contract Documents.

The Subcontractor shall be liable to the Contractor and the Owner for all loss, cost and expense attributable to any acts of commission or omission by the Subcontractor, its employees and agents resulting from the failure to comply therewith, including, but not limited to, any fines, penalties or corrective measures.

**12.5 PATENTS.** Except as otherwise provided by the Contract Documents, the Subcontractor shall pay all royalties and license fees which may be due on the inclusion of any patented materials in the Subcontractor's Work. The Subcontractor shall defend all suits for claims for infringement of any patent rights arising out of the Subcontractor's Work, which may be brought against the Contractor or Owner, and shall be liable to the Contractor and Owner for all loss, including all costs, expenses, and attorney's fees.

## ARTICLE 13

### INSURANCE

**13.1 SUBCONTRACTOR'S INSURANCE.** Prior to start of the Subcontractor's Work, the Subcontractor shall procure for the Subcontractor's Work and maintain in force Worker's Compensation Insurance, Employer's Liability Insurance, Comprehensive General Liability Insurance and all insurance required of the Contractor under the Contract Documents except as follows:


The Contractor, Owner and Architect shall be named as additional insureds on each of these policies except for Worker's Compensation.

This insurance shall include contractual liability insurance covering the Subcontractor's obligations under Article 12.

**13.2 MINIMUM LIMITS OF LIABILITY.** The Subcontractor's Comprehensive General and Automobile Liability Insurance, as required by Article 13.1, shall be written with limits of liability not less than the following:

A. Comprehensive General Liability including completed operations

   1. Bodily Injury   $_____Each Occurrence

                    $_____Aggregate

   2. Property Damage  $_____Each Occurrence

                    $_____Aggregate

B. Comprehensive Automobile Liability

   1. Bodily Injury   $_____Each Person

                    $_____Each Occurrence

   2. Property Damage  $_____Each Occurrence

**13.3 NUMBER OF POLICIES.** Comprehensive General Liability Insurance and other liability insurance may be arranged under a single policy for the full limits required or by a combination of underlying policies with the balance provided by an Excess or Umbrella Liability Policy.

**13.4 CANCELLATION, RENEWAL OR MODIFICATION.** The Subcontractor shall maintain in effect all insurance coverage required under this Agreement at the Subcontractor's sole expense and with insurance companies acceptable to the Contractor.

All insurance policies shall contain a provision that the coverages afforded thereunder shall not be cancelled or not renewed, nor restrictive modifications added, until at least thirty (30) days prior written notice has been given to the Contractor unless otherwise specifically required in the Contract Documents.

Certificates of Insurance, or certified copies of policies acceptable to the Contractor shall be filed with the Contractor prior to the commencement of the Subcontractor's Work.

In the event the Subcontractor fails to obtain or maintain any insurance coverage required under this Agreement, the Contractor may purchase such coverage and charge the expense thereof to the Subcontractor, or ter-. minate this Agreement.

**13.5 WAIVER OF RIGHTS.** The Contractor and Subcontractor waive all rights against each other and the Owner, the Architect, separate contractors, and all other subcontractors for loss or damage to the extent covered by Builder's Risk or any other property or equipment insurance,

except such rights as they may have to the proceeds of such insurance; provided, however, that such waiver shall not extend to the acts of the Architect listed in Article 12.3.

Upon written request of the Subcontractor, the Contractor shall provide the Subcontractor with a copy of the Builder's Risk policy of insurance or any other property or equipment insurance in force for the Project and procured by the Contractor. The Subcontractor shall satisfy itself as to the existence and extent of such insurance prior to commencement of the Subcontractor's Work.

If the Owner or Contractor have not purchased Builder's Risk insurance for the full insurable value of the Subcontractor's Work less a reasonable deductible, then Subcontractor may procure such insurance as will protect the interests of the Subcontractor, its subcontractors and their subcontractors in the Work, and, by appropriate Subcontract Change Order, the cost of such additional insurance shall be reimbursed to the Subcontractor.

If not covered under the Builder's Risk policy of insurance or any other property or equipment insurance required by the Contract Documents, the Subcontractor shall procure and maintain at the Subcontractor's own expense property and equipment insurance for portions of the Subcontractor's Work stored off the site or in transit, when such portions of the Subcontractor's Work are to be included in an application for payment under Article 5.

**13.6 ENDORSEMENT.** If the policies of insurance referred to in this Article require an endorsement to provide for continued coverage where there is a waiver of subrogation, the owners of such policies will cause them to be so endorsed.

## ARTICLE 14

## ARBITRATION

**14.1 AGREEMENT TO ARBITRATE.** All claims, disputes and matters in question arising out of, or relating to, this Agreement or the breach thereof, except for claims which have been waived by the making or acceptance of final payment, and the claims described in Article 14.7, shall be decided by arbitration in accordance with the Construction Industry Arbitration Rules of the American Arbitration Association then in effect unless the parties mutually agree otherwise. This agreement to arbitrate shall be specifically enforceable under the prevailing arbitration law.

**14.2 NOTICE OF DEMAND.** Notice of the demand for arbitration shall be filed in writing with the other party to this Agreement and with the American Arbitration

Association. The demand for arbitration shall be made within a reasonable time after written notice of the claim, dispute or other matter in question has been given, and in no event shall it be made after the date of final acceptance of the Work by the Owner or when institution of legal or equitable proceedings based on such claim, dispute or other matter in question would be barred by the applicable statute of limitations, whichever shall first occur. The location of the arbitration proceedings shall be the city of the Contractor's headquarters or _____

**14.3 AWARD.** The award rendered by the arbitrator(s) shall be final and judgment may be entered upon it in accordance with applicable law in any court having jurisdiction.

**14.4 WORK CONTINUATION AND PAYMENT.** Unless otherwise agreed in writing, the Subcontractor shall carry on the Work and maintain the Schedule of Work pending arbitration, and, if so, the Contractor shall continue to make payments in accordance with this Agreement.

**14.5 NO LIMITATION OF RIGHTS OR REMEDIES.** Nothing in this Article shall limit any rights or remedies not expressly waived by the Subcontractor which the Subcontractor may have under lien laws or payment bonds.

**14.6 SAME ARBITRATORS.** To the extent not prohibited by their contracts with others, the claims and disputes of the Owner, Contractor, Subcontractor and other subcontractors involving a common question of fact or law shall be heard by the same arbitrator(s) in a single proceeding.

**14.7 EXCEPTIONS.** This agreement to arbitrate shall not apply to any claim:
  (a) of contribution or indemnity asserted by one party to this Agreement against the other party and arising out of an action brought in a state or federal court or in arbitration by a person who is under no obligation to arbitrate the subject matter of such action with either of the parties hereto; or does not consent to such arbitration; or
  (b) asserted by the Subcontractor against the Contractor if the Contractor asserts said claim, either in whole or part, against the Owner and the contract between the Contractor and Owner does not provide for binding arbitration, or does so provide but the two arbitration proceedings are not consolidated, or the Contractor and Owner have not subsequently agreed to arbitrate said claim, in either case of which the parties hereto shall so notify each other either before or after demand for arbitration is made.

In any dispute arising over the application of this Article 14.7, the question of arbitrability shall be decided by the appropriate court and not by arbitration.

## ARTICLE 15

### CONTRACT INTERPRETATION

**15.1 INCONSISTENCIES AND OMISSIONS.** Should inconsistencies or omissions appear in the Contract Documents, it shall be the duty of the Subcontractor to so notify the Contractor in writing within three (3) working days of the Subcontractor's discovery thereof. Upon receipt of said notice, the Contractor shall instruct the Subcontractor as to the measures to be taken and the Subcontractor shall comply with the Contractor's instructions.

**15.2 LAW AND EFFECT.** This Agreement shall be governed by the law of the state of _____

**15.3 SEVERABILITY AND WAIVER.** The partial or complete invalidity of any one or more provisions of this Agreement shall not affect the validity or continuing force and effect of any other provision. The failure of either party hereto to insist, in any one or more instances, upon the performance of any of the terms, covenants or conditions of this Agreement, or to exercise any right herein, shall not be construed as a waiver or relinquishment of such term, covenant, condition or right as respects further performance.

**15.4 ATTORNEY'S FEES.** Should either party employ an attorney to institute suit or demand arbitration to enforce any of the provisions hereof, to protect its interest in any matter arising under this Agreement, or to collect damages for the breach of the Agreement or to recover on a surety bond given by a party under this Agreement, the prevailing party shall be entitled to recover reasonable attorney's fees, costs, charges, and expenses expended or incurred therein.

**15.5 TITLES.** The titles given to the Articles of this Agreement are for ease of reference only and shall not be relied upon or cited for any other purpose.

**15.6 ENTIRE AGREEMENT.** This Agreement is solely for the benefit of the signatories hereto and represents the entire and integrated agreement between the parties hereto and supercedes all prior negotiations, representations, or agreements, either written or oral.

## ARTICLE 16

## SPECIAL PROVISIONS

**16.1 PRECEDENCE.** It is understood the work to be performed under this Agreement, including the terms and conditions thereof, is as described in Articles 1 thru 16 herein together with the following Special Provisions, which are intended to complement same. However, in the event of any inconsistency, these Special Provisions shall govern.

**16.2 SCOPE OF WORK.** All work necessary or incidental to complete the _____

_____

_____

_____

Work for the Project in strict accordance with the Contract Documents and as more particularly, though not exclusively, specified in: _____

_____

_____

_____

_____

_____

with the following additions or deletions:

**16.3 COMMON TEMPORARY SERVICES.** The following "Project" common temporary services and/or facilities are for use of all project personnel and shall be furnished as herein below noted:

By this subcontractor;

By others;

**16.4 OTHER SPECIAL PROVISIONS.** (Insert here any special provisions required by this subcontract.)

**16.5 CONTRACT DOCUMENTS.** (List applicable contract documents including specifications, drawings, addenda, modifications and exercised alternates. Identify with general description, sheet numbers and latest date including revisions.)

IN WITNESS WHEREOF, the parties hereto have executed this Agreement under seal, the day and year first above written.

_____  _____
Subcontractor          Contractor


By _____  By _____
  (Title)            (Title)

# Appendix E

# AGC METHOD FOR CHARGING EQUIPMENT OWNERSHIP COSTS

## EXPLANATION OF ITEMS OF EQUIPMENT OWNERSHIP EXPENSE

### Contractors' Annual Equipment Expense

Annual equipment expense as treated in the accompanying schedule* embraces those items that cannot ordinarily be determined with exactitude for a specific project. It does not include loading, shipping, erecting, operating or dismantling, nor does it include fuel, lubricants, supplies, wages or transportation of operating crews or any of the contractor's general expense of doing business. Minor or field repairs are not included because they are generally regarded as operating costs. The annual equipment expense is composed of but six items, which are as follows: (1) depreciation, (2) major repairs and overhauling, (3) interest on the investment, (4) storage, incidentals and equipment overhead, (5) insurance, and (6) taxes.

These six items are expressed as percentages of the capital investment, and the capital investment is considered as the cost of the machine, plus the expense of freight to the contractor's initial unloading or receiving point, plus the cost of assembling and making ready for use. Any similar expenses incurred after the initial setup are not included in the investment.

---

* See Figure 13.2.

This appendix is reproduced with permission of the Associated General Contractors of America. It is quoted directly from *Contractors Equipment Ownership Expense*, AGC, 1974.

## Depreciation

Depreciation rates are based on operation under the wear and tear of average job conditions, where the personnel of operating crews frequently changes, where methods of handling even under favorable field conditions are not uniform, and where it is frequently impracticable to protect machines from the weather. These rates are figured by the straight-line method by which a uniform percentage is charged off each year during the life of the equipment. No salvage or scrap value is considered, as in contractors' equipment it is usually insignificant.

**The depreciation rates used represent the average experience and are to be viewed from that point, and are therefore subject to adjustment in accordance with the individual experience. They are calculated by contemplating one shift per day use.**

The straight-line depreciation method used here is considered the most practical for a presentation of this kind. It is deemed sufficiently accurate for estimating and distributing equipment expense.

For purposes other than the determination of the cost of owning and maintaining equipment for several depreciation methods allowed by the Internal Revenue Code of 1954 should be examined carefully to determine which method best meets individual requirements.

## Major Repairs and Overhauling

Major or shop repairs include those items of heavy repair which usually keep a machine idle for an extended period in contrast with minor or field repairs which entail comparatively little delay and which are necessary to keep the machine in operation. Such repairs include overhauling, painting, and maintenance at the contractor's shop or yard, but do not include rebuilding.

## Interest on the Investment

Interest is computed from the contractor's average cost of obtaining money for the purchase of equipment. This cost is composed of interest on short-time loans, or monthly equipment notes held by the manufacturer or distributor. The quoted rate of interest on loans is not the actual cost to the contractor. Interest is immediately deducted and a considerable portion of the money borrowed is usually left on deposit, which considerably increases his cost of borrowing money. Seven per cent is taken as an average rate, against which is credited two per cent for interest on depreciation reserves and other possible daily bank balances, leaving an average net interest charge of *five percent.* This figure is constant for all items.

## Storage, Incidentals, and Equipment Overhead

Storage and incidentals include the expense of storing equipment between successive jobs or operations. This expense is made up of such items as rentals and maintenance of storage warehouse or yard, wages of watchman, and such direct overhead as may be involved in providing the storage facilities and handling the equipment in and out of storage. None of the contractor's general expense of doing business is included. The average expense of storage and incidentals is *3.5 percent.*

## Insurance

Insurance includes the average cost of premiums on general policies covering the usual insurable risks, including fire and theft, and in the case of marine equipment, includes the premium on marine insurance. It does not include any contingency allowance to cover uninsurable losses. These uninsurable losses, including the surplus value of equipment beyond policy coverage can be covered only by setting up a contingency item in the job cost estimate, and charging them as a cost of the work.

The amount necessary to cover insurance is taken as *1 percent* throughout the useful life of the equipment and is a constant in the schedule.

## Taxes

Taxes include the average personal taxes on assessed valuation of equipment and corporation taxes on capital value of equipment, but do not include state or United States income taxes, which are more properly items of general expense. The amount included for this expense is *1.5 percent for all items.*

By reason of the mechanical requirements of printing, those expenses that are identical for all items have been combined into one column in the schedule. They total 15 percent for inland marine equipment, 16 percent for coastwise marine equipment, and 11 percent for all other.

## Expense per Working Month

The annual equipment expense explained in the preceding paragraphs must be recovered by a contractor from work that he performs. To do this it is necessary to establish for each item a monthly charge of such amount that when multiplied by the average number of working months, it will yield a revenue equal to the annual expense. In other words, the monthly charge for any item equals the annual expense divided by the average number of months during which the machine is in use. It varies extremely, depending on the type of equipment, the climate, nature of the work, business conditions, and other factors.

The average number of working months for each item of the schedule is based on average climatic conditions for the nation as a whole, such as usually prevail in the central states, upon normal business conditions, and upon normal seasonal fluctuations of the industry. Where the average number of working months shown does not fit the experience of a specific contractor, the contractor should substitute the number indicated by his own experience.

The expense rates per working month are based on a single operating shift of 8 hours per day. *Where the equipment is used on double or triple shifts, an additional charge of 50 percent of the single shift rate for each such additional shift of 8 hours should be made.* Charges for use in excess of one shift but less than a full additional shift should be made at a rate proportional to the extra shift rate. The monthly rate is not subject to deductions for Sundays or holidays and should be charged for the full calendar period elapsing between shipments to and from the job.

## Expense per Working Hour

When the expense per working month is reduced to an hourly basis, the monthly rate must be divided by the number of hours normally worked in a monthly period. On the average, *22 days per month are considered working days.* Consequently the monthly rate should be divided by *176 to arrive at the hourly rate.*

## Equipment Used Underground

Certain items in this schedule are used exclusively on tunnel, shaft, and other underground work. In such cases appropriate allowances have been made for the unusually high costs of depreciation, overhauling, and major repairs. When other items of equipment, not so designated in the manual, are used underground, the cost of ownership should be raised to allow for the above-mentioned increased costs.

## Caution—Important

**It is intended that the data carried in the schedule (See Figure 13.2) will be applied in accordance with the principles set forth in the Foreword and Explanation of Items. The schedule is intended to reflect the average experience with respect to many variable factors. It is believed that this schedule, applied in accordance with the explanatory matter with the contemplated adjustments, can serve many useful purposes.**

## Charges for Equipment

It has been set forth herein that the purpose of this schedule is to reflect the average expense to a contractor owning and operating his own equipment on his own contracts. The charges that a contractor is justified in making under such circumstances are to be distinguished from those charges that are justifiable where the contractor may lease or rent a machine to others.

Three classes of equipment charges are generally recognized in the construction industry. The first applies to the contractor's own lump sum or unit price work where no separate payment for the use of equipment is involved. This rate, under the policy of most companies, is placed as nearly as possible at the actual cost of ownership and maintenance. It is the same type of rate as the average rates shown in the schedule, though in the individual case it may be greater or less than the schedule rate.

The second type is that used in cost-plus-a-fee or force account contracts, where the client-owner pays a specified charge for machines used on this project. If field repairs are specified as a cost of the work under such contracts, this rate is generally identical with the one already mentioned; otherwise it is a higher rate and includes the cost of such repairs.

In computing equipment charges to client-owners, some companies add a service charge comparable to the "ready to serve" charge of various public utilities and representing compensation for providing efficient labor-saving equipment. This charge, which can reasonably be from 5 to 10 percent of the total equipment expense, depending upon the size and nature of the project and the duration of the period that it is needed on the job, is either added directly into the equipment charge or carried in the construction fee, depending upon the policy of the contractor.

**The third type of rate is that used by contractors in renting equipment for use on work that they themselves are not performing. This rate differs from the others in that it should include an addition for overhead and profit, similar to that which a regular equipment renting company must charge.**

When a contractor leases equipment to others, he is engaging not in construction, but in the business of equipment renting. This business has its own peculiar risks, and intangible costs that one without special experience is likely to underestimate. Determination of rates for such a business does not involve questions of construction service and can probably be discussed best by specialists in the renting business.

With respect to renting, the rates should be based on release of equipment in as good condition as when sent to a job, allowing, of course, for a reasonable amount of ordinary wear and tear.

## Equipment Rental Agreement

Unless a properly drawn lease is used for the renting of equipment, a number of misunderstandings are liable to arise and cloud the transactions. Some of these may pertain to payment of freight, duration of the rental period, liability of the parties, terms of payment, and other matters that can cause trouble when not clearly defined. The AGC has developed and approved a form of lease (AGC equipment rental agreement) which is suitable for renting most construction machines and which, with modifications, can be made suitable for marine equipment.

## NOTE

Column 3, "Interest, Taxes, Storage, Insurance" cost per cent is constant throughout the schedule. The component parts are: Interest, 5 per cent; Taxes, 1.5 per cent; Storage and Incidentals, 3.5 per cent; Insurance, 1 per cent.

# Appendix F

# Small Office Building Input Data

# FIGURE F.1 Precedence Network for Office Building Project

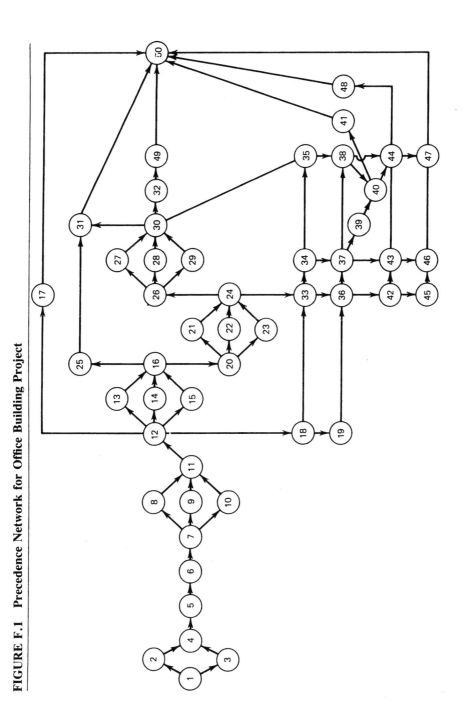

**Figure F.2   Activity model input for office building project.**

```
NEW FILE NAME STARTING IN MARCH
CONSTRUCT SMALL OFFICE BLDG.
1 4 500 'MOVE IN' 2 3
2 2 0 'SURVEY BLDG-SEWER LINES' 4
3 2 480 'ROUGH GRADE' 4
4 7 14480 'EXCAVATION' 5
5 12 42000 'BLDG. FOUNDATION' 6
6 4 9000 'POUR SLAB ON GRADE' 7
7 15 40000 'BUILD RETAIN WALLS-COLS CORES' 8 9 10
8 2 4800 'ROUGH IN MECH-ELECT' 11
9 12 6000 'FORM FLOOR-1' 11
10 2 1800 'ROUGH IN PLUMBLING' 11
11 4 6000 'POUR FLOOR-1' 12
12 10 14000 'BUILD COL-CORES-2' 13 14 15 17 18
13 2 4800 'ROUGH IN MECH-ELECT-2' 16
14 12 12000 'FORM FLOOR-2' 16
15 2 1800 'ROUGH IN PLUMBING-2' 16
16 4 6000 'POUR FLOOR-2' 20 25
17 20 60000 'MECH/ELECT PLANT' 50
18 7 50000 'HVAC-ELECT-1' 19 33
19 6 16000 'MAJOR PLUMBING-1' 36
20 10 14000 'BUILD COL-CORES-3' 21 22 23
21 2 4800 'ROUGH IN MECH-ELECT-3' 24
22 12 12000 'FORM FLOOR-3' 24
23 2 1800 'ROUGH IN PLUMBING' 24
24 4 6000 'POUR FLOOR-3' 26 33
25 25 12500 'ELEVATOR GAR-1' 31
26 10 14000 'BUILD COL CORES-ROOF' 27 28 29
27 3 4800 'ROUGH IN MECH-ELECT-ROOF' 30
28 12 12000 'FORM ROOF' 30
29 2 1600 'ROUGH IN PLUMBING-ROOF' 30
30 4 6000 'POUR ROOF' 31 32 35
31 25 12500 'ELEVATOR 2-3' 50
32 9 15000 'ROOFING SYSTEM' 49
33 12 50000 'HVAC-ELECT-2' 34 36
34 12 50000 'HVAC-ELECT-3' 35 37
35 12 50000 'HVAC-ELECT-ROOF' 38
36 8 16000 'MAJOR PLUMBING-2' 37 42
37 8 16000 'MAJOR PLUMBING-3' 38 39 43
38 8 16000 'MAJOR PLUMBING-ROOF' 40 44
39 1C 4800 'GLAZING 1-2' 40
40 5 2400 'GLAZING-3' 41 44
41 20 9000 'MASONRY-FACT BRICK' 50
42 15 30000 'ACC CEILING + INT PART.-1' 43 45
43 15 30000 'ACC CEILING + INT PART.-2' 44 46
44 15 30000 'ACC CEILING + INT PART.-3' 47 48
45 10 21000 'INTERIOR DECORATING-1' 46
```

**Figure F.2   (cont.)**

```
46 10 21000 'INTERIOR DECORATING-2' 47
47 10 21000 'INTERIOR DECORATING-3' 50
48 2 0 'TEST MECH-ELECT SYSTEM' 50
49 10 15000 'DRIVEWAYS,WALKS,ETC.' 50
50 5 0 'CLEAN UP + MOVE OUT'
```

# REFERENCES

*Accounting Program Manual,* Construction Data Control, Inc., Norcross, Ga., 1983.

Adamczyk, James P., "Project Planning and Scheduling with the GTICES-PROMAX Program," *Masters Special Topic,* School of Civil Engineering, Georgia Institute of Technology, Atlanta, 1982.

Adrian, James J., *Construction Accounting: Financial, Managerial, Auditing, and Tax,* Reston, Va.: Reston, 1979.

American Institute of Certified Public Accountants, *Construction Contractors—Audit and Accounting Guide,* prepared by the Construction Contractor Guide Committee, New York, 1981.

Antill, James M., and Ronald W. Woodhead, *Critical Path Methods in Construction Practice,* 2nd ed., New York: Wiley, 1970.

The Associated General Contractors of America, *Cost Control and CPM in Construction,* AGC, Washington, D.C., 1968.

The Associated General Contractors of America, *CPM in Construction, A Manual for General Contractors,* AGC, Washington, D.C., 1965.

Au, T., and E. Parti, "Building Construction Games—General Descriptions," *Journal of the Construction Division,* ASCE, Vol. **95,** No. CO1, Proc. Paper 6645 (July 1969), pp. 1–9.

Au, T., R. Bostleman, and E. Parti, "Construction Management Game—Deterministic Model," *Journal of the Construction Division,* ASCE, Vol. **95,** No. CO1, Proc. Paper 6647 (July 1969), pp. 25–38.

Benedetti, Eloy, "Comparative Financial Analysis of Construction Companies Using Ratio Analysis." Unpublished special problem, Georgia Institute of Technology, 1979.

Blasi, Luigi, "Equipment Control Systems." Unpublished special problem, Georgia Institute of Technology, 1984.

Bonny, John B., and Joseph P. Frein, *Handbook of Construction Management and Organization,* New York: Van Nostrand Reinhold, 1973.

Brigham, Eugene F., *Financial Management Theory and Practice,* Hinsdale, Ill.: Dryden Press, 1979.

*Building Construction Cost Data,* Robert Snow Means, Kingston, Mass. Published annually.

*Building Estimator's Reference Book,* 18th ed., Frank R. Walker, Chicago, 1973.

Burman, Peter J., *Precedence Networks for Project Planning and Control,* London: McGraw-Hill, 1972.

*Caterpillar Performance Handbook,* 9th ed., Caterpillar Tractor Company, Peoria, Ill., 1979.

Clough, Richard H., *Construction Contracting,* 3d ed., New York: Wiley, 1975.

Collier, Keith, *Fundamentals of Construction Estimating and Cost Accounting,* Englewood Cliffs, N.J.: Prentice-Hall, 1974.

*Contractors Equipment Ownership Expense,* Associated General Contractors of America, Washington, D.C., 1974.

"Contractor's Job Cost System," *Hadley Service Bulletin No. 194,* April 1956.

Coombs, William E., and William J. Palmer, *Construction Accounting and Financial Management,* New York: McGraw-Hill, 1977.

*Credit Reports* (individual subscription), Building Construction Division, Dun and Bradstreet, New York.

Dabbas, M. A., and D. W. Halpin, "Integrated Project and Process Management," *Journal of the Construction Division*, American Society of Civil Engineers, Vol. **108**, No. CO3, Transaction Paper 17304 (September 1982).

Dallavia, Louis, *Estimating General Construction Costs*, 2nd ed., New York: F. W. Dodge, 1957.

Davis, J. Gordon, "Keeping Project Costs in Line," *Machine Design*, Vol. **47**, No. 29, Dec. 11, 1975, pp. 128–33.

Davis, J. Gordon and Daniel W. Halpin, "Productivity Improvement in Project Planning and Scheduling, *Proceedings, American Society of Civil Engineering Workshop on Civil Engineering Productivity*, St. Louis, Mo., September 26–28, 1983.

*Dodge Manual for Construction Pricing and Scheduling*, New York: McGraw-Hill Information Systems, Published annually.

Dopuch, Nicholas, Jacob Birnberg and Joel Demski, *Cost Accounting*, New York: Harcourt Brace Jovanovich, 1974.

Douglas, James, *Construction Equipment Policy*, New York: McGraw-Hill, 1975.

*Engineering News Record*, New York: McGraw-Hill. Published weekly.

*Financial Ratios for Electrical Contractors*, Operation Overhead, National Electrical Contractors Association (NECA). Published periodically.

Fondahl, John W., and Ricardo R. Bacarreza, *Construction Contract Mark-Up Related to Forecasted Cash Flow*, Technical Report No. 161, Stanford University, Department of Civil Engineering, The Construction Institute, Stanford, Calif. (November 1972).

Gibb, Thomas Wilson, Jr., *Building Construction in the Southeastern United States*. Report presented to the School of Civil Engineering, Georgia Institute of Technology, Atlanta, 1975.

Graham, R. G., and C. F. Gray, *Business Games Handbook*, New York: American Management Association, Inc., 1969.

Halpin, D. W., "Constructo—An Interactive Gaming Environment," *Journal of the Construction Division*, American Society of Civil Engineers, Vol. **102**, No. CO1, Proc. Paper 11969 (March 1976), pp. 145–156.

Halpin, D. W. and R. W. Woodhead, *Construction Management*, New York: Wiley, 1980.

Halpin, D. W. and R. W. Woodhead, *Constructo—A Heuristic Game for Construction Management*, Urbana, Ill.: University of Illinois Press, 1973.

Halpin, D. W., K. O. Reiff, and R. W. Woodhead, "Ein Baustellenbezogenes Planspiel," *Bau + Bau industrie*, Dusseldorf, Germany: Werner Verlag, Nov. 1971.

Harris, Robert B., *Precedence and Arrow Networking Techniques for Construction*. New York: Wiley, 1978.

Hoyos, J. E., "Analysis of Equipment Replacement Decision." Unpublished special problem, Georgia Institute of Technology, 1982.

"Introducing the Construction Management and Accounting System for IBM System/34," International Business Machines Corporation, General Systems Div., Atlanta, 1978.

Jackson, I. J., III and M. H. Gilliam, *Financial Management for Contractors*, New York: The Fails Institute, McGraw-Hill, 1981.

Kibbee, J. E., et al., *Management Games*, New York: Reinhold Book Corporation, 1961.

Kollaritsch, Felix P., *Operating and Financial Ratios of Ohio Highway Contractors*, Bureau of Business Research Monograph No. 118, College of Commerce and Administration, Ohio State University, Columbus, 1966.

Kollaritsch, Felix P., *Job Order Systems for Highway-Heavy Contractors*, College of Administrative Science, Ohio State University, Columbus, 1974.

Lang, Hans J., and Michael Decoursey, *Profitability Accounting and Bidding Strategy for Engineering and Construction Management,* New York: Van Nostrand Reinhold, 1983.

Mott, Charles H., *Accounting and Financial Management for Construction,* New York: Wiley-Interscience, 1981.

Neil, James N., *Construction Cost Estimating for Construction Control,* Englewood Cliffs, N.J.: Prentice-Hall, 1982.

Neil, James N., "A System for Integrated Project Management," *Proceedings, ASCE Specialty Conference on Current Practice in Estimating and Cost Control,* Austin, Texas, April 1983.

Park, William R., *Construction Bidding for Profit,* New York: Wiley, 1979.

Peurifoy, Robert L., *Estimating Construction Costs,* 3rd ed., New York: McGraw-Hill, 1975.

Pulver, Harry E., *Construction Estimates and Costs,* 4th ed., New York: McGraw-Hill, 1969.

*RMA Financial Ratios for Commercial Contractors,* New York: Robert Morris Associates. Published annually.

Royer, King, *The Construction Manager,* Englewood Cliffs, N.J.: Prentice-Hall, 1974.

Simpson, J. M., "The Construction Cost Accounting Practices of Highway Contractors in the Southeastern United States." Unpublshed special problem, Georgia Institute of Technology, 1975.

Teicholz, Paul, "Requirements of Construction Company Cost System," *Journal of the Construction Division,* ASCE, Vol. **100,** No. CO3, Proc. Paper 10786, (September 1974), pp. 255–263.

Teicholz, Paul, "Labor Cost Control," *Journal of the Construction Division,* ASCE, Vol. **100,** No. CO4, Proc. Paper 11020 (December 1974), pp. 561–570.

*Uniform Construction Index,* Construction Specifications Institute, Washington, D.C., 1974.

U.S. Congress Joint Committee on Taxation, *General Explanation of the Economic Recovery Tax Act of 1981,* H.R. 4242, 97th Congress, Public Law 97-34, December 1981.

Van Horne, James C., *Financial Management and Policy,* Englewood Cliffs, N.J.: Prentice-Hall, 1980.

Wager, Oliver E., Jr., "Construction Cost Accounting Practice—At a Glance." Unpublished special problem, Georgia Institute of Technology, August, 1974.

# Index